Ágnes Veszelszki

Digilect

Knowledge & Information

Studies in Information Science

Ágnes Veszelszki

Digilect

The Impact of Infocommunication Technology on Language

DE GRUYTER
SAUR

ISBN 978-3-11-049990-2
e-ISBN (PDF) 978-3-11-049911-7
e-ISBN (EPUB) 978-3-11-049713-7

Library of Congress Cataloging-in-Publication Data
A CIP catalog record for this book has been applied for at the Library of Congress.

Bibliographic information published by the Deutsche Nationalbibliothek
The Deutsche Nationalbibliothek lists this publication in the Deutsche Nationalbibliografie; detailed bibliographic data are available on the Internet at http://dnb.dnb.de.

Typesetting: Konvertus, Haarlem
Printing and binding: CPI books GmbH, Leck
♾ Printed on acid-free paper
Printed in Germany

www.degruyter.com

Contents

Acknowledgements

The analyses in this book were performed between 2006 and 2015. As the topic is tightly related to technology, the examined material quickly becomes obsolete and therefore demands continuous updating. Before the book went to press, the author did her best to actualise its content but the changes that occurred during the manuscript-to-publication time were out of her reach. Therefore, the author appreciates the kind reader's critical mind and understanding.

I am grateful to my theme leader, Prof. Borbála Keszler for supporting me all along my work, for continuously providing me with precious advices, and for improving the quality of the book with her expert comments. I am also thankful for her assistance, and for saying "everything will be fine" when I needed it most.

I would like to thank Prof. Géza Balázs for being my mentor for more than fifteen years now, for encouraging me to carry on scientific research, and for constantly giving me opportunities to demonstrate my aptitude.

I am thankful to Judit Bóna and Árpád Zimányi for their useful advices given to me as external examiners, as well as to the members of the PhD Thesis Defence Committee (namely: Prof. em. István Szathmári, Erzsébet Heltainé Nagy, Prof. em. Janusz Bańczerowski, and Judit Balogh) for their work.

I am grateful to the editor of the Series "Information and Knowledge" Prof. Wolfgang Stock and the Hungarian academician Prof. Tibor Frank for encouraging me to publish the book in English. The book was funded by the Scientific Book and Journal Publishing Programme of the Hungarian Academy of Sciences.

I wish to express my thanks to the translator, Levente Nagy, for his thoroughness and extensive background work while translating the text from Hungarian to English.

I am indebted to all my teachers for being a source of inspiration and motivation.

I also want to say thanks to my fellow PhD students, my students and my friends, as well as to those who completed my survey questionnaires, sent photos, or offered their exercise books and private correspondence, for contributing to this book.

Special thanks to my partner, Zoltán Szalai. Words cannot describe how lucky I am to have him in my life.

Finally, if I wanted to write down all the things I owe to my parents, these acknowledgements would be longer than the book itself. So I have to confine myself to saying:

I gratefully dedicate this book to my parents.

DOI 10.1515/9783110499117-202

1 Introduction: Infocommunications and Digilect

1.1 Objectives

The high degree of internet penetration and its social (and linguistic) effects evidently influence how people, and especially the highly susceptible younger generations, use language (see *Annex 1*). The primary aim of the book was not to identify the characteristic features of the digital language variety (this has already been done by several works in both Hungarian and other languages) but to examine how digital communication affects the language of other mediums of communication: orality, handwritten texts, digitally created but not digitally perceived, that is printed texts, including in particular advertisements (which quickly respond to linguistic change). Naturally, the book presents the characteristics of the digital language variety but only to give a framework to the impact analysis.

On the basis of my examinations conducted on elementary and secondary school students, it is reasonable to assume that a generation has grown up which is unable (unwilling) to distinguish between conceptually oral but medially written and "traditional" literacy. The book also touches on the phenomenon that infocommunications (including especially multimedial and hyperlinked texts which are replacing unimedial linear texts) is making changes in reading comprehension strategies.

The methodology applied included corpus analysis (linguistic examination of several thousand pages of notes and private letters written by secondary school and university students, comparison of these texts with instant messages, and analysis of advertisements), questionnaire survey, observation and field work (observation of oral conversations and utterances).

It is important to document changes in progress and thus direct attention to potential outcomes. The current linguistic change is different from previous ones primarily in its speed and form of spreading, and it not only brings innovative grammatical forms and writing/spelling solutions but may also have far-reaching cultural and educational consequences in the long run.

In view of this, the book aims to document this unfolding linguistic change, to present and describe the new language variety (touching on the topics of written and spoken language and visuality), and to examine the effects of infocommunications on language.

DOI 10.1515/9783110499117-001

1.2 Structure of the book

The book comprises six main conceptual units (chapters).

The first, introductory chapter discusses the relationship between infocommunications and digilect, and presents the methodology applied. The methodological subchapters (1.1. Objectives, 1.2. Structure of the book, 1.5. Methodology, 1.6. Corpuses used for the analysis) embrace two other subchapters which provide an overview and give reasons for the choice of topic. The present introductory chapter outlines the most important views on digital communication, with special regard to the relationship of writing and reading comprehension, and to social networks, and then discusses various opinions about the dichotomy of orality and literacy.

The second chapter is titled "Digilect as a language variety". It introduces a new linguistic term, the *digilect*, which fits into the theoretical system of other *lects* (language varieties). An attempt is made to establish the linguistic status of digilect. The second chapter is concluded with an overview of the most important text types relating to digilect (e-mails, internet forums, blog and vlog posts, tweets, message wall posts, comments, online chat messages and instant messages, text and multimedia messages).

Chapter 3 is the longest one. It deals with the characteristic features of digilect, examined from the following four main aspects: pragmatic-textological, lexical, grammatical and formal (technical). The most remarkable pragmatic features of digilect (Subchapter 3.1) were the relationships between participants, perspective, thematic progression, the distinct role of the phatic function, usage of T and V forms, and metatexts. The vocabulary of digital communication (Subchapter 3.2) is characterized by foreign language influence, new words and collocations, IT vocabulary, expletives (fillers), slang elements and taboo breaking. The grammatical aspect of the digital language variety received particular attention (Subchapter 3.3). This part deals with character-saving techniques and neologisms, which were examined by type of word formation (morphological derivation, composition, other types of word formation). The part on morphological derivation includes a case study which describes the productivity of the suffixes *-(V)l* and *-(V)z* through new verbs created in digilect. A notable grammatical feature of digilect is the use of impersonal structures. Since the relevant literature mentions the influence of spoken language as one of the most typical syntactic characteristic of digital communication, the subchapter briefly summarizes the grammatical approaches of spoken language illustrating the most relevant statements with examples. The description of syntactic characteristics involves two other case studies as well: one on the topic-repeating structure as a spoken language element in digilect texts, and another on the grammaticalization of the discourse marker *asszem* 'I mean'. Of the formal/technical characteristics of digilect (Subchapter 3.4), pseudo-phonetic spelling, lowercase and uppercase writing, and punctuation marks were discussed in more detail. Subchapter 3.5 elaborates on the relationship between digilect and visuality (in particular on emoticons, which are considered as the key characteristic feature of digilect, as well as on image-text relationships.)

Digilect and its effects are demonstrated through the findings of two questionnaire surveys in Chapter 4. The first survey (Subchapter 4.1) was carried out on elementary school students in 2008 and focused on the differences between typewriting and "traditional" handwriting. The second survey (Subchapter 4.2) was conducted in 2010 and involved a large number of respondents. It principally focused on the effects of digital text types on (hand)written and spoken language forms (such as: internet-specific expressions, abbreviations, word forms, use of emoticons etc.) The survey was primarily aimed at collecting qualitative data through open-ended questions, and only secondarily at collecting quantitative data.

Chapter 5 basically worked with corpus analysis and concentrated on the impact of digilect on other types of text (oral conversations, handwritten notes, school test papers, dialogue letters and advertisements).

Summary, conclusions and directions for further research are provided in the sixth chapter. The book ends with literature, index and annexes.

1.3 Digital communication

The most important results of the digital revolution is that interpersonal communication became independent from place and time (German: Entortung, Entzeitlichung),[1] and became faster (Schlobinski 2009: 6). Heijnk lists the following main characteristics of the network, or as he put it, the *here-now-everything-medium* (German: das Jetzt-Hier-Alles-Medium; Heijnk 2002: 16–18):

- enables sharing up-to-date contents (which therefore rapidly become outdated);
- interactive (users can change its content);
- offers theoretically unlimited storage capacity, that is unlimited space for information;
- multimedial;
- interconnected through hyperlinks;

1 Most of the key notions are given in German language, too—as a complementary information.

- global (in theory, it is available everywhere);
- multidirectional and its audience can be determined on the basis of user interface (1 – 1, 1 – n, n – 1, n – n);
- offers a (relatively) permanent form for communicating information: the ideal website is up-to-date, dynamic, multimedial, well-structured, continuously accessible and editable, and has external and internal links.

Of the general and linguistically relevant social characteristics of the internet era, the most commonly mentioned feature is that the amount of information is growing at an incredible pace ("*knowledge explosion*", Pscheida 2007: 9, 27). The expression *information explosion* refers to the thought-provoking (if not alarming) phenomenon that every year humanity regenerates, that is doubles the volume of information generated until that year throughout its entire history. As of 2012, every day 2.5 billion gigabytes of data were „created in a variety of forms, such as social media posts, information gathered in sensors and medical devices, videos and transaction records" (W4). The notion *big data*, which first appeared in a scientific study in 2003 but became widely used only in 2008 (cf. Boellstorff 2013), refers to the phenomenon that the volume of data out there will eventually become so vast and complex that it will be impossible to process and receive it through simple, conventional, manual data processing methods and applications.[2]

Due to its vast, imperceivable and unmanageable character, this information flood creates both a sense of insecurity (Pscheida 2007: 5, 28), and a sense of confidence that all information is available online. However, in the *a'la carte culture* (Origgi 2006: 202) one has to know how to search and select information. The person who is capable of this is called *homo surfiens* by Meinel and Sack (2004). "Knowing how to search" is the simplest definition of the type of knowledge which is a key sociological notion in the 2000s: *media competence*. Information society faces the paradoxical situation that although collective knowledge is accumulating, individual knowledge is shrinking due to the limited capacity for receiving information and the unmanageable and unstructured nature of the knowledge (Pscheida 2007: 29). Knowledge-based society demands continuous flexibility, life-long learning and conscious metacognition from the individual.

2 For example the website http://www.worldometers.info provides real-time information, among others, on the number of new books published in a year or the number of newspaper circulated, mobile phones sold, e-mails and tweets sent on a day.

Several sociological studies (for a summary see: McCrindle–Wolfinger 2010) deal with the generation whose members were born into or have grown up in the digital age (the generation theory—the existence of generations itself—is a highly discussed scientifical topic; cf. Balázs–Veszelszki eds. 2016; Neuland ed. 2012). The generation born in the 1980s are commonly referred to as Generation X, and those born in the 1990s as Generation Y. According to a survey there is a third term for this demographical cohort: *Generation C* (or *Gen C* for short). It is not their age but their online behaviour that connects the members of Generation C. The letter *C* in the term refers to the basic unit of internet communication: *content*, but many other expressions can be linked to it, such as: constant connectivity, collaboration, change, curiosity and co-creation. Similarly to young people in general, Gen C places critical importance on group identity (i. e. belonging to the same "virtual tribe"), and on shaping and expressing opinion in social media. Their influence depends on what and how frequently they share. Gen C recognizes creativity and participation as basic values (W3).

"Printed books, which have been the dominant medium of communication, are now challenged by various electric and electronic mediums," says Hungarian philosopher Kristóf Nyíri (2004). Digitalized information can be reproduced, stored in a compressed form, quickly forwarded to great distances and further edited on other computers without any loss of information (Hodel 2010: 25). The primary form of digital information is immaterial, though it can be made material by printing. It is up to the user to decide on the manifestation of the master text: whether they read it on screen or in print (McCloud 2006: 14–15).

An often mentioned buzz word of the here-and-now culture is *telepresence* (German: Fernanwesenheit). A compact definition of this term is provided by Hrachovec (2002: 115): "Telepresence is the art of enabling social proximity despite geographical or temporal distances through the integration of computers, audio-visual, and tele-communicative technologies." Proximity and distance has gained a new meaning in the internet world: "Everything is equally near or—if you like—far away" (Karácsony 2002: 128).

The notion of *digital homeless* (digital vagabonds) refers to a new trend: more and more young people get rid of their movable and immovable property to be free from social ties and live their lives exclusively through digital assets. With the development of information technology, more and more physical objects (data storage devices for music or films, photo albums, television) can be replaced with immaterial electronic means of storing the same information. This trend is supported by statistics as well: the Association of American Publishers reported that ebook sales tripled in only one year between 2008 and 2009, while the sales of printed books falls year by year. Similarly, the sales revenue of online music stores multiplied by four, while CD sales dropped by half in five years.

Digital goods are now accessible anytime anywhere thanks to the increasing capacity of computers and data storage devices (and the cloud technology). Followers of "the cult of less" sell their belongings on the internet; they only keep consumer electronic goods and some clothes. The greatest danger in the life of digital vagabonds is loss of data, that is when one of their HDD breaks down and all information stored on it are gone. To avoid this, they back up their data and rely on the services of data recovery companies. Data loss to some people may be really disastrous. Researchers say that in the long run a possible outcome of this trend is the ability to make a digital copy of our minds ("mind uploading"), but for the time being this remains fiction.

Today's technology is for today's needs. Unlike e-mails, online forums or website contents, the messages and comments posted on Facebook or Twitter cannot be traced back. (Theoretically and technically, of course, it is possible but in practice the enormous amount of information makes it impossible to retrieve data older than two or three days.) András Karácsony (2002) considers the increased importance of the *present* as a general characteristic of modern world ("in the time-consciousness of our modern world [...] critical importance is attached to the dimension of the present", 2002: 136), but the internet makes this phenomenon much more spectacular. And what does "present" mean in this context? To cite András Karácsony again, it means "the period of time when future becomes past. Naturally, this can be a moment but a much longer period as well. [...] It has more to do with our experience about the present, so it relates to subjective time perception" (Karácsony 2002: 136).

In the age of infocommunications we use *new search techniques*: while in the past we consulted a dictionary, lexicon, encyclopaedia or technical book to find specific information, but today we rely on web search engines such as Google. According to Hodel (2010: 26), this new searching method is a hybrid form: our competence used for the search (background knowledge, performance of the search) is still rooted in the analogue age but faces the challenges of the digital age. Background knowledge is essential for evaluating and ranking the results of the search. In knowledge representation and information retrieval three ways of behaviour can be differentiated: searching, browsing and retrieving (Peters 2009: 287–293).

Internet users have accumulated a vast amount of *collective knowledge*, especially on Wikipedia. Wiki is the most characteristic structure of web 2.0 (its most popular representative is Wikipedia): it means an online knowledge base which users can both read and edit. The independent wiki articles are interconnected through hyperlinks. One of the biggest advantage of wikis, these freely accessible and editable knowledge bases (but we may also mention ebooks), is not their (nearly) unlimited storage capacity but their searchability and interactivity (Harnad 2006a: 16).

The traditional *author-editor-reader relationship*, which is linked to the Gutenberg Galaxy, is changing: it becomes more direct and more dialogical as the reader becomes co-author and co-editor of the texts. The essence of web 2.0 is interactivity: these online services are based on a community which produces all the content. Traditionally, the publication of a book involves several professionals whose names appear in the book (author, publisher, editor, reviewer, proofreader), whereas on Wikipedia articles are created by many anonymous authors.

The main criticism of such community edited pages relates to their authenticity and *reliability*. A 2005 study in the journal *Nature* reported on the following experiment: randomly selected articles of the English Wikipedia and Encyclopaedia Britannica were assessed through blind peer-review. The comparison showed that there were no significant differences between the accuracy of the online and the printed encyclopaedia. According to reviewers' ratings Encyclopaedia Britannica contains 2.92 and Wikipedia's 3.86 errors per article (Giles 2005; Terdiman 2005). In 2007 a similar study was published by the German journal *Stern*. The German Scientific Information Service (Wissenschaftlicher Informationsdienst, WIND) was invited to carry out a comparative examination of articles. In the framework of this, the researchers selected 50 random articles from Wikipedia and Brockhaus encyclopaedia which consists of 15 volumes. In 43 out of the 50 cases, Wikipedia performed better (it achieved 1.7 GPA compared to Brockhaus's 2.7 GPA on a five-point scale where 1 GPA was best). Surprisingly, Wikipedia out performed Brockhaus in the most important aspect, the accuracy of facts as well (Becher–Becher 2011).

That being said, online knowledge is not always reliable: the author, the sources and the recency are usually unknown, and the information is often uncontrolled which means that anyone can publish anything. The sources of knowledge have changed so internet users have to develop a skill for the critical evaluation of information, as part of their new media competence.

Not only the reliability of texts is changing as word-processing and online publishing is gaining ground, but also the *manner in which texts are written and read*. Looking at the manner of writing we will find the following: Although computers have the same keyboards as typewriters, the text is not put directly on the paper; its protoform is kept on the computer. This phenomenon is referred to as "the triumph of the abstract over the concrete" (Benczik 2001: 104). As digital texts have no material form, they permit untraceable deletion, insertion and non-linear editing (Balázs 2009). The former linear way of reading has also changed: digital texts containing hyperlinks are perceived through circular reading.

Hyperlinks are used to connect individual nodes (texts, images) which give the web-like character of internet texts. *Hypertext*, according to Ted Nelson (1974, 1987) who coined the term, is a nonsequential (multimedial) text made up of written material or images so complexly connected to each other that it is impossible to print it on paper. In the definition of hypertext, the Hungarian linguist S. János Petőfi emphasizes the aspect of non-linear organization: "the text is structured in a way that its parts, which are separated from each other for some reason, are linked together through relationships with 'non-linear organization'" (Petőfi 1995). Therefore, hypertext is a typically digital text with a network-like structure in which the relationships between the individual elements are created by hyperlinks making it non-linear. This type of text through its hyperreferences offers an option to and expects interactivity from the reader. Objects (hubs) linked to each other and to the text through hyperlinks, that is the information units themselves, can be texts, images and videos (Storrer 1997: 121). These are interlinked by electronic references, i. e. links, which make up the non-linear network-like structure (Frindte 1999b: 44).

In hypertext readers have to make a decision every time they arrive at a node: users either enter them or pass them by. Users reading for information "continuously encounter so many extra references that they become unable to check them so they lose control over them. [...] There is a constant fight for linearity, for chronology as a principle of sorting" (Karácsony 2002: 135). Hypertextuality has two inherent problems for individuals: one is the feeling of being "lost in hyperspace", that is the loss of orientation, and the other is the "cognitive overhead" or overload on users facing information flood and constant pressure to make decisions (Pscheida 2007: 49).

Hyperlinks make it less clear what a text is; the traditional linearity of texts in books (beginning on the first page and ending on the last one) is suspended in hypertexts and "reading comprehension never really ends for the reader" (Karácsony 2002: 135). This *patchwork-like information structure* gives rise to a new way of thinking: conventional linear categories (before–after, real–unreal) either disappear or gain new meaning. Frindte (1999b: 46) says that this cognitive change makes the category of "development" empty as well; the continuity of linearity is being replaced by the movement around a center. In this system it is more reasonable to talk about inter-subjectivity instead of objectivity (1999b: 49).

In the history of communication, humans have developed newer and *newer forms of keeping in touch* with each other. Frindte (1999a: 11–12) evokes Zygmunt Bauman's metaphors about each period. Bauman says that the main message (or rather the medium of message) of the modern era was photo paper. Photo paper can be used only once, and the information recorded on it cannot be changed anymore. By contrast, videotape as the message medium of the post-modern era can be replayed and overwritten anytime, it does not record information permanently. While the keyword of the modern period was *production*, for the post-modern period it was *recycling*. Frindte completes Bauman's metaphors with his own ones: he thinks that the post-modern era would be better described by the expression *"information highway"* or Jean Baudrillard's (1987: 20) expression *"the promiscuity of networks"* (on further metaphors cf. Pléh 2010).

The timeline of the development of internet usage is wittily illustrated by the following instance of internet folklore:

Example 1: Timeline of the development of internet usage in internet folklore[3]

1998- tudod mi az az internet? ('Have you heard about the Internet?')
2000- na bekötötték nálam a tárcsázós netet!:) ('Imagine, I've got dial-up internet access at last.:)')
2002- van tevéd a teveclubon? ('Do you have a camel at the CamelClub online game?')
2004- csetelünk? ('Have time to chat?')
2006- megadod az msn címed? felveszlek (:('Send me your MSN address and I'll add you to my partner list (:')
2008- küldtem egy meghívót az iwiwre meg bejelöltelek myvip-en! ('I invited you to iwiw and marked you as friend on myvip')
2009- hallod, mi az a facebook? ('Man, what's this Facebook?')
2011- tee, figyu lájkoltam a képedet fészen, azt amelyiken be van jelölve az xy is. de képz a geci tanár rajtakapott hogy facéztam a telómmal órán és el is vette, de küldtem neki farmville-n cuccokat, szóval ma visszaadta :D ('hey, look, I liked your photo on Facebook, the one where xy is tagged. But guess what, the bloody teacher caught me while hanging on Facebook and took my phone away. But I've sent him some stuff on Farmville so he's returned it today. :D')

Having a look at *digital genres* we will find that asynchronous forms have lost significance and synchronous forms have gained ground (e. g. a possible order of the change: e-mail, online forum > IM, blog > tweet, Facebook post). Note, however, that in this pluralistic process older genres do not disappear, they only lose importance and surrender some of their functions to newer genres. The pressure of time forces the creation of increasingly shorter forms. Between 2003 and 2005 we could say that "text message is the new postcard",[4] but according to recent observations (the surveys presented in Chapter 4) today good wishes are no longer sent in text messages but increasingly posted on the message walls of

---

3 Hungarian examples are provided with the original spelling. Examples are translated to English only if their meaning is important to understand what they illustrate. They are also explained where necessary. The sources of the examples are always indicated.

4 This linguistic structure was popularized in Hungary by the animated motion picture "Wall-E" released in 2008. The film gives an open social criticism by depicting dull citizens who follow the fashion as they are told over the loudspeaker. "Try blue, it's the new red!" says the loudspeaker, and everyone changes their dress from the uniform red to blue. This expression ("X is the new Y") has become the phrasal template (or "snowclone") of expressions describing new global trends.

social networking websites. Long narratives (lengthy genres) become silent and are replaced with multi-voice, subjective micro-narratives which choose topics from everyday life (Frindte 1999a: 13–14).

Forms related to communication technologies are subject to *conventionalization*: for example, text messages with many abbreviations and special linguistic forms have become conventional so text messages with standard spelling and regular forms now seem odd, that is "marked" (they may indicate that the sender rarely uses text messages or in fact wishes to express a more formal relationship with the receiver).

The most commonly mentioned linguistic features of digital change are the involvement of images, the substantial effect of spoken language, and the phenomenon of spontaneous literacy (Echtzeitschriftlichkeit; cf. Schlobinski 2009: 6). The role of visuality is much more significant today than at any time in the history of communication: this is clearly shown by the graphic interface of computer programmes, and the visual design of websites, the use of smileys (see: Chapter 3.5).

1.3.1 Text composition and reading comprehension

The connection between literacy and "high culture" has long been subject to dispute in the philosophy of science. One side holds that the emergence of literacy detached social relations as literate societies lack the spontaneity and "warmth" of orality and therefore people living in literacy-based cultures have separated from each other both psychologically and emotionally (Lakoff 1982: 257). Today, however, orality seems to gain ground against literacy so these "values" are making their way back to society. The age of literacy was followed by the age of post-literacy (Benczik 2001: 241) which, however, does not entail the disappearance, only the loss of absolute power and dominance of literacy. The other side argues that the linearity of writing, the visual pattern of successive sentences is closely related to the discovery of linear-logical thinking, temporal sequentiality and causal relationships (Lakoff 1982: 258). McLuhan presents a similar argument: The development of alphabetic writing created the paradigm of the so-called fixed point of view—lasting chronology, uniform spatial perspective, consideration of the subject of a thought from a fix perspective—, but this paradigm

(Independently from me, the same snowclone was used by Marc McCrindle and Emily Wolfinger who wrote in the November 2010 issue of the Transsylvanian journal *Korunk* the following in the context of generation changes: "Thirty is the new twenty-one!" [2010: 13–14].)

gained absolute power only in the age of printing, i. e. the age of smooth and silent reading (McLuhan 1962, 1964; as interpreted by Nyíri 1998: 15–16). The relationship between linearity and scholarship, however, is not so simple. Olson (1977) adopts a central position in the debate by saying: both cognitive styles have their pros and cons (for example, members of oral-based cultures have better short-term memory and a different relation to story-telling).

Humanity has never had so many mediums to keep in touch with others than it has today. Our everyday life is mediatized (Krotz 2001 referred to by Höflich 2003a: 7). This diversity forces users to make decisions several times a day: how to reach a given person, how the selected medium (phone call, text message, e-mail, instant messaging, blog, formal letter) influences the text as a product, what communication principles must (should/may) be observed in the selected medium (German: Kommunikette; cf. Höflich 2003a: 9).

Electronic literacy is a fundamental requirement of information society (cf. Knorr–Jakobs eds. 1997: 2; Szépe 1997). As we saw above, hypertextuality demands new production and reception strategies from both readers and authors. The subchapter discussing this topic only briefly refers to the issue of functional illiteracy (which means the ability to read shorter texts but inability to read longer ones). Only reference is made here to the matter of media competence (its existence and necessity) (cf. Hug 2010).

Relying on the findings of usability examinations, Heijnk (2002) provided a list of advices for users wishing to produce online texts. Most importantly, he recommended keeping such texts short. An experiment conducted on 81 respondents (Heijnk 2002: 57) showed that people read 25 % slower on a computer screen than on paper, so a text of the same length and difficulty is read for five minutes instead of four on screen. Though it may appear paradoxical, texts can be made shorter without losing any word. The solution lies in the hypertext structure: the text has to be split into parts and connected with hyperlinks.

1.3.2 The social network

The web-like structure, however, does not only characterize hypertexts but social relations as well. Social networks are social structures which are made up of relationships between individuals or organizations ranging from casual acquaintances to long-standing friendships and family ties. The expression was first used by J. A. Barnes in 1954. In general, social networks comprise 150 persons at the most and 124 persons on average (Hill–Dunbar 2002). The evolutionary psychologist Robin Dunbar currently uses Facebook to tests the accuracy of the "social brain" hypothesis (which assumes that the size of social networks is relative to the neocortex volume of our brain; cf. Dunbar 2002) and find out whether social networking websites have changed the group size that has been regarded as optimal.

The analysis of social networks play an important role in modern sociology and anthropology (Balaskó–Balázs–Kovács 2010; Balázs 2007b). Social networks have become increasingly popular parts of the internet: they enable users to search, exchange messages and share other information in systems built on relationships between friends, business partners and family members. The Hungarian-American physicist Albert-László Barabási's research findings clearly support one of the most well-known principle of communication theory, namely the global village theory, as well as the rule of six degrees of separation (Barabási 2002, 2016).

The scientific theory has a literary antecedent: the idea first appeared in short story *Chain-Links* written by Hungarian author Frigyes Karinthy in 1929 (published in the volume "Everything is Different"). The relevant paragraph is cited here: "One of us suggested performing the following experiment to prove that the population of the Earth is closer together now than they have ever been before. We should select any person from the 1.5 billion inhabitants of the Earth—anyone, anywhere at all. He bet us that, using no more than five individuals, one of whom is a personal acquaintance, he could contact the selected individual using nothing except the network of personal acquaintances" (Karinthy 1929).[5]

Karinthy believed that number of acquaintances grows exponentially with the number of links in the chain, so only a few links are necessary to reach anyone on the planet (Barabási 2002). Karinthy's intuition that any two persons can get into contact through no more than five other persons (six links) is the first printed version of the scientific concept which we now know as "six degrees of separation". The small world phenomenon is not only characteristic to society but to the major part of complex networks ranging from the World Wide Web to biological cells (Barabási 2005). The six degrees of separation holds that anyone can contact any other person in the world through a chain of acquaintances in a maximum of six steps.

In 2006 the American magazine *Time* awarded the title "Person of the Year" to the builders of the new information highway, the authors of commonly created and shared contents, that is to the average internet user (W1). In other words, according to *Time*, the person of the year in 2006 was the everyday internet user who had achieved an unprecedented level of community and cooperation between people; the citizen of the "new digital democracy" who had created

5 Translated from Hungarian and annotated by Ádám Makkai, edited by Enikő Jankó. https://djjr-courses.wdfiles.com/local—files/soc180 %3Akarinthy-chain-links/Karinthy-Chain-Links_1929.pdf [16. 04. 2015]

things like the "cosmic knowledge base" of Wikipedia, the multi-million channel video sharing website YouTube, the (in the meantime outdated) online metropolis MySpace, or the social media site Facebook.

Image 1: Time magazine's Person of the Year in 2006

The new global communication also produced a new elite: the *netocracy* as termed by the American tech magazine *Wired* in 1990. The term, which is a blend of *internet* and *aristocracy*, refers to the global upper-class whose power is based on its technological advantage and networking skills. (The notion was used by Swedish philosophers Alexander Bard and Jan Söderqvist [2002] in their book *Netocracy – The New Power Elite and Life After Capitalism.*)

1.4 On the verges of orality and literacy

The examination of the language of new mediums once again raised the issues of language varieties, literacy and orality. There are four approaches to the dichotomy of written and spoken language, which were summarized by Gáspár Edgar Onca (2006: 1–7) as follows (the well-known representatives of the given approach are indicated in brackets): 1. written language is simply the graphic representation of language (see: Aristotle, Plato, Bloomfield); 2. written language serves as a model for language (see: Leibniz); 3. the primacy of spoken language over the written word (see: Ferdinand de Saussure);[6] 4. spoken and written language have completely different concepts (see: Koch and Oesterreicher).

6 The secondary, derivative role of written language in relation to spoken language is explained by Biber (1988: 6) by quoting among others Sapir (1921: 19–20): "writing is »visual speech symbolism«", Bloomfield (1933: 21): "writing is not language, but merely a way of recording language by visible marks", Fillmore (1981: 153): "written communication is 'derivative of the face-to-face conversational norm"; Aronoff (1985: 28): "the undoubtedly correct observation that spoken language is »true« language, while written language is an artifact".

In *Phaedrus* Plato compares writing to painting. He states that „the painter's products stand before us as though they were alive: but if you question them, they maintain a most majestic silence. It is the same with written words: they seem to talk to you as though they were intelligent, but if you ask them anything about what they say, from a desire to be instructed, they go on telling you just the same thing for ever" (Plato 1925;[7] cf. Thaler 2003: 8–9).

This almost 2500-year-old argument seems to be challenged, if not overruled, by the communication technology of the 21st century which makes it possible for physically remote people to communicate in writing directly and interactively (continuously taking turns).

The *dichotomy of literacy and orality* has always been a debated issue in language philosophy: the Great Divide theory holds that writing and speech are two rigidly separated entities which are in binary opposition (e. g. Knoop 1983; Günther 1988: 17). According to a more widely accepted view, the two entities form a continuum with many in-between categories[8] (cf. Benczik 2001). Spoken and written language characteristics are specifically combined in internet communication where users simultaneously exchange messages in writing but as if they were having an oral conversation (Gósy 1999).

Lakoff (1982) advocates the traditional binary opposition when he enlists the characteristic features of written and spoken communication, but he was ahead of his time when he observed that the purest forms must be the two end-points of a continuum. The two communication types use *different strategies*: while spoken language is more spontaneous and direct, written language is rather planned, organized and non-spontaneous. Lakoff argued that in the society of the late 20th century ideal human communication started to shift from a literacy-based model to one based on oral discourse (1982: 240). He saw the reason for this in the benefits and drawbacks of literacy and orality: spontaneous discourse is characterized by immediacy, emotional directness but also the lack of clarity, slip of the tongue, hesitation and repetition. By contrast, planned discourse avoids these pitfalls but it also lacks the warmth, closeness and vividness of oral conversations (1982: 242).

---

7 The full text: „Socrates: Writing, Phaedrus, has this strange quality, and is very like painting; for the creatures of painting stand like living beings, but if one asks them a question, they preserve a solemn silence. And so it is with written words; you might think they spoke as if they had intelligence, but if you question them, wishing to know about their sayings, they always say only one and the same thing" (Plato 1925. Plato in Twelve Volumes. Vol. 9. Translated by Harold N. Fowler. Cambridge, MA: Harvard University Press; London: William Heinemann Ltd.: 275d).

8 Biber (1988) tried to override the continuum approach with quantitative research: he compared different dimensions on a large corpus of English texts. The text types were categorized by the various co-occurrences of the dimensions.

Lakoff pointed out that the shift, or more appropriately cultural revolution, which can be traced back to both social and technical reasons, had already started (1982: 243). The shift is clearly indicated by the fact that planned communication is no longer considered ideal. Lakoff (1982: 243–245) illustrates this turn the following way: up until the middle of the 20th century dialogues in literary works had been characterized by written language as the authors did not strive to imitate spoken language; from the middle of the 20th century, however, more and more dialogues resembled stylized orality, real dialogues, the representation of spontaneous conversations: novels, screenplays and dramas were no longer free from mistakes, hesitations, slips of the tongue and other types of speech error. Naturally, this shift is only catalyzed by the emergence of "new" (in 1982 primarily visual and audiovisual) technology as it questions the information storage and transmission function of writing.

A different, more diachronic approach is reflected in how Walter J. Ong divided *communication history* into periods. In Ong's system the pre-literate period was characterized by *primary orality* (Ong 1982: 136); the emergence of writing marked the beginning of the period of *primary literacy*; and the period of *secondary literacy* started with the spread of voice transmitting and recording devices. "It was the development of sound recording that made it possible to record, reproduce without any alteration, analyze and reflect on spoken language" (Benczik 2002: 103).

Based on the works of the philosopher Kristóf Nyíri, the Hungarian linguist Géza Balázs (2005a: 39) proposed the term *secondary literacy* for the language of digital communication which fits to the system of Walter J. Ong. The term suggests that although they are written in form, digital texts primarily show the characteristics of spoken language, which is true. Although it is not acoustic but visual and is not completed by body language,[9] it is similar to spoken language in that it demands fast thinking and instant verbalization (response); it is also less disciplined, less arranged and more intuitive than traditional written text, and it deviates more from writing norms. In the framework of literacy internet discourse "follow dominantly typographical patterns, so it is represented in the form of written texts" (Benczik 2002: 105)—it converges to the rules of spoken language. Another term related to this type of text production is *new spoken language* which is primarily characterized by transitionality: "a form of writing and communication [...], speech synthesis which converges to but is not identical with spontaneous spoken language" (Balázs 2005a: 40). In other words, the digital revolution made it possible to connect the distance-bridging property of writing with interactivity which is traditionally attributed to spoken language. Among others, this is what has loosened the (long-disputed) dichotomy of literacy and orality.

9 For this function emoticons are used (with certain limitations). See Chapter 3.5 for more details.

The most important characteristics of orality and literacy are summarized in Table 1 below. "Spoken language and written language as such do not exist, they are only the manifestation of different communication practices", says Dürscheid (2006: 25) on the basis of Fiehler (2000: 100). The table is intended to present the basic differences between the two forms but please note that such characteristics are only valid to prototypical cases (for detailed explanation and exceptions see: Dürscheid 2006: 27–34). It also limits the scope of the table that the pairs of characteristics are only valid to the situation of our modern age and to alphabetic writing systems. The traditional dichotomy (which in this case only contains the "pure cases", that is the two end-points of the scale) was completed in the table with the characteristics of secondary literacy (written-spoken language).

The sharp binary opposition—on the basis of Koch and Oesterreicher (1985 and 1994)—is often mitigated with the categories of conceptual and medial literacy and orality (Table 2).[10] Koch and Oesterreicher found that the terms "spoken" and "written" have at least two meanings, so they distinguished them by medium and concept and created the above-mentioned four categories. The distinction by medium simply refers to the medium of communication, that is whether the text is produced phonetically or graphically (1994: 587). The distinction is dichotomous.[11] By contrast, the distinction by concept refers to the modality, the way of expression of the utterance, and is expressed on a scale. The literacy-pole is characterized by temporal and spatial distance, publicity, unknown partner, lack of emotions, independence from situation and action, few reference to the origo, few opportunities for the recipient to cooperate, monologic composition, planned character and fix topic parameter values (Koch–Oesterreicher 1994: 588; cf. Onea 2006: 5); the orality-pole is characterized by the opposite of these values. The two types of orality and literacy are not strongly correlated (Dürscheid 2006: 44), there are special cross-cases as well (e. g. medially written but conceptually spoken texts).

While before the appearance of internet text production (primarily instant messaging) conceptual orality was definitely linked to medial orality (e. g. records/minutes: speeches were recorded in writing; screenplays: the text to be performed orally or read out loud was written down), with online instant messaging this relationship has disappeared: conceptual orality can exist without being preceded or followed in time by medial orality. Instant messaging is the first in the history of communication to use writing for simultaneous, direct and

10 Christa Dürscheid (2016) criticizes the long term reception of the Koch–Oesterreicher-model from a media-linguistic point of view.

11 Sign language which is neither oral nor written is disregarded by the model. For further information about the deficiencies of the model, potential ambiguity and the changed communication conditions see: Dürscheid 2006: 50–53.

situation-bound verbal communication (Storrer 2001). In the Post-Gutenberg Galaxy online writing and reading return to the original state: orality is revived and communication is nearly as fast as thinking (Harnad 2006a: 16).

Table 1: Characteristic features of written and written-spoken language
(Sources of the aspects applied in the table: Dürscheid 2006: 25–27; Lakoff 1982; Knoop 1983; Günther 1988: 11; Hoffmann 1998: 6; McCarthy 1998: 78; Krauthamer 1999: 2; Biber 1988: 5, Chafe–Danielewitz 1987: 2)

spoken form	written form	written-spoken form
not restricted to instrument	restricted to instrument	restricted to instrument
restricted in space and time	less restricted in space and time	less restricted in space and time
sender and receiver in the same place (co-presence)	sender and receiver normally in different places	sender and receiver may be in different places
synchronous	asynchronous	quasi-synchronous/asynchronous
momentary	recorded, material, more stable and lasting (may be used later in the same form)	momentary but recordable (through saving, printing, multiplication)
acoustic	visual	visual
phoneme	letter	letter
accompanied by: body language, such as facial expressions, gestures etc.	accompanied by: visual (typographical) elements, such as segmentation, layout	accompanied by: visual (typographical) elements, such as emoticons
characterized by suprasegmental features	indirectly characterized by suprasegmental features	indirectly characterized by suprasegmental features (emoticons)
continuous, voice continuity	comprises sequential, discrete elements	sequential
temporal	spatial	temporal and spatial
deixis	no deixis, rather: anaphors and cataphors	deixis is possible
implicit	explicit	more implicit
specific	abstract	more specific
quick thinking, instant verbalization	time to think	relatively short time to think (see: IM)

Table 1: (continued)

spoken form	written form	written-spoken form
less strict, less developed structure	more logical, decent and elaborate structure	logical but less strict structure which is closer to orality
more intuitive, less regard for norms	more conscious, more regard for norms	less regard for written language norms
several types of feedback	limited or delayed feedback	text-dependent feedback: e-mails: delayed but quick; IM: instant
instant correction, completion, repetition	no repetitions	delayed repetitions
innovative	conservative	innovative
mainly dialogical	monological	more dialogical

Table 2: Conceptual and medial literacy and orality
(Source: complemented and altered by Ágnes Veszelszki on the basis of Koch–Oesterreicher 1994: 588; Dürscheid 1999: 17–30, 2006: 45)

		conceptual written ←——→ **conceptual oral**
medial written	**non-electronic**	legal texts; scientific paper; postcard; minutes; screenplay
	electronic	official e-mail; forum; private e-mail; message wall post; tweet; IM
medial oral		scientific pres.; sermon; radio interview; talk show; informal conversation

Though earlier she often referred to the concept of Koch and Oesterreicher, in her recent works Dürscheid (2016) calls it outdated and over-referenced.

Kilian[12] (2001: 76) thinks that the written form of spoken vernacular is only language used creatively by part of a speech community. As a particularly large proportion of internet users are (were) university students, states Kilian, this is especially true to the educated class of society which knows how to distinguish between conceptually oral but medially written language from medial-conceptual written language. By the 2010s, this statement seems a bit obsolete: on the one hand, the group of mobile phone and internet users is no longer restricted to university

12 The author's general remarks in the study are primarily related to the German language and society but they can be generalized in view of the entire study.

students but includes younger and older generations as well; on the other hand, it can be assumed that a generation has grown up which is unable (unwilling) to distinguish between conceptually oral but medially written and "traditional" literacy.

As a conclusion, we may point out the following: Electronic texts can be considered *hybrid texts* from several aspects (similarly: Beutner 2002: 105; Ferrara–Brunner–Whittemore 1991: 10; Frehner 2008: 26–27). Using the term "hybrid" seems even more appropriate if we consider that electronic texts are typically multimedial ones, so they may incorporate images, videos or sounds (by contrast, conventional written texts are based on verbal and iconographical elements). Electronic written texts are very often based on oral concepts so they exhibit the conceptual features of orality; in fact, they may be considered as the written mimesis of conceptual orality (Kilian 2001: 69) on the grounds that electronic media are characterized by interactivity. This duality is reflected in the term *oraliterality* (German: Oraliteralität; Döring 1997: 290). As opposed to written communication in the traditional sense, which builds on the dissociation (i. e. the spatial and temporal distance) of writer, text and reader,[13] electronic communication often features synchronous forms[14] (temporal proximity), and sometimes (though less often) even spatial proximity.[15] Furthermore, Eckkramer and Eder (2000: 269, 272) consider electronic texts as hybrid forms because they are generated by combining different conventions.

1.5 Methodology

Possible methods for examining electronic communication and its impact are summarized in *Table 3*.

Eckkramer and Eder (2000: 21–22) developed a special method known as "contrastive cyber-textlinguistics" (kontrastive Cybertextlinguistik). This area of linguistics is concerned with the texts of a society which communicates both through traditional printed (so-called typographical) media and through computers. It covers every dimension of the interaction between different medium spaces and also involves practical and literary texts into the analysis. The authors examined and compared among others dating ads, recipes and job advertisements in English, German, Spanish and French to document the changes of conventions associated with different text types. I also rely on this methodology in my examinations in the field of internet linguistics.

---

13 A European book from the 16th century, for example, can be studied in Asia in the 21st century.
14 For example: instant messaging.
15 It is not at all unusual that people in the same room exchange texts or instant messages if other forms of communication is not possible at the moment.

Table 3: Methods of examining internet communication and the effects of internet on non-digital texts
(Source: Köhler 1999: 19, modified by Ágnes Veszelszki)

Method type	Method group	Methods
reactive	polling	electronic survey sent in e-mail, online or paper-based survey...
	interview	F2F interview; interview in e-mail or a virtual world, through instant messaging or phone
	experiment	psychological, psycholinguistic experiment...
semi-reactive	observation of participants	observation of group dynamics as participant, participation in chat rooms or internet forums...
non-reactive	observation	observation of chat rooms, internet forums, blogs, social networking websites...
	content analysis	analysis of URL structures, hyperlinks, social networking website dynamics...
	text analysis	(qualitative and quantitative) analysis of online texts, examination/comparison of non-electronic texts...

Three main methods were applied for this book. First (though it is not a real method but the basic principle of scientific work) I looked through the relevant literature in Hungarian, German and English to present the theoretical background of the topic, taking into consideration and, where necessary, giving a detailed summary of empirical research findings.

Second, I performed text analyses: my own collection of texts from the field of digital communication (mainly: instant and text messages, blog and message wall posts), handwritten texts (mainly: school notes, test papers, notes passed in class), and, to a lesser extent, traditional printed texts, more specifically advertisements.

Third, to complete the other two methods I also conducted paper-based and online surveys.

By combining the three methods, I tried to describe, as fully as possible, the linguistic characteristics of digital communication in the beginning of the 21st century.

1.6 Corpuses used for the analysis

The analysis of electronic texts was carried out on instant and text messages, blog and forum posts, and messages shared on the message wall of Facebook and the Hungarian social networking website iwiw. Most message wall posts and comments were also saved as image.[16] The corpus of texts collected from various sources runs to several hundreds of pages (as the corpus is continuously expanding and the texts are stored in different formats, its size cannot be determined more exactly). The share of text messages is remarkable: there are nearly 10,000 of them in the corpus, all of which were collected between 2000 and 2012. The case studies included in the book (concerning iwiw message wall posts, netologisms, the topic-repeating pronoun, and the grammaticalization of the expression *asszem*) are all based on independent corpuses suitable for quantitative analysis.

The analysis of handwritten texts was performed on the exercise book notes prepared by the students of Bibó István Grammar School in Kiskunhalas, Hungary (the high school I graduated from) between 2007 and 2009. The corpus consisted of altogether 2409 pages (115 exercise books and 45 final examination notes). This corpus was completed with the lecture and seminar notes of university students studying at Eötvös Loránd University, Budapest, which also accounted for hundreds of pages.

In the exercise books of high school students I found 29 pages of private letters, typically dialogue letters sent secretly between students in class. This corpus of private letters was later completed with 18 other dialogue letters.

In addition to the notes, test papers and letters, I also examined advertisements in printed press and on posters/billboards in search for elements of the digilect. Currently, the corpus contains 76 images of ads featuring some kind of digilectic characteristics but the collection is continuously growing.

One of the most difficult challenges of the research was to build a spoken language corpus. In this case, no actual systematic corpus building could be carried out as it would have required the collection and processing of an enormous volume of audio files recorded between 2010 and 2011, which would have clearly gone beyond the framework of the book. For this reason I had to rely on my own observations, as well as on those supplied to me by my colleagues and friends between 2008 and 2011. Furthermore, the speech corpus[17] was also extended with the relevant observations of the students of Eötvös Loránd University, recorded in linguistic diaries between 2010 and 2011 (naturally, the diaries were only used if their author had consented to the scientific utilisation of the information).

16 The images are available in the *Annex*.

17 The spoken language corpus can be found in the *Annex 10* of the book.

2 Digilect as Language Variety

2.1 Introducing the concept of digilect

German authors dealing with this topic more or less agree that communication on the web has special, genre-independent linguistic features (see for example: Androutsopoulos–Ziegler 2003: 252; Dürscheid 2005: 94). This observation is not only true to German but to the English language as well. Christina Haas and Pamela Takayoshi (et al. 2011), researchers of Kent State University, argue that online chat does not (only) mean substandard language use or masses of mistakes but it is rather an independent variety of formal English which operates with special linguistic tools. The researchers examined the chat messages of university students in search of characteristics which deviate from standard written English. They concluded that nonstandard written linguistic features are customary characteristics of the language of online chatting. The language of instant messaging is informal, explicit and playful, it uses both abbreviated and extended forms, and it prefers meaning to form and social relationships to content. The researchers also stressed that communication technologies used by young people (such as chat programmes, blogs and social networking sites) are primarily based on writing (typing). Even mobile phones can be used for writing (mainly text messages, but instant messaging is also possible). The research team plans to extend its research to texts posted on Facebook to find out whether the popular social networking website share any common characteristic with online chatting in terms of language usage.

In his 2001 book David Crystal, one of the first monographers in the field of internet linguistics, termed this language variety as *netspeak*. Later, in his book published in 2008 (Crystal 2008: 13), he lists a couple of other terms for the digital language variety, including textese, slanguage, new hi-tech lingo and hybrid shorthand. He goes so far as to call people who speak this language variety as bilingual. Schlobinski (2000, cited by: Frehner 2008: 27) opposes this view saying that internet text types are so heterogeneous and they integrate the features of so many text types that it is impossible to bring them all under a single umbrella like *netspeak*. In my view, this heterogeneity and integrative nature is exactly what characterizes digital communication. Heterogeneity is also highlighted by other authors, which is reflected in the terms coined by them:

Ferrara, Brunner and Whittemore (1991) use the expression *written interactive register*, and they consider the increasingly dominant presence of dialogical elements in this new register as one of its most important characteristics.

DOI 10.1515/9783110499117-002

Eckkramer and Eder (2000: 22, 266) argue that the shift of medium also brings about linguistic change and results in a new, digital language register (2000: 273). For this phenomenon they introduced the notions *virtual* or *digital textuality* (German: virtuelle Textualität, digitale Textualität).

As far as the relevant Hungarian literature is concerned, Géza Balázs (2004, 2005a) uses the term *secondary literacy* in several of his works (cf. Dittmann 2001: 10). Furthermore, this new linguistic mode of existence is termed *symbolic literacy* (2005) and *new spoken language* (2004a, 2004b) by Zoltán Bódi, and *virtual literacy* (2003) by Nikoletta Érsok.

Digilect[18] (Hungarian: digilektus) is my own term for referring to the language use of computer-mediated communication (CMC) in a wider sense. Digilect is a new language variety with specific features that are not characteristic to communication transmitted by other mediums. Point 2.1.1 below discusses the evolution of the concept of digilect and its (hypothetical) status in linguistics.

2.1.1 The system of lects (language varieties)

It is a commonly accepted principle in sociolinguistics that language exists in varieties (Kiss 2002: 74; Wardhaugh 1995: 49, 102). There is no widespread consensus, however, on the definition and classification of these varieties.

Different authors provide different definitions for the term *lect*, which is believed to have been coined by Charles-James Bailey. Some define it as a language variety with specific phonetic, vocabulary and structural characteristics, "whose speakers mutually and easily understand each other" (Kálmán–Trón 2005: 29). Peter Trudgill gives a somewhat broader definition: "it is a neutral expression which may refer to any variety of language—dialect, phonetic variant, sociolect, stylistic variant or register—which linguists wish to discuss as a separate entity" (Trudgill 1997: 63).

According to the classification of Árpád Sebestyén (1988: 117–118), the (Hungarian) language is divided into three major layers: normative varieties (literary and standard language), regional varieties (vernacular language, including various dialects) and social varieties (including jargon: technical and hobby terminology, age-related variants and argot). According to Imre Wacha (1992: 89–90) varieties can be classified from a demand-related, a regional-geographical, an occupation-related and a social layer related aspect. Ronald Wardhaugh (1995: 115–130) distinguishes between regional and social varieties. Peter Trudgill (1997) not only mentions social (sociolects) and regional (dialects) varieties, but also adds the category genderlect (i. e. the gender-specific language variety). Trudgill also introduces the theory of the social dialect continuum (with the basic concepts of basilect, mesolect and acrolect). In his book on sociolinguistics, Jenő Kiss (2002: 74) deals with the following three main categories: standard (subcategories: spoken and written standard registers), social varieties (group-specific and technical languages), and regional variants (dialects). László Kálmán and Viktor Trón (2005: 30) distinguish between ethnic, regional and social language varieties, the latter including *idiolect* (an individual's distinctive use of language).

Janusz Bańczerowski (2000 and 2008: 20) uses similar classification, although his terminology differs from the previously mentioned ones. Bańczerowski also considers idiolect as the language of a specific individual, but he terms the language of any community or group of people as *common/collective polilect*. The polilect of an ethnic community is an *ethnolect* which has specific type-variants in his system: natiolect (the national language); sociolect (the polilect of groups, social layers/classes, viewed from a social perspective) and dialect (certain variants of specific ethnolects, viewed from a regional perspective). *Table 4* below summarizes the various types of classifications:

Table 4: Classifications of lects

Author [year of publication (in Hungarian)]	Variants
Árpád Sebestyén (1988)	normative varieties (literary and standard language) regional varieties (vernacular language, including various dialects) social varieties (including jargon: technical and hobby terminology, age-related variants and argot).
Imre Wacha (1992)	from a demand-based aspect: literary language, standard, vernacular from a regional-geographical aspect: dialects from an occupation-related aspect: technical languages from a social layer related aspect: layer and group specific languages
Ronald Wardhaugh (1995)	regional social
Peter Trudgill (1997)	sociolect = social language variety dialect = regional language variety genderlect = gender-specific language variety
Jenő Kiss (2002)	standard (written and spoken) social (group-specific and technical languages) regional variants (dialects)
László Kálmán and Viktor Trón (2005, 2007)	ethnic territorial social + idiolect

Table 4: (Continued)

Author [year of publication (in Hungarian)]	Variants
Janusz Bańczerowski (2000, 2008)	idiolect (a specific individual's language) common/collective polilect (the language of any community or group of people) – ethnolect (the polilect of an ethnic community) – natiolect (national language, a special category of ethnolects) – sociolect (the polilect of groups, social layers/classes, viewed from a social perspective) – dialect (certain variants of specific ethnolects, viewed from a regional perspective).

2.1.2 The status of digilect in linguistics

Digilect is certainly a language variety, if we take (for example) Trudgill's definition: "it is a neutral expression which may refer to any variety of language—dialect, phonetic variant, sociolect, stylistic variant or register—which linguists wish to discuss as a separate entity" (Trudgill 1997: 63). It is not so obvious, however, where to place it in the system of language varieties: depending on the approach we take, it can be considered as a sociolect, a mediolect or even a stylistic layer (register).

2.1.2.1 Digilect as sociolect

Sociolect is a language variety which may be linked to a certain social group. Digilect may be regarded as some kind of sociolect as it is primarily used by younger generations—though not exclusively, so it cannot be equated with school slang. We cannot attribute digilect to "computer users" as a specific social group either, as this category would be too broad.

This language variety has more to do with the medium of communication, i. e. its occurrence is determined by the device on which it is used: most language users opt for the expressions of digilect when writing on a computer. From this perspective, digilect can be taken as a mediolect, that is a lect characterized by the medium.

2.1.2.2 Löffler's mediolect and the digilect

Except for the system developed by Jenő Kiss in 2002, all the models described above ignore the medium of communication. This aspect, however, is taken into account in a German sociolinguistic book published more than 30 years ago in 1985: Heinrich Löffler designed his model to describe linguistic reality (German: Sprachwirklichkeitsmodell). The model looks at language varieties from six main aspects. Idiolect, which refers to an individual's use of language, represents an independent category. In addition to the traditional distinction between areal (dialect) and social (sociolect) varieties, Löffler categorized language varieties by function (functiolect), situation, style and text types (situolect), sex (sexolect) and age, as well as medium (mediolect). According to Löffler, these categories are governed by extra-linguistic aspects but the specific lects are distinguished from each other by linguistic characteristics (Löffler 1985: 87–175; *Figure 1*).

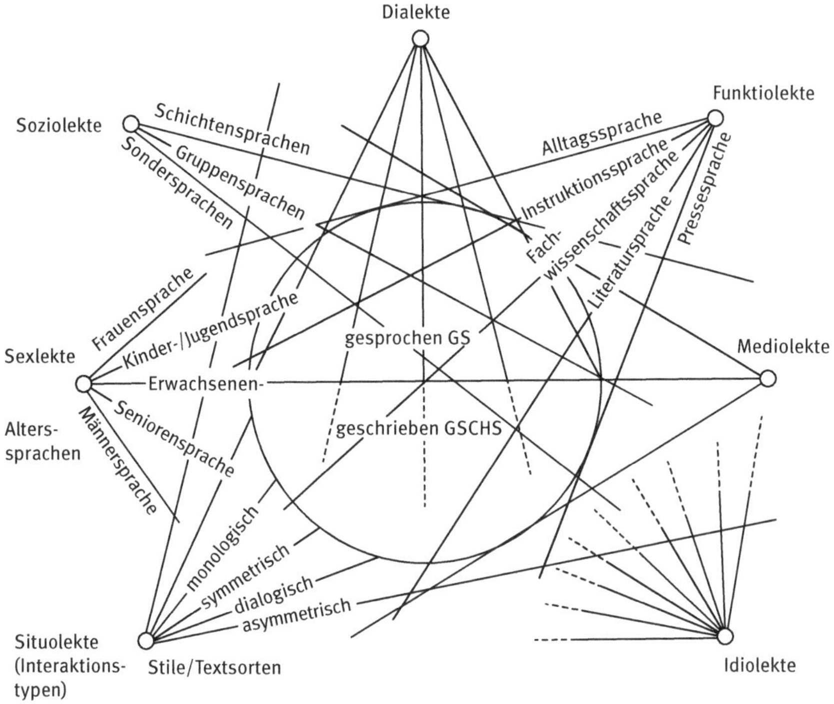

Figure 1: Löffler's sociolinguistic model
(Source: Löffler 1985: 87)

Even this model would offer a suitable place for a new type of language variety whose characteristics are specific to digital communication, i. e. for the digilect. The question *May digilect be considered as a lect?* is answered by Trudgill's already cited definition--that is any category of language variety established for analytical purposes can be considered as a lect. David Crystal (2008: 179) uses the term *electronic discourse* for the language variety used in computer-mediated communication, particularly in e-mails, online chatting and text messages.

Due to its wide availability and usage, the internet has an impact on language use. New digital devices have been created (mobile phones, computers with internet access, and their combination: the smartphones) which not only influence the form of messages but also their other characteristics, including linguistic ones.

Recent cognitive schools teach that linguistic signs are not based on objective reality but on human experience. Reality is continuous so its linguistic representation also has to be so. As the linguistic categorization of phenomena in our world is also based on human experience, subjective borderlines become blurred. According to the advocates of cognitive semantics, it is primarily the prototype and stereotype theory that makes it possible to build linguistic categories (Bańczerowski 2008: 7–13). Although language is one and undivided (Bańczerowski 2008: 88), we can establish different categories to process and analyze it even if we know that these categories cannot be strictly separated from each other.

Löffler's system operates with strong dichotomies which we may find reasonable to relax a bit. Obviously, there are no such strict borderlines between the categories in Figure 2, still they can be used as a guideline. Digilect, therefore, can be considered as a type of mediolect (medium specific language variety).

In relation to Figure 2, we should reiterate the already described theory of Koch and Oesterreicher (1994): the matrix comprised by conceptual and medial literacy and orality.

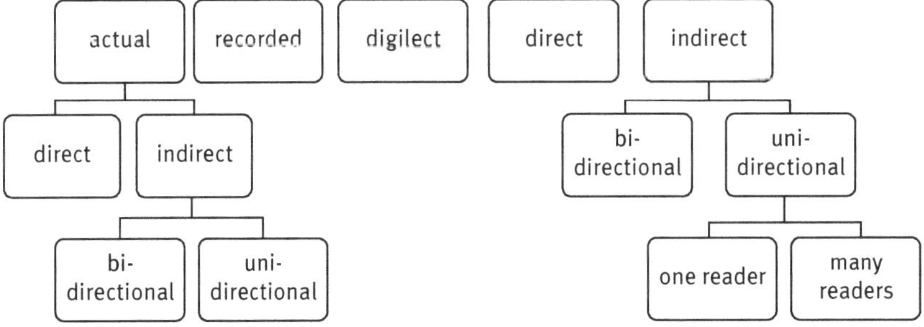

Figure 2: Löffler's classification of mediolect, built on strict dichotomies
(Source: based on Löffler 1985: 91 and 98, complemented with the category of digilect)

2.1.2.3 Digilect as a stylistic layer (register)

As a further option, digilect may be regarded as a new, special stylistic layer as well. Both Trudgill and Wardhaugh define stylistic layers (registers) as language varieties which are not based on social or regional features (Trudgill 1997: 63; Wardhaugh 1995: 48).

If we add to this István Szathmári's definition of stylistic layers—"a constituent system of language"; "functionally different modes of communication whose historic evolution was influenced by social and cultural circumstances" (Szathmári 2004: 198)—, the digilect could also be regarded as a stylistic layer.

2.1.3 Again on the status of digilect in linguistics

Digilect, however, has merely twenty or twenty-five years of history to look back on. Since it is technically determined, it quickly responds to the development of information technology, therefore it is also *constantly changing*. Let me illustrate this quick adaptation/modification through two examples: while between 2000–2005 text messages sent from mobile phones had the biggest impact on language use, by 2015 this technical device and therefore the communication genre itself have lost a lot of their significance. In Hungary the social networking website Facebook became widely used in 2009 which today strongly influences changes in language, more specifically in the vocabulary (consider for instance the new Hungarian word *lájkol* 'to like' and its family of words).

In the light of the above considerations, the status of digilect in linguistics may be resolved in the following ways: it can be considered as a new stylistic layer (register); we can create the category of *sociomediolect*, a new provisional category taking place somewhere between sociolect and mediolect; or—as the most feasible option—due to its novelty, continuous change and instability, we refrain from categorising it for the time being and we only accept its occurrence and record its effects.

2.2 Genres and text types of digilect

How can we draw a line between digilect text types and other types of text? For this we have to use the term *tertiary medium* as described by several authors (cf. Frehner 2008: 29; Dittmann 2001: 29; Schmitz 1995: 9). According to the technical criteria of communication, we can differentiate between primary, secondary and tertiary media (which apply to texts as well). Primary media means that no technical means are required to communicate, as in the case of a personal, face-to-face conversation. Secondary media require a technical means on the side of production but not on the side of reception, while tertiary media demand technical means on both sides. The text types of digital, computer-mediated communication (CMC) are evidently linked to tertiary media. This approach is also supported by the fact that "the examination of linguistic communication and information processes are impossible without taking the specific medial environment into consideration" (Ziegler 2002: 11).

The most typical text types (genres)[19] of digilect are the following: e-mails, posts and comments on internet forums, blog and vlog posts, tweets, online chat texts, posts and related comments on the message wall of social networking websites. As the detailed description of these text types is beside the point of this book, only the most important characteristics of them will be presented later below. Internet-based text genres are embedded to each other, their relationship is characterized by fluidity, and it is difficult to make clear distinctions between them.

Gábor Tolcsvai Nagy (2001: 331–333)—relying heavily on the system of Barbara Sandig (1972)—summarized the aspects of text typologies using a formal system of criteria as follows: spoken/written, monological/dialogical, spontaneous/planned, and traditional/non-traditional. As opposed to static models, Gülich developed a model which adopts a dynamic approach (Tolcsvai Nagy 2001: 334). According to Gülich, the purpose of classifying text types is the following: "differentiating between text types is important for the participants in communication, and [...] knowledge about the characteristics of different text types form part of the participants' every day knowledge" (Gülich 1986: 18). According to this dynamic model which reflects a more cognitive approach, text types are not part of objective reality but are "constructed during communicative interactions" (Gülich 1986: 189). Thus, according to the prototype theory, participants in communication continuously compare their ideas about the given text type to the prototype and try to place them in the network, therefore participants constantly modify their image of text types and modify their linguistic acts accordingly (Tolcsvai Nagy 2001: 335). In the light of this, Tolcsvai Nagy (2001: 336–338) suggests the following examination aspects: extent of explicitness/inclusion, markedness of point of view, characteristics of communication scenes, typical features of the semantic structure of

19 The terms *text type* and *text genre* are usually used interchangeably, as overlapping concepts in the relevant literature, which tradition is followed in this book as well (cf. Ziegler 2002: 10). The subdiscipline of Textsortenlinguistik (in German for text type linguistics) and the usability of internet-based texts in linguistic analyses are covered in detail by Ziegler (2002).

the text, subfeatures of the general semantic structure of the text typical to the text type, and style.

As it is outside the scope of this book to provide the typology of text types, only the points shared by the above-mentioned two theories will be borrowed to describe the text types used in digital communication. The static model will provide us with a dichotomous approach, while the dynamic model will give us a critical view on the "methodological self-assuredness" of the static model (Tolcsvai Nagy 2001: 336). According to the prototype theory, text types cannot be clearly assigned to any features so it is better to view them on a scale between the two extremities.

Text types are sorted by five main aspects: asynchronous/synchronous, spontaneous/planned, restricted/unrestricted length, private/public, anonymous/not anonymous. Asynchronous forms of communication do not allow for direct interaction (though they show traces of synchronicity, i. e. the potential for coming closer to real-time communication; cf. Frehner 2008: 16), so it is reasonable to call them *quasi-synchronous* as well. Spontaneity, that is formulating messages in an ad-hoc manner, is related to constant presence and availability (cf. Frehner 2008: 30). According to observations, the maximum allowed number of characters clearly influences the formulation of texts: such length constraints force/inspire users to use abbreviations or reformulate their message.[20] Text types with unrestricted length are more narrative than those with fixed length (Frehner 2008: 31). The last two pairs of aspect are interrelated but they do not fully overlap with each other. The public/private distinction refers to the number of participants in communication: public texts are accessible practically to anyone, while private forms of communication have a restricted number of recipients. This dimension influences the topic, style and formal characteristics of messages. Anonymity is also a relative category: no electronic text type assures complete anonymity because they all assume at least a nickname (or can be traced back by IP address). By contrast, in the case of non-anonymous texts, authors not only disclose their true name but some of their personal data as well.

20 This is supported by the findings of the internet survey detailed in Chapter 4.2. Respondents stated that they adjust the form of their message to the maximum length allowed: "*I consciously avoid abbreviations, it's rather that I formulate my message in a more compact and straightforward way*"; "*I rarely use abbreviations, I prefer finding shorter synonymous words or reformulate my message*"; "*I never use abbreviations in text messages, I would sooner reformulate my message if it's too long*".

Naturally, the categories may overlap with each other. *Table 5* only presents prototypical cases.

Table 5: Digilect genres sorted by their characteristics in a continuum

asynchronous	e-mail, sms, blog, tweet, post-comment, chat	*synchronous*
planned	blog, e-mail, sms, forum, post-comment, tweet, chat	*spontaneous*
unrestricted length	chat, blog, e-mail, forum, post-comment, tweet, sms	*restricted length*
public	blog, forum, tweet, post-comment, chat, sms	*private*
anonymous	forum, blog, tweet, post-comment, chat, sms, e-mail	*non anonymous*

2.2.1 E-mail

Electronic mail or e-mail is a much faster way to convey messages than traditional ("snail") mail, and it also allows for sending attachments (including images, formatted documents, audio and video files etc.).

The formality of e-mails is mainly determined by the offline (i. e. not internet-based) relationship between the sender and the recipient. Accordingly, we can differentiate between private (informal) and official (formal) e-mails. Private e-mails, on the one hand, are characterized by conversational, indirect style and breaches of the standard rules of letter composition; official e-mails, on the other hand, pay more attention to the standard (see: Érsok 2007: 36–37; where Érsok establishes the interim category of peer e-mails). The e-mail address itself plays in important role in distinguishing between formal and informal e-mails: official letters are not recommended to be sent from private address (i. e. address that contains an indecipherable name or a nickname) The rules governing official e-mails are almost similar to those applying to traditional paper-based letters, including letter structure (missing subject is a breach against internet manners, the so-called netiquette), formal style and handling (e-mails must be filed as well).

Based on the number of addressees, we can differentiate between personal/ unique e-mails and e-mails sent to multiple addressees. Logically, the former is sent to a single and the latter to several recipients. E-mails can be sent to mailing lists as well.

Composition by inserting comments is also a characteristic tool of e-mails and particularly *mail cycles* (where the same partners exchange multiple mails in the same topic for several days). When this type of composition technique is used, the addressee does not write its reply above or below the sender's email but insert it directly into the email body (sometimes only those parts of the original e-mail are left which need to be answered). To make reading easier different font formats (normal, boldface, italics, bold italics) or colours are used. This technique is also used by Stevan Harnad (2006b) in his Scholarly Skywriting. This term refers to a scientific method which combines topic-threaded web archives (such as forums or newsgroups) and e-mails with multiple addressees. Texts relating to the same topic can be read ("skyreading") and commented ("skywriting") by all online participants using, in particular, the quote/commentary tool which enables users to copy part of the text and add a comment to it. Quotations and comments can be multiply embedded, representing a unique form of hyperlinking. Harnad believes that this technique combines the advantages of oral and written communication in the digital age: it merges the synchronous, interactive character of orality with the permanence of writing in a new form: commenters can enter into a dialogue with the text.

A thorough and exhaustive analysis of emails is available in the book edited by Ziegler and Dürscheid (Ziegler–Dürscheid eds. 2002).

2.2.2 Internet forum

An internet forum is an asynchronous, written form of group communication in which comments center on a given topic. Forums are used by people with similar interests—who are not necessarily present at the same time in the forum—posting comments to a central topic, typically under nicknames. Forum comments are displayed in reverse chronological order (the most recent one appears first) and they can be traced back.

As Érsok (2007: 58) observed, internet forums—though they share certain linguistic characteristics with spoken language—feature carefully composed texts because the asynchronous nature of the forum relieves the pressure on users to be quick and they can always correct or improve their comment before posting it.

Forums are still used but their importance seems to have declined. Comments added to online news articles operate similarly: most news portals enable their readers to comment on their articles, but as opposed to forums, here comments are not only posted under nicknames but increasingly under real names too (typically by logging in with a Facebook or other social networking site profile).

2.2.3 Blog, vlog and tweet

Blogs are private or organizational websites where short entries are posted periodically. The entries (posts) represent a unique type of digital texts. The term *blog* itself was created by cutting off the first two phonemes of the expression *weblog*. Bloggers often use URLs to direct their readers to other websites, including other blogs, and they even comment on the posts of other bloggers (Bőgel 2006; Gagyi 2008).

If the blogger adds (usually homemade) videos to its blog posts, such blog is referred to as *video blog* (shorter forms: *vidlog, vlog*). Thus, *vlog* is a form of blog in which video content dominate over textual content. “We’re going from being media consumers to media makers”, said a vlogger to the *Wired* magazine (W5) about the process which runs parallel to the spreading of multimedia devices and broad band internet access. “Anyone can make videos at home, and instead of keeping them in the family archives they can share them online with millions of viewers. The quality of such videos range from home made recordings to professionally produced films.”

Blogs and vlogs cover a wide range of topics: you can find accounts and video recordings about grandpa’s birthday, about the vlogger’s everyday life and about dance performances, but you can also get cooking lessons from talented housewives. Vloggers say that this way people can deal with topics which are not covered by traditional media.

Blogs were classified by Bertalan Csala (2009) by (multi)mediality (text-only, image-only and URL collecting blog), authorship (e-media, blog-newspaper, personal blog, social blog) and theme (blog comics, blog novel, music blog, technical blog etc.). The list of blogs in the first category must be definitely completed with video blogs and blogs featuring mixed mediality. The thematic classification calls for some rearrangement as well: it would be reasonable to remove blog comics and blog novels from this category and put them into the category of mixed-genre literary blogs (together with poetic blogs for instance), and the thematic categories could be further extended, for example with the category of gastronomy-related blogs. (There are two types of gastronomy-related blogs: gastroblogs and recipe blogs. *Gastroblogs*, in my understanding, are articles published in the internet in the field of gastronomy, typically centered around a subtopic, such as cocoa rolls, tea or gastronomic experiences in a foreign country. Gastroblogs usually, but not always, post recipes as well but they are much more dominated by personal experiences and subjective commentaries. Gastroblogs are often illustrated by high quality and even artistic photos. As opposed to this, *recipe blogs*, as the name suggests, deal with recipes sorted primarily by meal. Recipe blogs are not dominated by subjective remarks, though they sometimes occur to complement the given recipe. Photos here are usually home-made, serving practical, illustrative purposes, rather than artistic or decorative purposes.)

Twitter is an online social networking and microblogging service which enables users to send and read short messages called “*tweets*” (this word, like *blog* and *vlog*, was also borrowed by Hungarian in an unchanged form).

The length of tweets are limited to 140 characters, which is even less than the length permitted by text messages, and therefore is only sufficient for making mini comments. Users' latest tweets are displayed on their profile but are also automatically shared with their followers. It is up to the user posting the tweet to determine who will see it (friends only or all subscribed followers). Unwanted messages, including spam, and annoying users can be ignored. Twitter had a peculiar German variant, the twitkrit.de, where users comment on miniature texts with almost literary quality. In Hungary Twitter is not as relevant in public communication as in the US or Germany—while North American companies and stars post their messages on Twitter, the Hungarian celebrities are using Facebook for this purpose.

2.2.4 Social media, social networking websites: posts and comments

Since its introduction in 2006, the term *web 2.0* has become a common buzzword in information science. It differs so from the so-called web 1.0 technologically structurally and socially: the former object-oriented approach is replaced by a more individual-oriented network (Peters 2009: 15). In the web 2.0 era, professional content developers (journalists, web designers and developers etc.) are no longer the only persons who can produce and share digital contents in the internet, but practically any computer user: anyone can be an author in their own way. On social networking websites—as notable examples of the idea of the web 2.0—social groups organise themselves in a new way. From the anonymous online presence of the old days, users now appear in social media with their true names and have a digital identity. Increasingly often, we may have the impression that users' online and offline identities differ from each other, or at least they do not fully overlap anymore. This phenomenon is not only related to the online construction of the ideal self, but also to the intersubjectivity factor mentioned by Brubaker and Vertesi (2010). This means that the identity created on social networking websites is not isolated but is a result of intersubjective relations: friends contribute to it with comments, likes, photos and tags (*collaborative authorship of the self*), though, users can choose what type of virtual identity building content is published on their profile. The conscious process of identity creation (sharing content which strengthens and hiding content which weakens the image of the ideal self) becomes particularly evident when the person dies (cf. Veszelszki–Parapatics 2014, 2016). Thus, though their online activity, others (partners, relatives, friends) also contribute to how a user's personality is presented. The content generated by such activity can be modified: approved, highlighted, made public or private, or deleted. Users can decide whether something has happened or not—or at least whether it has an online record or not. In the offline world, however, users do not have the option to cancel a mistake (by hitting the backspace or delete key, or pressing the cancel button). Users pursue conscious self-presentation activities: presence and visibility is important for professional reasons (as well). *Self-representation* is a form of impression management; it is an instinctive endeavour to influence—preferably unnoticeably—others' opinion about us with all available means. Quoting Röttgers' variation of Descartes' proposition: *I am seen, therefore I am* (Röttgers 2009: 91), or even: *I share, therefore I am*. Social media provides an excellent environment for this.

Social media is probably dominated by younger generations, but the share of users aged 65 or above has also been continuously increasing. Certain analysts say that, if Facebook survived and developed according to current trends, by 2060 (other estimates say 2130) more Facebook-profiles would belong to deceased people than to living ones (Condliffe 2013). According to the tendency in late 2013 and early 2014, more and more youngsters, teenagers leave (or not even register themselves on) Facebook, partly due to the increased presence of advertisements and the loss of magic (registration of older members of the community, parents, grandparents, teachers etc.) which may terminate the dominance of the website in the market (Alter 2014).

Social media has different definitions emphasising different aspects (e. g.: Kaplan–Haenlein 2010; H. Kietzmann–Hermkens 2011). Instead of discussing these definitions in detail, it is sufficient here to summarize their common points: Social media is the sum of web-based devices or websites which are built on user participation, community interaction and user generated content (images, texts, videos or multimedia). User generated content can be shared, discussed and modified by others and it can be co-created with others. Relevant keywords here are the following: participation, openness, communication, community and interconnectedness (Fehér 2012: 7); as well as "permanent online status" induced by social media, and constant availability for communication (Pólya 2012: 37; Veszelszki ed. 2013).

In fact, social media is an umbrella term which covers social networks (such as LinkedIn, Facebook or MySpace), social bookmarking services (such as Del.icio.us), social news websites (such as Digg), content share sites (such as YouTube or Flickr), wikis (the most famous being Wikipedia), virtual worlds and games (such as Second Life or World of Warcraft). Thus, social networks are only one form of social media, though certain websites and service providers cannot be placed into clear categories: attempts to classify them have resulted in many overlapping categories.

The terms *social networking sites (SNSs)* and *online social networks (OSNs)* refer to web-based services which enable individuals to create (semi)public profiles within a system and share them with users they want to keep in touch with (the nature and basis of such relationships may vary by website) (Boyd–Ellison 2007).

Wikipedia lists nearly 200 well-known and commonly used social networking sites (W1; see also: Krishnamurthy–Wills 2008; Weller–Meckel–Stahl eds. 2013). The most popular social networking websites used for finding friends and recommending content is evidently Facebook in the European and North American market, but the list continues with Google+ and the Hungarian iwiw (now closed). Social networks are the microblog service provider Twitter, the photo and video messaging service Snapchat, the photo sharing sites Flickr and Instagram, the video sharing site YouTube, the virtual bookmarking site Pinterest, the local search and discovers service Foursquare/Swarm, and the professional networking site LinkedIn as well. In 2014 a new actor made its debut under the name Ello, which claimed to be an advert-free social networking site (it did not gain enough users).

Iwiw, which is short for "international who is who" was once the most popular social networking website in Hungary with a vast number of users. It flourished between 2006 and 2009, but it was discontinued in 2014.

Today the largest social networking website with the highest number of users in Hungary is *Facebook*. It is frequently referred to in common talk, it is the largest marketing platform, its Messenger application is a highly popular IM tool, and the portal itself is synonymous with the Internet for many users. Despite the fact that originally, back in 2004, it was created specifically for university students, Facebook shows the symptoms of aging: its users are increasingly dominated by middle-aged and older people. As a result, younger generations, especially teenagers (the so-called Millennials) are eager to find new forms of communication which are less known and used by their parents, grandparents and teachers.

One such relatively new mobile application is *Snapchat* (established in 2011), an image messaging service which serves to capture and share micro-moments. It is chiefly used to share edited and manipulated images with friends who can view such images ("snaps") for 1 to 10 seconds (as specified by the sender) before becoming inaccessible. The time limit can be bypassed by taking screenshots, though the system will notify the sender of a snap has been saved. Naturally, snaps are not permanently deleted from the system; they only become inaccessible for the recipient. From a communication point of view, Snapchat introduced two innovations: on the one hand, it changed users' attitude to time, as when in the continuous flow of shared images users have only a few seconds to view a picture, they are expected to fully focus on such picture in this short period of time. On the other hand, Snapchat accounts reveal that users do not really process but just skim through the snaps shared with them, and they rather flood others with their own snaps (in five minutes users can share up to 40 differently filtered selfies about themselves: those where they look perfect are sent to more distant friends, while closer friends may get not so advantageous photos too). The service is primarily used by teenagers; this is why it is called "The King of Millennial Micro-Moments". Similarly to Facebook, Snapchat may also face aging as older generations start to discover it and—consequently—younger users turn away from it.

Most social networking websites feature a message wall function which enables users to post texts, photos, videos or any user activity (the timeline function for example displays when you change your profile, use an application, join a group or wish to attend an event announced on the website).

The message wall/timeline is the place for messages users want to share with multiple people, as their friends (and even a friend's friends) can see messages posted here, depending on profile settings. More personal things are recommended to be disclosed in private messages. On Facebook all published messages can be deleted and even modified later on.

Image 2–3: Screenshots from the Facebook message wall (27th October 2010; 5th May 2016)[21]

Messages posted this way (including texts, images, audio and video messages) can be commented by others. Users may also recommend things to each other by clicking on the "like" button. Interestingly, this English word was borrowed by Hungarian by nativizing its pronunciation to match Hungarian spelling rules (*lájk*).

Example 2: Popular folklore about Facebook

Facebook is like a jail. You sit around, waste time, write on walls and get poked by guys you don't really know.

---

21 I blurred the pictures and names due to data security.

2.2.5 Online chat and instant messaging (IM)

"Online chat is real time communication through text messages" (Nyíri 2006). The texts generated through online chatting carries the characteristics of both literacy and orality. "Due to its informal style, [...] spoken language idioms and conversational phrases, [chat communication] is functionally closer to orality; however, it takes place in writing" (Érsok 2007: 5). Written discourse is closer to the rules of spoken language, in fact, it can almost be regarded as the written form of spontaneous speech (Bódi 2004a: 45).

Chatting can take place publicly (e. g. in a virtual chat room where even hundreds of users are available for conversation—this is called *chatosphere* by Harnad [2006a: 16]) or privately (e. g. through Google Talk, which is integrated in the Gmail service, through Skype or through the Messenger service of Facebook). This two types are often distinguished from each other (online chatting vs. instant messaging, cf. Table 6; Juhász 2010, 2011), and the major difference between them is in the application used and the number of users present at a time. From a pragmatic point of view, it should be highlighted that conversation partners in instant messaging know each other from offline life, so the topic between them can be their common background or friends; while in the case of online chatting this relationship between the partners is usually restricted to the virtual world. This explains why users log in with their real names for instant messaging and with nicknames for online chatting. This distinction, however, is irrelevant from the perspective of the present linguistic analysis, so the term online chatting shall be understood in a more general way, so as to include instant messaging as well.

While chatting, users can also send files to each other and most chat systems allow them to make video calls as well. But why do users chat in writing if they could do so through online voice or video calls? There may be several reasons for this: first they do not need to control their non-verbal communication this way, second they can correct and even delete their responses before sending them, third others nearby cannot overhear what they are saying. As a special case of online chatting, two users may send chat messages to each other while engaged in an online voice call with each other (according to my observations, when it comes to information, data or names, the participants in communication prefer written forms even while having a voice conversation. However, if they do not want others to overhear their conversations, they write down certain parts or whole sentences instead of uttering them.)

Further literature about examinations on chat communication in German and English: Thaler (on internet relay chat in relation to the French language, 2003); Dittmann (in relation to the German and French languages, 2001); Kilian (2001); Beißwenger (2001); Storrer (2001); Fix (2001, from a psychological aspect); Schneider et al. (2005); Teplan (2005).

Table 6: Differences of online chatting and instant messaging (IM)

Aspect	**Online chat**	**IM**
intimacy	public	private
number of participants	up to several thousand users, "chatosphere" (Harnad 2006a: 16)	typically two users (less typically multiple users,"conferences")
programme	virtual chat room, online chat sites, e. g.: chat.hu	for example: Windows Live Messenger (MSN), Google Talk, Skype, Facebook Chat
aim	typically entertainment, less typically collection of information	private life related aim (relationship oriented IM); work-related aim (information oriented IM)
familiarity	usually only in the online, virtual world	typically in offline life as well
name usage	pseudoname, nickname	real name
topics	general topics, often: sexual hints	common background, common friends
greetings	saying hello is important	lengthy procedure of saying goodbye
organization of turn takings	delayed reactions, parallel speaking	overlapping, interruptions
segmentation of turns	one thought, one turn; thoughts are rarely interrupted by hitting the enter key	hitting the enter key instead of using punctuation marks (aim: maintain partner's attention)

2.2.6 Text message and multimedia message

As text messages sent on mobile phones are created spontaneously, with an aim to be quick and short, I consider them (together with the gradually disappearing technology of multimedia messages)[22] as a digital genre as well (even though, in the standard case, they are not generated on computers as the other digital genres listed above). Kristóf Nyíri (2006) mentions a hypothesis in cognitive philosophy which assumes that mobile phones operate as an external, connected part of our minds, in fact, they are becoming a component in our mechanism of thinking. "People, who otherwise seldom resort to writing, send text messages to each other even though they could use the same medium (the mobile phone) to initiate oral conversation. The popularity of text messages go against the obvious trend that writing is disappearing as a form of communication, and this is surprising even if these text messages are really short" (Benczik 2001: 209).

22 Multimedia messages were analyzed from a linguistic point of view by Frehner (2008: 203–233), who also described MMS-related photo and audio content.

Image 4: The cover of Crystal's book: *txtng*

SMS is an acronym for Short Message Service.[23] The service allows users to send messages of up to 160 characters between two mobile phones or a computer and a mobile phone (and vice versa). The first text message was probably sent in 1992 (Höflich 2003a: 11). Text messages have various names in Hungarian, including: életjel, emese, esem, esemes, esi, írok, írjál, levél, masszázs, message, mese, rövid szöveges üzenet, rszü, rtk, segédmunkás, semes, simze, SM, SMS, sömi, sömö, sömös, sömösö, söms, sösmös, susmus, SusMuS, szöszmösz, szöveges, szöveges üzenet, text, üzcsi, üzenet, üzi, virtuális galamb... (the list can be continued with English variants: short messaging, short mail, SMSing, person-to-person messaging, mobile messaging, wireless messaging, text messaging, texting, txtng; see: Crystal 2008: 6).

Image 5: Ed McLachlan's caricature of text messages (Source: Crystal 2008: 1)

23 Further literature on (Hungarian) text messages: Balázs ed. 2002; Balázs 2003, 2004, 2005a, 2005b, 2006, 2007a, 2007b, 2011; Érsok 2004; Frehner 2008; Tokaji–Adamkó 2006; Veszelszki 2002, 2005a, 2005c és 2007, 2011c; Zimányi 2005.

According to my former survey (Veszelszki 2006), the main aims of text messaging is keeping in touch, amusement, quick sharing of information, receiving news, doing official administration, participation in SMS lotteries, sending good wishes, replacement of phone conversations (according to a respondent: "if [the other party] will not pick up the phone"), and recording information ("recorded information will last"). This is supported by Frehner's examination (2008: 29).

Typing on older cell (not smart) phones was peculiar because characters are allocated to the keypad buttons in alphabetical order. *Homonumeric words* or *textonyms* are words which have the same numeric combination when typed on a mobile phone keypad. This way the numeric combination 726 on a mobile phone could mean *pam*, *ram*, *sam* or *ran*, depending on which number key is pressed more than once (Crystal 2008: 187). When using the so-called T9 system (text on nine keys; cf. Frehner 2008: 13), keys did not have to be pressed multiple times to reach a certain word because the system suggested possible words on the basis of keys already pressed. It would be interesting to see how this technique, which is also known as predictive texting, influences the manner and result of text composition. Smart phones have a QWERTY keypad.

According to David Crystal 2008: 187), the language of text messages featuring abbreviations is informally referred to as *text speak* (though sometimes, when used in a more general sense, the term refers to texts generated on mobile phones, irrespective of the existence of abbreviations). Some consider textese as an independent language (Kroker 2001: 145; „Do you speak SMS?": Lauer 2001: 19; termed as: "Handysch": Schmidt 2001: 9). Textese is of course not a new language but it may be considered as part of digilect.

Text messages are increasingly less common, they seem to be gradually replaced by free instant messaging services which require internet access.

Text messaging was a very important form of communication in the early 2000s as it enabled users to communicate in writing without internet access (Balázs 2006, 2007a; Veszelszki 2006). However, with the spread of mobile internet and wifi hotspots (and as these technologies become cheaper) it lost much of its importance. Today (in 2017) it is only used to quickly send very brief information (such as *"I'll be late 10 minutes."*) and becomes more and more unique as a rare alternative of instant messaging. For the very same reason, the above described SMS language, which operates with many abbreviations (Bieswanger 2007), has also lost some of its significance: earlier users tried to compress as much information into the maximum 160 characters of a text message as possible—even by violating spelling norms—, by now this conversational function has been taken over by social media (chiefly by the Messenger app of Facebook which is used both for emailing and instant messaging purposes). Thus, in 2017 it is pointless to talk about SMS language as the use of abbreviations no longer characterizes the language of younger or older generations. In instant messaging, abbreviations are used (increasingly rarely, as I have observed) in connection with more traditional, codified forms (e. g. *h* for 'hogy' (that), *m* for 'mert' (because), *fb* for 'Facebook'). Creative abbreviations used in texting in the early 2000s (e. g. *5let* for 'ötlet' [idea; öt 'five'], *jo8* for 'jó éjt' [good night; 8 pronounced as eight ~ éjt]; cf. Veszelszki 2007) today mainly express linguistic playfulness and have a stylistic function.

This trend which goes against inventing and using creative abbreviations is only strengthened by the predictive text feature of mobile phones which corrects misspelled forms on the basis of a built-in vocabulary (sometimes causing misunderstanding or funny situations by automatically replacing words with irrelevant ones.) As another new and increasingly accurate feature on mobile phones, speech to text apps (already available in Hungarian) convert live speech to written text ready to be sent (after optional editing/correction) to the recipient as a text message or e-mail. The technology giant Google predicts that voice controlled devices will dominate the market in the future (their attempt in this filed with Google Glass, though, has turned out to be a failure due to ergonomic reasons; cf. Veszelszki 2016a).

2.2.7 The hashtag

Although we may often have the impression that new technology creates new text forms (genres, text types) as well, most apparently new digital genres have their antecedents in the offline—or even in the online—world (cf. traditional letter vs. e-mail, diary vs. blog, internet forum vs. chains of comments in social media). As an exclusion from this rule, *hashtags* can only work fully in a digital context as they link different content with hyperlinks (naturally, they have a stylistically marked use in printed advertisements, posters, shop windows etc, but their connecting function detailed below can only be fulfilled in electronic communication).

A hashtag is a type of label or metadata tag which is associated with several social networking and micro-blogging websites. It makes it easier for users to find content in the same topic. A hashtag can be created by inserting a # (number sign) before an expression without spaces. Because of its widespread use, the word *hashtag* was added to the Oxford English Dictionary in 2014. In Hungarian the word *hashtag* refers to both the label market with a hashtag (number sign) and the sign itself.

The digital usage of # can be linked to programming languages, and it gained its present thematic connecting function as early as in the first Internet Relay Chat (IRC) conversations. Hashtags were first used in the Twitter microblogging service in 2007. As it became highly popular on Twitter, the tagging system was taken over by Instagram, Facebook and other social networking websites. Now hashtags appear in e-mails, internet forums, instant messaging and even in live conversations. The expression has not only cropped up in orality in a verbalised form, but it even influenced nonverbal communication: the sign is sometimes shown with the fingers (Parker 2011; Kamer 2013).

As regards its position, the tag can occur anywhere in a text, though prototypically it takes place at the end of the text. Using multiple hashtags in the middle of the text makes comprehension quite difficult (*Example 3*). Certain tags most likely go hand in hand with others. Co-occurring, multiple hashtags are called in this paper *hashtag chains* (cf. Veszelszki 2016b). Analysing various contents shared on Instagram (Cohen 2015), the highest interaction is triggered by 11 or more hashtags, so on this platform the more hashtag someone uses, the more efficient he or she will be in building community. Hashtag chains typically relate to the most popular topics of amateur photography (travel, touristic sights; food and drink; fashion; selfie 'self-picture' and work), and their structure prototypically follows the user name, text, hashtags order.

Example 3: Multiple hashtags under a photo (Instagram, 2015)

Nagyon #cuki #aranyos #cica #macska #kiscica! Annyira #love #szeretem!
(So #cute #lovely #pussycat #kitty #moggie! I so #love her!)
It's a classic mille feuille but with a twist...the flavors of bananas!
#recipeoftheday #syntaghthshmeras #millefeuille #napoleon #dessert #sweets
#deliciousrecipes #recipes #food #foodlover #foodblogger #foodtricks
#instafood #instagood #instadaily #instalover #instamood #foodpics #bananas.

The most typical picture-text pattern is the following (Veszelszki 2016b): the picture is followed by non-linked text which can be taken as a subtitle and then – as further textual content – the hashtags. Subtitles can be a single word, a sentence, a poem, a quote, and sometimes, though relatively rarely, even a very long text. If the post is shared on Instagram, subtitles sometimes appear as hashtags. This may take the specific form that instead of a series of independent hashtags all syntactic units or all words of a phrase or sentence are marked with a # sign. Sometimes more important elements and keywords are emphasized as hashtags within the subtitle text (and this may be further combined with @ signs marking persons; *Table 7*).

Searching on a hashtag will list all content with the given hashtag on the social networking website. With this sorting and searching function, however, hashtags not only connect contents, thematic blocs but also users having similar fields of interests. It is often used as an instrument by social movements (for example, with the hashtag *#bringbackourgirls* Nigerian students directed attention to the fact that 270 girls were kidnapped from their school in 2014). Hashtags are important for social sciences too, as they make it possible to search opinions from different individuals about an event, then sort and analyse them. Marketing professionals have also recognised the potential in hashtags: they can use these labels to measure the effectiveness of promotion in social media (Ting et al. 2015: 16).

Table 7: Photo subtitling patterns on Instagram (cf. Veszelszki 2016c)

Pattern	**Examples**
no text, only image	–
only hashtags	*#awesome #time #sunday #bff #love #music #followback #Delhi #sdamarket*
only non-linked text	[No occurrence in the corpus, as only hashtagged pictures were collected. Still, there may be one example: *Christmas*]
non-linked text + independent hashtags (the most typical form)	*Found this old photo from over 25 years ago. Scary how quickly time passes! #time #passingtime #teenager #teenageyears*
hashtags within a text (keywords are emphasised by hashtagging)	*Our latest #mashreads book selection is #FatesandFuries by Lauren Groff, a love story told from two very disparate #perspectives. Based off of the themes of the #book, for this week's MashReads social challenge, we want to see examples of juxtapositions you find in the wild such as light and dark or two contrasting textures. Be sure to tag your photos with #mashreads to enter. #NYC #PhotoChallenge (Image: @tylertronson)*
coherent text with hash-tags + list of extra hashtags	*#cloudporn with #lighthouse at old #port of #Chania #Crete #Creta #Greece #travelgram #instatravel*
hashtagged words in expressions, collocations	*#sweet #dream #my #time #crazy #time*
each word in a text is hashtagged	*#life #is #like #a #car #if #you #keep #going #in #reverse #you #will #not #get #to #your #destination #but #if #you #speed #up #and #do #not #take #the #time #to #enjoy #your #surroundings #you #might #get #lost*

"Hashtags are deictic, indexical—yet what they point to is themselves, their own dual role in ongoing discourse" (Rambukkana 2015). The metacommentary written after the hash can be used to indicate the user's attitude toward the hash-tagged content (an apparently serious text can be made funny or ironic by adding the right label to it; cf. Parker 2011). For example: "*I can't decide which series to watch this evening. #firstworldproblems*". The expression „*first world problem*" is a self-ironic reflection on the fact that for the user the mentioned issue is the greatest problem in life. In its broader sense, the hashtag may be taken as the marker of epistemological modality in relation to the text.

Hashtags should be used in a relatively constant form, in line with their original function of promoting searching and connecting contents. Examples of this fixed form include: *#mik* (acronym of Hungarian Instagram Community), *#mik_gasztro, #yolo* (you only live once), *#swag, #sundaymorning; #bestoftheday, #picoftheday, #pictureoftheday, #worstnightmare, #grexit*. The same need for stability may motivate the usage of joker suffixes, such as *-porn* and *-gram* (deriving from the word Instagram): *coffee porn, foodporn, wordporn, poemsporn, lobster porn; latergram, travelgram, foodstagram*; as well as prefixes, such as *insta-* (also deriving from Instagram): *instafood, instatravel, instagood, instamood, instadaily, instasize, instahun, instanight* etc. Hashtags are also bound by certain syntactic rules, as exemplified by the following tags which are often used by the Instagram community: *#throwbackthursday* (for short: *#tbt*) and *#flashbackfriday* (for short: *#fbf*) indicate that users share older (childhood) photos of themselves on Thursdays and Fridays. Similarly, the following example refer to pragmatic bondedness: *#followfriday* (*#ff*) by which every Friday users recommend to their followers something that they think is worth following. These examples partly justify the idea that they are bound by lexical rules as well: creating hashtags by replacing the day name, such as **flashbackmonday* or **throwbacksaturday,* result in an odd effect, making the tag stylistically marked.

So we could see that hashtags are keywords, some sort of metadata, a new minimalistic text type which users attach to their texts (or pictures) to contextualise their content (Veszelszki 2016b). Tags have a great role in finding multimedial digital content: as titles contain only minimal information, tags provide additional verbal information to search engines and help open up new access routes to content.

3 Characteristic Features of Digilect

The most important features of this novel way of writing: phonetic spelling, the influence of spoken language, abbreviations, the influence of the English language, new collocations, new words, grammaticalization, the use of emoticons, interweaving thematic threads (a new type of intertextuality in hyperlinked texts), and increased role of visuality.

"The diversity in personal electronic communication usually exhibits a strong level of creativity, if and to the extent that it supports the creation of new linguistic forms. Key components of this linguistic world include the creation of new word-forms and new meanings, the frequent use of slang words, hypocorism, chattering (joking), obscenity, conciseness, and the usage of emoticons and fresh loan words (usually of English origin) adapted to Hungarian phonetics. On the other hand, it has also developed a conventional component which restrains creativity", says Gábor Tolcsvai Nagy (2004: 201) summarizing the most important linguistic characteristics of digital communication.

In the following, digilect will be described from pragmatic-textological, lexical and grammatical-formal aspects, summarizing the characteristics of each aspect and considering the following basic principle: "Drawing demarcation lines between linguistic components may only serve practical research purposes as the purpose of the analysis always determines the limits of the research" (Bańczerowski 2008: 77).

3.1 The pragmatic-textological characteristics of digilect

Pragmatics—as defined by functional and cognitive linguistics (see: Tátrai 2005; Verschueren 1995, 1999)—is an approach which focuses on linguistic activity and considers language use as an activity, a function (cf. Veszelszki 2015a).

3.1.1 Relationship types between participants of digital communication

There are four main types of relationships between participants of digital communication (as in other forms of communication) (Figure 3). The relationship of the parties involved in private messaging or exchanging e-mails resembles to a private conversation (Figure 3.1). Technology enables users to send the same message to several parties at the same time (Figure 3.2). Bulk e-mails—similarly to bulk mails or text messages—are sent to multiple specific recipients. Forum, message wall and

DOI 10.1515/9783110499117-003

blog posts, tweets, however, are sent to non-specific recipients (i. e. to anyone) or to a group of persons linked by specific interest. Messages sent to mailing lists are usually put into the "many-to-one" category (Figure 3.3), though they also fit to the next category as any participant on the mailing list can send e-mail to the others (they are not always passive recipients). Several users communicate (often at the same time) with several others on public chat rooms, and the internet forum as a text type also belongs to this category (Figure 3.4, though the messages posted on such forums rather belong to the "one-to-many" category mentioned above).

Figure 3: Relationships between participants (with examples)

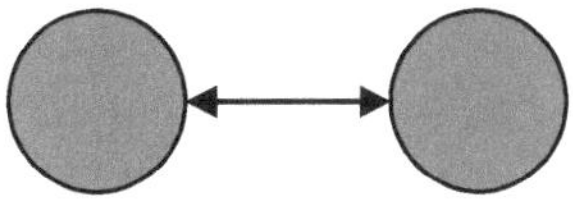

Figure 3.1: One-to-one:
private online chatting, e-mail

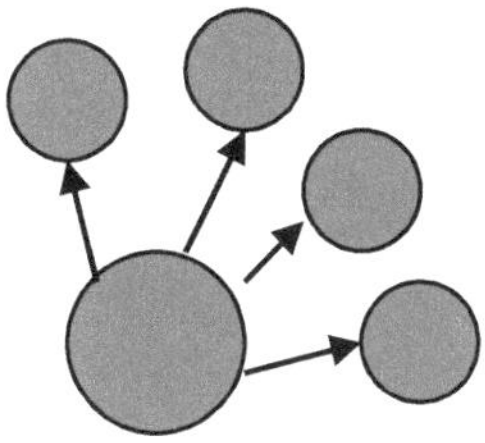

Figure 3.2: One-to-many:
e-mail with multiple addressees, forum post, comment, message wall post, blog post, tweet (+text messages with multiple addressees)

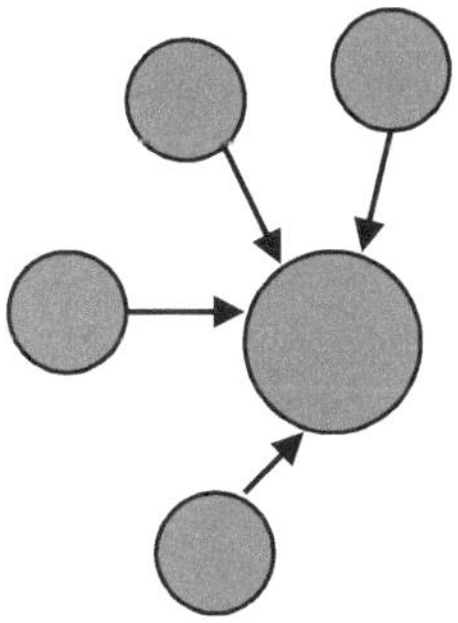

Figure 3.3: Many-to-one:
mailing list

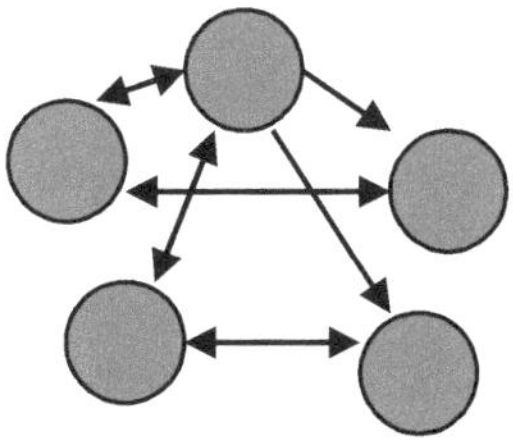

Figure 3.4: Many-to-many:
jointly used blog, internet forum, (public) online chatting with multiple participants

3.1.2 Perspective

"Perspective [...] plays a decisive role both in the structure of semantic relations of the sentence(s) comprising the utterance and in the interpretation of the utterance as a cognitive unit" (Tátrai 2005: 207). According to the functional cognitive approach, in order to understand narratives the recipient sets up a mental model in which "the position (perspective) from which the speaker (narrator) views the things (actors and events) of the story world plays an essential role" (Tátrai 2005: 207). The reference center, which is of paramount importance, is determined by the speaker's person and spatial and temporal position. The reference center is characterized by egocentrism (the first person singular): "the central person is the one who makes the utterance, while the second person singular (addressee) and the third person singular (other person mentioned) are defined in relation to this central person" (Tátrai 2005: 220; cf. Langacker 1991). In certain cases the deictic center may be transferred, this is called "deictic projection" (Lyons 1977/1989: 579).

The reference center of the major part of digilect-related utterances examined are egocentric. Certain Facebook and Twitter users, however, use a specific perspective when posting messages: they relate the events which happened to them from an external point of view (third person singular). For example:

Example 4: External perspective (Facebook post, 2010)

[name] *saw a ground-to-ground rainbow.*

3.1.3 Thematic progression

As regard topic switches, instant messaging is characterized by leaping turn-takings and interweaving thematic plots. This is illustrated by the two conversation excerpts in *Example 5*. The first conversation is characterized more by linear progression—a term used by Daneš (1970, 1974)—and the second one by recursion where the participants revisit topics introduced at the beginning even at the end of the dialogue.

To maintain attention, participants not only hit Enter (i. e. send their messages) when they have finished a thought or utterance, but after smaller chunks as well. This was noted by Nikoletta Érsok who, for this very reason, distinguishes between contributions (Hungarian: *hozzászólás*) and turns (Hungarian: *forduló*). In her terminology, contribution is the formal basic unit of instant messaging, that is the unit of communication "created (typed and sent by hitting the Enter key) as a private act" (2007: 94). By contrast, the term *turn* in instant messaging is reserved by Érsok for the totality of those contributions "which were created by the same user and constitute a closed content-conceptual-emotional-pragmatic unit" (2007: 95).

Example 5: Excerpt from two instant messaging conversations from the aspect of thematic progression (2009)

(topic switches are indented, participants are marked with different letters: A + B, C + D)

Example 5.1: Conversation #1: relatively linear thematic progression

[...] B: well
B: it's a serious decision
        A: ok, it's about time to call my man (husband)
B: to leave everything behind
        B: okay
        B: as for me, I'll have a shower
                B: i'm travelling tomorrow
                B: we'll leave in the afternoon
                B: have a good weekend
                A: have a good time
                B: enjoy the season
                        A: i didn't interrupt you
                        A: did i?
                B: talk to you on tuesday
                        B: nooo :D
                                A: i'll make pie tomorrow
                        B: just you were silent
                                B: wow, easter pie?
                                A: i'll tell you about it on tuesday
                                B: cool
                                B: i would call in too
                                        B: saturday dinner was great
                                A: come
                                        B: i like your pictures
                                        B: your place looks so cosy
                                        A: great to hear this
                                        A: i like your things
                                                A: kiss
                                                B: kisses

Example 5.2: Conversation #2: recursive, cyclical thematic progression

C: hi there
C: already up?
C: we've just got up too
C: im already cooking
C: but doing my stuff too
D: hi

D: only half-awake
D: im working from bed
C: you corrected your webpage
C: i envy you
D: [name] dashed over to [name]
C: im also doing the laundry
D: i did, yes!
D: only that i couldnt insert page-break
C: what're you working on?
D: so it stays this way
C: but at least its new and fits the season:)
D: ive started to do the translation
C: i see you've sent me sg
D: [name]
D: can you help me at the weekend??
C: of course
C: when will you come
D: thurs morning
C: cos we're doing the corrections until saturday evening
D: oh no correcting what?
D: on sat I'll go to [name]
C: Hungarian, more than 280 test papers
D: to a birthday party
D: more than 280, uhhh
C: friday is also admission exam for the sixth class
D: for me its language exam
D: with [name]
C: how bout monday?
C: then ill dry to stay at home
C: t
D: monday?
D: ok, ill try to stay
C: we'll arrange this later
D: fine
D: i have to do this translation..
C: okay
C: ill check the [name]
D: okay, thx
C: kisses (puszka)
D: kisses (puszurák)

3.1.4 The distinctive role of the phatic function

Linguistic rituals (e. g. saying hello, goodbye or thanks) play a distinctive role in spontaneous digital texts, such as instant messages. On the one hand, this emphatic role is marked by the lengthy process of saying goodbye (in *Example 6*, ten out of the total 44 turns are dedicated to this act).

Example 6: Saying goodbye (instant messaging text, 2010)

A: *that'll be great, but now I've gotta go to work*
A: *thank you so much for your help, and sorry for being so unreliable*
B: *okay, take care*
B: *kiss*
A: *you too!*
B: *so am I these days unfortunately*
B: *kisssses*
A: *well, we'll fix this issue then:9*
A: *kisses bye*
B: *kissesss*

On the other hand, the emphatic role of this function is also indicated by the wide range of linguistic elements created for this purpose (Hungarian examples: agreement: *okker, oksa, okk, oké* [all variants of the English *OK*], *ja, jah, ya* [variants of the universal word of agreement *ya*]; greeting: *cyóka* [variant of *szia* 'hi'], *pápá* [variant of *pá* 'bye'], *pusszancs, puszika, pucc, cupp* [variants of *puszi* 'kiss']; saying thanks: *thx* [taken from English], *köcike, köcce, köszi, köszke, köx* [variants of *köszönöm* 'thank you']).

Jó8pusz [*jó* + *éjt* + *pusz* 'good night kiss'] plays with the similar pronunciation of the Hungarian *éjt* 'night' and the English *eight*. A similar example can be found in German as well: *guten8* which can be considered as an oronym [*gute* + *Nacht* 'good night' > *guten+acht* (acht in German = eight)]. The following examples are part of the same family of words: *pukk*; *pucc*; *pu$*; *PU§*; *pub*; *puc*; *PUKK*; *pumm*; *puq*; *pus*; *pux*. They all stem from the root *puszi* 'kiss', more specifically its shortened form *pusz*, but the last *sz* letter replaced with another consonant or symbol. The eagerness to avoid customary forms is also apparent in the examples *ppuc* and *zsup* which are the reverse forms of the words *cupp* 'smooch' and *pusz*, respectively. In fact, any word starting in *p* can replace *puszi* in this function (saying goodbye), whether meaningful (e. g.: *pápua* 'Papua'; *pipa* 'pipe'; *piszok:* 'dirt'; *pitypang!* 'dandelion'; *plüssss* 'plush'; *pöcök* 'peg'; *pötty* 'dot'; *puli* 'Hungarian sheepdog'; *puliszka* 'maize porridge'; *pumpa* 'pump'; *pußta* 'lowlands' [intentionally misspelled]; *puszedli* 'macaroon') or gibberish (*pacat*; *paccs*; *plökk*; *pötyi*; *pulo*; *puszmancs*; *pusszer*).

Another family gathers around the word *csók* 'kiss' based on similar sounding: *csoki* 'chocolate', *csook* [where the double *o* indicates the same *o:* sound marked by *ó* in the original stem], *tsok, tsók, tshók* [where *ts* indicates the same *tʃ* sound marked by *cs* in the original stem], and *tsokiskex* 'chocolate biscuit'. Variants using numbers or symbols include *99* and * [the latter depicts kissing lips]. The slang word *cső* [derived from the Italian *ciao* but used both for saying hello and goodbye] plays with the homonymous *cső* 'tube' to create the following variants *csőtészta* 'maccaroni', *csőváz* 'tubular frame', *csövi* 'from the tube', *légcső* 'windpipe' and *lépcső* 'stairs'.

Digilect is constantly changing which is demonstrated well by the fact that it quickly responds to linguistic trends. Though it had already appeared in the 2001 Hungarian film *Üvegtigris*, the expression *Száábúsz* [*szervusz* 'hello' pronounced in a special way by one of the characters in the film] only became fashionable in 2006 when the second part of the movie was released. This form of greeting and its variants appeared almost instantly in text messages: *Száábbúússzzz*; *Xáábuux-ttokk*; *xábbuuxxx*.

Anna Zsigó (2010) noticed that saying goodbye has reduced importance in instant messaging which is just the opposite than the trend in text messages: "instead of saying goodbye the parties [only] express their intention to take leave", e. g.: *now, I'm leaving; I'm off; I'm busy; I've gotta go*. These expressions clearly close the conversation.

Bańczerowski (2008: 285–336) distinguishes between two levels of communication: the information level carries the propositional elements of the utterance, while the meta-information level refers, among others, to the relationship between the speaker and the recipient. Signs with phatic function relate to the meta-level (*Example 7*).

Example 7: Dominance of the phatic function (instant messaging text, 2011)

A: [*C name*] ♥ ***you here?***
B: *íXDDDDDDDD*
B: ***ciao***
C: *I'm here* ***say it*** dear :D
A: *dear xD*
B: *and he's really here XD*
A: *I'm flattered*
B: *oooh*
B: *where did I fall into this*
B: *yesterday* [*name*] *missed you much :D*
B: *and so did I*
C: *xd*
B: *why didn't you come? :D*
A: ***but really***

A: [*C name*]
A: *why didnt u come?*
C: *cos I didnt want to see your ugly faces :D*
A: *„!„*
B: *thanks*[*C name*]
B: *how nice^^*

3.1.5 Meta-themes

Meta-thematic references are those parts of a conversation which refer to the medium of the conversation (e. g. to the former Google Talk instant messenger in which the dialogue in *Example 8* took place).

Example 8: Meta-thematic reference using GoogleTalk (instant messaging text, 2010)

A: *I didn't know you can write here too*
[B: *I'm looking at the things you posted on iwiw*]
B: *you can :)*
A: *I learn something everyday*

3.2 The lexical features of digilect

Examining English and German e-mails and text messages, Frehner (2008: 118–21) described the use of discourse markers, interjections, response forms, regionalisms, slang expressions, spoken language words/structures (e. g.: *ya*, *none of my business*) and taboo expletives. According to Kilian (2001: 64) the use of conversational phrases (e. g. the German *achwas*) and regionalisms indicates the effort to transform orality into writing (there are non-lexical signs as well, such as using capitals and overusing characters to indicate intonation in writing).

As far as the Hungarian language is concerned, the vocabulary of digilect is characterized by the following five factors:

1. foreign (mainly English) influence;
2. new words and collocations;
3. IT vocabulary;
4. "expletives";
5. slang, taboo-breakers, individual- or group-specific linguistic elements.

3.2.1 Foreign (mainly English) influence

The substantial influence of the English language on Hungarian digilect is evident if we consider that the major part of internet-specific acronyms used in instant messaging (Fix 2001: 59; Teplan 2005: 51–53) have been borrowed from English. For example: *OMG* (= Oh, my God), *LOL* (= laugh out loud[ly]), *Thx* (= Thanks).

Hungarian users often write down English loan words as pronounced in English (*like* > *lájk*, *love* > *láv*), and even attach Hungarian suffixes to them (*like* > *lájkol* [*lájk+ol*], *love* > *lávol* [*láv+ol*], *lávcsi* [*láv+csi*]; *chat* > *cset* > *csetel* [*cset+el*] [where *-ol/-el* is a verb-forming suffix, *-csi* is a diminutive suffix]). They sometimes treat the original written English form as if it was spelled in Hungarian (*file* > *filé* [fi-le:]; *Facebook* > *face* [fɒ-tsɛ] or *fácse* [fa:-tʃɛ] which is the playful version of *face* written as if it was pronounced as an Italian word).

Example 9: Usage of the word *fész* (Facebook post and comments, 2010)

A: *Hej. Vagy? Milyen órád van?*	A: *Hey. U here? What class do you have now?*
B: *Electro áh fos*	B: *Electro ah shit*
A: *Magyar:(de j perc azt win;)*	A: *Hungarian:(but j mins and win;)*
A: *5^^*	A: *5^^*
B: *Kenyed: d*	B: *U got it: d*
A: *Szünet*	A: *Break*
A: [név]*éknak nem sikerül* ***fész***.	A: [name] *and his friends have troubles with* ***fész***.
B: *Bénák. Nyakas üzeni tevagyaz gáboorka.*	B: *They're lame. Nyakas says you're the man, gáboorka.*
A: *Ehh?:-) Szállodám lesz:(cső* [név]*;)*	A: *What?:-) Next class is Tourism :(CU* [name]*;)*

3.2.2 New words and collocations, new meanings

Digilect abounds in newly created words. Here are just a few examples: *csörgizik* 'call on the phone', *MSN-ezik* 'uses MSN messenger' (already outdated), *iwiwezik* 'uses the iwiw social networking website' (already outdated), *csetel* 'chats', *fotosoppol* 'edits images with Photoshop'.

New forms of communication always give rise to new words too. Facebook, the international social networking website became popular in Hungary in 2009 and the user community soon came up with several alternatives for its name: *face* [fɒ-tsɛ] where the English word *face* is taken as if it was a Hungarian word; the vulgar *faszbúk* [*fasz* 'dick' + *búk* = *book* written with Hungarian pronounciation]; the calque *arckönyv* [*arc* 'face' + *könyv* 'book'] and the phonetically adapted form *fészbúk*. New words were created to refer to Facebook related activities as well: *fészel* 'be on Facebook'; *lájkol* 'use the "Like" button in Facebook' (spelling variant: *like-ol*); *posztol* 'post a comment'.

New words are discussed in detail in Chapter 3.3.2 in relation to the grammatical features of digilect.

New lexical elements not only include words (e. g.: *fon* 'phone', *tel* 'phone', *mob* 'cell phone', *telcsi* 'phone', *üzi* 'message') but also word + noun collocations, such as: *lóg a neten* 'surf the internet' [literally: "hang on the internet"], *felmegy a netre* 'go online' [literally: "go up to the internet"], *dob egy sms-t/mailt/mélt* 'drop a text message/e-mail' [literally: "throw a text message/e-mail"]; *felmegy MSN-re* 'log in to MSN' [literally: "go up to MSN"]; *leesik a netről* 'suddently go offline' [literally: "fall off from the internet"]; *felszalad FB-ra* 'quickly check something on Facebook' [literally: "run up to Facebook"], *tol egy szmájlit* 'send a smiley' [literally: "push a smiley"].

3.2.3 IT vocabulary

The vocabulary of information technology, that is technical terminology once used only by IT professionals, is now gaining ground in standard Hungarian too. Digilect, which relates to digital media, is rich in such words and phrases.

For example: *szkennel* 'scan' [verb], *wlan* 'WLAN (wireless local area network)', *letöltés* 'download', *fájl* 'file', *program* 'computer programme', *alkalmazás ~ applikáció* 'application', *júzer* 'user'; *GPS* 'GPS (Global Positioning System)'; *Bluetooth* 'Bluetooth (wireless communications system)'; *vírusirtó* 'antivirus'; *Space* 'Space key'; *klikkel* 'click with the mouse'; *csatolmány* 'e-mail attachment'.

IT vocabulary also influences slang. For example the word *bebootol*, which stems from the English *boot*, literally means that the computer's operating system is up and running ('booting up'), and figuratively that somebody is 'waking up or regaining consciousness'; cf. Veszelszki ed. 2012).

3.2.4 Expletives

The term *expletive* (*töltelékszó*) is quite a problematic one (Keszler 1983: 171, 1985): there is no widely accepted definition for the term, and it is also doubtful whether in certain cases a word can have absolutely no communication function at all. Even if we assume the existence of expletives, it is still difficult to distinguish them from other words whose communicative function is to play for time, maintain personal relationship, express attitude, promote cohesion or which are used like particles.

According to Runkehl–Schlobinski–Siever (1998: 101) the lexical characteristics of internet communication in German language include interjections (Interjektionen), onomatopoeia, tag-questions (e. g. the German *ok*, *gell* or *ne*), and hesitation or pause particles (*ähm*, *naja*, *achja*, *tja*). Expletives are often used in a special way in writing. Certain words in the following example (*hallod*, *ohh*, *ooh*, *mi*, *ó*) may be deemed to have phatic communication function (Image 6).[24]

Image 6: Using expletives (Facebook message wall, 2010)

3.2.5 Individual and group-specific linguistic elements. Slang and taboo-breakers in digilect

Digital (private) texts may also contain idiolectic elements (*elslattyogok fürdeni* ['I'll have a shower' where *elslattyogok* is an onomatopoeic expression describing

24 Expletives are also discussed in Chapter 3.3.5 on topic-repeating pronouns and Chapter 3.3.6 dealing with the grammaticalization of *asszem*.

movement], *mogat fosok* ['I'm brushing my teeth' where the first letters of the words are exchanged so that the second word gains a vulgar meaning]), elements which express a specific group identity or refer to common background knowledge, or elements taken from slang (*gyökérség* 'lameness' [literally: rootness], *beolt egy buxát* 'have sex with a girl' [literally: injects a vallet], *smárol* 'smooch', *kavar valakivel* 'flirt with somebody' [literally: stir with somebody], *tök gáz* 'completely lame' [literally: perfectly gas], *zúzzunk* 'let's go away/let's party' [let's crash]).

The slang verb *ad* 'give', used in the accusative case, is a relatively new linguistic element with so diverse meaning that it is impossible to describe it with one word: 'great, cool, enjoys it, likes it'. The following examples have been taken from a Facebook wall where the expression is used twice in the same post: *alig adtam a mai napot... Rómában* 'I was not amused by this day... in Rome'; *adom Zoltán Erika nénit nagyon* 'I really like the singer Erika Zoltán' (Image 7, 8). A common, almost ritual response to the structure *adom a valamit* 'I give something' [meaning "I like something"] is *kapom* 'I get something' [meaning "I like it too"] form. For example: "– *Adom a bulit.* 'I give this party' [I like this party.] – *Jaja, kapom én is!* 'I get it too' [I like it too.]".

Image 7: Use of slang elements: *adom, adta*... [Facebook message wall, 2010]

Image 8: Use of slang elements: *adta*...[Facebook message wall, 2010]

As digilect is close to spontaneous speech, users more easily give way to foul language than they probably would in the framework of traditional literacy (cf. Veszelszki 2011c). In terms of explicitness, taboo-breakers can be rated on a scale of 1 to 5 (where 1 is the most explicit).

1. dysphemism;
2. fully explicit form;
3. euphemism;
4. intentional omission (by graphical means);
5. usage of words which are non-stigmatized in themselves, but the context makes them taboo-breakers.

Dysphemism (i. e. the intentional use of coarse or derogatory words instead of neutral ones) is often the source of vulgar humour (Pusztai 2003; Berencsi 1996).

Example 10: Dysphemism as source of humour in digilect

The sleigh is ***screwed****, the reindeers have* ***fucked off****, I'm stuck up in a chimney with a stubbed toe, while the elves are* ***whoring****. This year I ain't go* ***fucking*** *nowhere!!! Regards: Santa... [nickname]*

*OnceUponATime,InAFarawayLand,BeyondManySeasAndMountains,ThereLivedAMan.NowThisManWokeUpOneDayAndSaid:Dammit,ILiveSo****Fucking****FarAway!* [text message, 2003]

Digilect texts contain swear words and other elements of foul language in fully explicit form, which may be explained by the fact that digilect is quite close to orality. Perhaps this is why respondents do not consider expressions like the following as taboo-breakers: *I'm so damn depressed*; *fuck*; *this day sucks like crazy*; *SHUT THE FUCK UP WILL U?-*; *I feel fucking awful.*

Euphemistic and inoffensive expressions constitute a unique category. Technically, euphemism is becoming a pictorial periphrasis: the number of euphemistic expressions for obscene swear words is almost endless (Zlinszky 1925; Kis 1991). The euphemistic effect in the following examples is given by the exchange of letters so that it is more difficult for outsiders to understand them: *ki van a faxom* [where the *sz* in the word *faszom* 'my dick' is replaced with *x*]; *kuma szünet* [where the *l* in the word *kula* 'shit' is replaced with an *m*]; *bazzegol* [where the form of the original word *bazdmegezik* 'repeatedly uses the word "fuck"' is distorted]; *transzveszcica* [where the *tita* part of the word *transzvesztita* 'transvestite' is altered to *cica* which means 'cat']. The expressions *durva(jó)volt, durvasok* may be considered as examples both for letter exchange and word exchange as the original word *kurva* 'bitch' is altered to an existing word *durva* 'rough' by changing only one letter. Taboo expressions are replaced with a non-taboo expression in the following examples: *megy a búsba* 'go to hell' [where *búsba* 'to a sorrow place' is a replace word for *picsába* 'to an ass/cunt' both indicating the destination of the movement]; *a csudába* '~blimey'; *1 rakás szerencsétlenség* 'a pile of misfortune; *1kalap-szamócát* 'a pile of strawberries' [where *szamócát* 'strawberries' replaces the word *szart* 'shit']; *elbénáztam* '~I lamed it up'; *tele van vele a hócipőm* and its variant *televanacsizmám* "I'm fed up with it' [literally: "my moonboot is full with it"]. Using English swear words, such as *damn* or *shit* (and its variant *shitság* described below), appears to be less taboo-breaking in a Hungarian context. Another euphemistic solution is to use childish variants (*elbambiz* 'spoil sg') or baby talk (*kaki* 'pooh').

Example 11: Maccaroni language and foul language

Gaz van,nem birom ezt a kisbetüs ***shitság****-ot olvasni,de muszaj.miért vagyok ilyen loutkitoowagen@es?mas mar reg abbahagyta volna...én küzdök tovább...köszi a tPT* [*shitság* is a blend of the English *shit* and the Hungarian noun-forming suffix *-ság*] [text message, 2002]

Due to its special nature, the form termed here as *intentional omission* (*elhallgatás*) is a separate category and not a sub-category of euphemism. Real intentional omissions are for example the elliptic *egy kalap sz* '~a piece of sh...'; *befagy a...!* '~my ... has frozen to the seat', or *Nagy kalappal* [literally: "I wish you a big hat of ...", where the word *shit* is only implied. The phrase is used to avoid wishing good luck, which is considered bad luck. Similar English phrase: "Break a leg!". A special form of intentional omission is when certain letters of the taboo word is replaced with dots or other symbols (e. g.: *ta.ony ido* [full word: *takony* 'snot']; *elb..tam a napot* [full word: *elbasztam* 'I fucked it']). Special characters include asterisk (*átb**ztak ezzel!* [full word: *átbasztak* 'I was fucked up']; *M**KOK* [full word: *MOCSKOK* 'creeps']), plus sign (*egy f++ volt* [full word: *fasz* 'dick']; *p+++a fáradt* [full word: *picsa* 'cunt']) or number sign (*sza#ok* [full word: *szarok* 'pieces of shit']), and even replacement letters, such as *q* which stands for *k+u* based on homonymy (*qrva idegesitö*; *qrva jo szam*; *qrvára el vagyok fáradva*; *qrwa korán* [original spelling: *kurva, kurvára* 'bitch, fucking']). Quite often, though not necessarily, the number of substitute characters match the number of letters omitted.

Example 12: Euphemism (intentional omission) with various characters

Qrwa *korán van!, alig látok még:-) ébresztö van ám!:-))) na puszi.* [full form: *Kurva* 'fucking'] [text message, 2001]

ez egy ***f++*** *volt,****p+++a*** *fáradt vagyok!mosolyod nélkül nincs jókedvem...*[full forms: *fasz* 'dick' and *picsa* 'cunt'] [text message, 2003]

Eddig csöd:MEXIKO-OLASZORSZ.1:0!FÉLIDÖ!Az olasz pasijaim egyszerüen ***sza#ok****...a masik meccsen ECUADORnak muszaj nyerni.Tudod!mint a szam!Sash:Ecuador...tütü* [full form: *szarok* 'pieces of shit'] [text message, 2003]

Háló!már mióta nem csipogtaПunk?most fexem,tanulka vót,mert ***elb..tam*** *a napot.vansajtó fotókiállítás+kellenenézni.nemtommeddigvan. veledOKminden?puli* [full form: *elbasztam* 'I fucked it'] [text message, 2004]

3.3 The grammatical features of digilect

Technicalization, as a result of rapid ICT development, influences and alters both "pure" orality and "traditional" literacy. New regularities and conventions can be observed and described *in situ nascendi* (Fiehler–Barden–Elstermann–Kraft 2004: 128). Christa Dürscheid (2002) identifies the following common characteristics of e-mails and text messages (Dürscheid's German examples are provided in brackets): reduced forms (*was ist das denn für ne einstellung!*; *umwerfend wars*), spoken language elements (*cool was?*), interjections (*OHH*), conversational/slang phrases (*Darauf hab ich echt keinen Bock!!*, *scheiße WOLL?*), phrases with dialectic origin (*Wir kucken grad fotos an*), phonetic spelling (*BUA*), and dominance of coordinate clauses over subordinate clauses.[25] These features are also characteristic of digilect in general, which also includes instant messaging, blog, forum etc. texts. (Clearly German-specific categories were omitted from the above list.)

Example 13: Reduced grammatical structures in Hungarian (excerpt from two instant messaging texts, 2009)

A: *book: Kati Iván: szerelem@szerelem.hu* [*love@love.hu*]
B: *:)*
B: *wow*
B: *that's just for you*

C: *anyway, how is your Arabic plan going?*
D: *I need 500 hours on the plane*
C: *how many do you have so far?*
D: *200*
C: *then you need some more patience*
D: *I'll surely stay for the summer*

In the following I will first briefly summarize the relevant stages of spoken language research, then present the characteristics of spoken language and try to apply them to our corpus of Hungarian digilect texts.

25 However, certain examinations (Keszler 1983: 191; Zimányi 2011) seem to contradict with this observation. Borbála Keszler was the first to call attention to characteristics which vary by text type (e. g.: monological and dialogic texts).

Spoken language, according to Scherer (1984: 147) on the basis of Schank and Schoenthal (1976: 7), is spontaneous speech formulated freely in a natural communication situation (and the definition is continued by structuralists as follows), so it is a language in the sense of language use (*parole*), rather than the linguistic system (*langue*). A speech can be considered spoken language if it was "not transcribed and not even thought over before delivery" (Steger 1965–66/1967:[26] 262 as cited in Keszler 1983: 166).

The (grammatical) problems of examining spoken language are linked to three factors: First, producing written texts requires linguistic consciousness, while spoken language is a less conscious activity due to its automaticity and relative ease. Second, written language is "objective" and permanent, while spoken language is auditive and elusive (also, this is why the compilation of spoken language corpuses is difficult). Third, traditionally, higher social rank is attributed to written texts than to spoken language (cf. Fiehler–Barden–Elstermann–Kraft 2004: 159).

From a linguistic-theoretical perspective, spoken language research started with the emergence of structuralism which advocated the primacy of spoken language over written language (for details, see: Onea 2006: 2–4), and from a technical perspective, with the development of sound recording and transcription (cf. Scherer 1984: 148). Schwitalla (2001: 896–903) provides a decent outline of German spoken language research, and his observations are also true to the development of Hungarian examinations in this field (as far as trends are considered).

Initially, in the 1960s, the subject of examinations was restricted to: sentence length, simple and complex sentences, subordinate and coordinate clauses, tense and mood, and the use and omission of conjunctions. Linguists examined these topics with statistical methods and focused on deviations from written language characteristics (deficit or difference perspective, Schwitalla 2001: 896; Deppermann–Fiehler–Spranz-Fogasy 2006: 5). Spoken language examinations called for an "abstinence from theory", that is the complete refusal of any theory and a commitment toward empirical evidence (as noted by Imo 2007b: 3).

In the 1970s attention shifted to the analysis of expressions referring to the relationship between the speaker and the listener: response, addressing, turn taking, "request-for-repetition particles" (in German: Rückversicherungspartikeln), formulaic introductory and closing structures, self-correction. It is worth noting that spoken language examinations initially concentrated on

26 Steger's text was published in the 1965–1966 Yearbook of Institut für deutsche Sprache, but the year of publication is 1967 (Steger 1965–66/1967).

monologues—following the traditions of written language analyses—and their attention shifted to dialogues only later (which became the foundation of discourse analysis).

Examinations were further refined by the involvement of text typology. In the 1980–90s research was primarily concerned with finding out the medium-related differences and characteristics of orality and literacy (Schwitalla 2001: 901). It was also during this period when the following linguistic features were described: code switching (Variantenwechsel), speech error, parallel syntactic editing, interruption and being interrupted, reformulation techniques (repetition, repetition with modification, concretization, explanation, summarization), the linguistic formulation of various speech intentions (e. g.: irony, joke, seriousness, solemnity), and the communicative effect of phonetic characteristics (among others: the importance of prosody, intonation and stress in the communication process).

The involvement of phonetic and psycholinguistic approaches gave new impetus to research: new problems included word order, ellipsis, and the beginning and end of sentences (presently, this direction is represented in Hungary by the speech research conferences, proceedings and the series *Beszédkutatás* 'Speech Research' of the Phonetic Research Group headed by Mária Gósy). The 2000s witnessed the emergence of the (radical) construction grammar (cf. Croft 2001; Imo 2007b: 11–12; Deppermann 2006) which tries to find out how grammar functions in speakers' mind from a cognitive perspective, and refuses the former methods of formal abstraction and description. Construction grammar sets out from the basic unit of *construction* which is any pairing of form and meaning, though *meaning* in this context not only involves semantic information but pragmatic and discourse function related information as well (cf. Croft 2001). According to the advocates of construction grammar, this method is suitable for describing the grammar of spoken language (Imo 2007b illustrates this through the German discourse marker *ich mein* 'I mean').

Another novel way of analysis is provided by interaction linguistics which derives its theory from interactional linguistic and ethnomethodological research and in practice deals with the empirical analysis of conversations primarily in everyday, secondarily in official settings, by combining syntax and prosody (Selting 2007). Interactional linguists view spoken language not as divergence from written language but as an independent system of language use (Selting 2007: 131; note they have only conducted empirical analysis on texts but have not presented any broad category, theory or terminology which could be applied to the analysis of other texts).

Though Reinhard Fiehler wrote the chapter on spoken language in the 7th edition of Duden Grammar (2005), his work relied on written grammars („Schriftlastigkeit", cf. Fiehler 2007: 297–298, 2006). In Fiehler's system the basic units of written language are letter, word, sentence and text, while the basic units of spoken language are phoneme, word, functional unit, turn and speech/conversation (2006: 24–5).[27]

Fiehler (2007: 302) points out that most examinations relating to spoken language focus on conspicuous and unique phenomena such as phonetic characteristics (phonemic merger and elision), words typically appearing in spoken language (discourse markers) and special syntactic structures (inverted word order). Further problematic fields that hinder the creation of a spoken language grammar include the following (Fiehler 2007: 306–308): 1. multimodality of the comprehension process (in written language comprehension is restricted to the verbal dimension, while in spoken language several channels are used in parallel); 2. the nature of the speech process (spoken language production is closely linked to interaction, feedback, and thus the line between product and production is blurred, unlike in written texts. This problem calls for a process oriented approach instead of the current product oriented one); 3. category building (a process oriented reinterpretation is needed both in the definition of concepts and terms); 4. delimitation of the subject (spoken language has two common definitions. According to the media-extensional definition, the only criterion for orality is the medium, while the prototypical-gradual definition follows the center-periphery model and distinguishes between typical and atypical texts). (Similar ideas are discussed in more detail here: Fiehler–Barden–Elstermann–Kraft 2004.)[28]

Even though I accept the approach of the above described linguistic schools that the basic concepts of spoken language grammar should not be defined in opposition to written language, still—as none of the schools have developed a comprehensive terminology or adaptable methodology[29] (and remedying this deficiency is outside the scope of the present book)—I use for the purposes of

---

27 The relevant literature also mentions the following units which correspond to the category of sentence in written language: communicative minimal unit, utterance, intonation unit and turn (for the complete list see: Fiehler 2007: 309).

28 Other remarkable instances of German spoken language research: Schank–Schoenthal 1976; Schlobinski 1997; Schwitalla 1997; Steger 1967; Deppermann–Fiehler–Spranz-Fogasy 2006; Ágel–Hennig 2007; Imo 2007a and 2007b; Eisenberg 2007. Hungarian works in this filed include: Szende 1973; Szalamin 1978; É. Kiss 1979 and Keszler 1983.

29 The authors of the German monography *Eigenschaften gesprochener Sprache* [Characteristics of Spoken Language] also mention the methodological difficulty that—for the time being—spoken language can only be analyzed in relation to and using the categories of written language analyses. Still, written language specific categories can only be adapted to spoken language examinations with certain restrictions (Fiehler–Barden–Elstermann–Kraft 2004: 158).

this analysis the terminology and methodology of traditional spoken language research.

Before identifying the properties of spoken language, we have to note that—similarly to written language—spoken language is not homogeneous but exists in varieties. Despite this, it is often homogenized and approximated to a prototype, i. e. linguists often focus on typical, highly characteristic "communication practices" and ignore non-prototypic elements (cf. Fiehler–Barden–Elstermann–Kraft 2004: 157).

The authors of the *Eigenschaften gesprochener Sprache* [Characteristics of Spoken Language] identified 15 functional units which are specific to spoken language (Fiehler–Barden–Elstermann–Kraft 2004: 213–220). According to Fiehler (2006: 28), these units are created by allocating functions to them during the communication process.[30] They can be used for Hungarian spoken language texts as well. The concept of reactive linguistic acts (German: reaktive sprachliche Handlungen) is an umbrella term for several different linguistic acts, and may be related to the basic unit of the speech act theory. Reactive evaluative positions (German: reaktive bewertende Stellungnahmen) may be considered as a sub-group of the previous category, still they are treated as a separate category due to their frequent occurrence and particular importance. This category includes evaluative remarks and comments (e. g.: *super* 'Superb!', *gut gemacht* 'Well done!', *Himmel* 'Heaven knows!', *Blödsinn* 'Rubbish!'), certain interjections (e. g.: *ach, ui*) and swear words (e. g.: *Verdammt noch mal.* '~Dammit!'). Expressions drawing and directing the attention of the listener (German: hörersteuernde Ausdrücke, e. g.: *Achtung* 'Attention! Watch out!', *Moment* 'Moment, please!') and exotheses (German: Exothesen, e. g.: *Was wollt ich doch sagen?* 'What did I want to say?') are also considered as independent functional units. In this system, elements, which are traditionally considered as clauses, are functional units which indicate conditions, consequences, cause, purpose etc. (German: Einheiten, mit denen Bedingungen, Folgen, Gründe, Zwecke, nähere Angaben etc. für/von etwas benannt werden). This hard-to-determine category includes, for instance the dependent clause of the following sentence: *Wenn die Rahmenbedingungen gleich bleiben, können wir das Ergebnis wiederholen.* 'If the conditions remain the same, we can repeat the result'. Forms of address (German: Adressierungen) is a much more definite category (e. g.: *Kinder, so kommen wir*

30 The functional units must be separated from those signs which belong to the formulation process, such as the German *äh*—or its Hungarian counterpart *ööö*—which signifies articulation. Other signs of utterance design and verbalization signify interruption, resumption, correction, reformulation, looking for a word, supplementation etc. (Fiehler 2006: 28).

nie zu einer Lösung. 'Boys, we'll never get this done this way.'; *Idiot, kannst du nicht aufpassen!* 'Idiot, can't you pay more attention?'). The self-identification (German: Selbstidentifizierungen) category contains self-introduction either in person or via phone, as well as self-reference, e. g. the speaker mentions his/her name or introduces him/herself in full. The category of imperatives influencing the discourse process (German: diskursprozessierende Imperative), in my opinion, is similar to the category of expressions drawing and directing the attention of the listener. The only difference is the grammatical form as this category uses imperatives (e. g.: *Sag mal...* 'Tell me', *Sieh mal alles hat zwei Seiten.* 'Look, everything has two sides.'). Operator-like discourse markers (German: Augmente)[31] resemble to modifiers or particles (e. g.: *nee* 'no', *nich wahr* 'isn't it?', *gell* 'isn't it?', *weißte* 'you see?', *siehste* 'you see?'). Modalizers (German: Modalisierungen) are postpositive units which slightly modify the meaning of what has been said (e. g.: *und so* 'and so', *irgendwie* 'somehow'). Text segmentation markers (German: Gliederungssignale) are for example *so* 'so', *also* 'therefore', *nun denn* 'and then'. Units forecasting the type of speech act (German: vorgreifende Verdeutlichung des Handlungstyps) are performative formulae (e. g.. *ich verspreche dir* 'I promise you', *ich warne dich* 'I warn you'). Functional units may also refer to the mental status of the next unit (German: vorgreifende Verdeutlichung des mentalen Status der folgenden Einheit; e. g.: *ich schätze* 'I estimate', *ich befürchte* 'I'm afraid'). Reference to reported speech or quotation (German: Ankündigung von Redewiedergaben) means that the speaker quotes somebody else's thoughts or utterances (e. g.: *Er rief uns zu, ob wir nicht noch zu ihm herüberkommen wollten.* 'He shouted to us asking if we wanted to go over to him.'). The broad category of operators correspond to Bańczerowski's *meta-operator* term (cf. Bańczerowski 2008). Topicalization and reference complex serve to identify a person, object or situation (e. g.: *Was der Großmutter ihr Haus ist, das ist letzte Nacht abgebrannt.* '~It was grandma's house—the one that burned down last night.').

The following functional units of the Fiehler–Barden–Elstermann–Kraft system (2004) were found in the digilect corpus (*Example 14*). By collecting texts and compiling a corpus in a systematic way, further research could reveal the statistic frequency of the above-mentioned functional units of spoken language in the digital communication corpus, and could compare the units with the data of a spoken language corpus.

31 The expression cannot be translated as *augment* in English because this term is already used for a morpheme expressing the perfect aspect in the Indo-European parent language.

Example 14: Functional units of the Fiehler–Barden–Elstermann–Kraft system (2004) in digilect

Example 14.1: Relative linguistic acts

– *ulamsa ?? xDDDD*
– ***igen*** ['yes'] *ez egy másik nyelvXDXDXDXDXDXD (félre nyomtam, egyébként unalmas akart lenni)* [Facebook pair of comments, 2010]

Jahjah [*'~yeppers'*] *am nelli ezt eccer néztük együtt de 8 x D:)* ♥ [Facebook comment, 2010]

:) ***naná*** [*'~but of course'*] *majd eggyütt nyomattyuk béres alexandrára :/ ;P* [Facebook comment, 2010]

Example 14.2: Reactive evaluative positions

Te nagyonhülyevagy ['You are so silly'] *xd...* [cset, 2010]

bakker [*'~dammit'*] *ezt de utáltam xD az egyik részbe volt egy csaj akit úgy hívtak h csütörtök..vagy péntek?xd* [Facebook message wall, 2010]

jézusom ['jeez'] *muszály ilyen képet rakni a kutyámról küldök neked másikat ami jobb is mint ez >D* [Facebook message wall, 2010]

– ***aztaa*** [*'~wow'*]*:O mennyire hasonlítotok* ***azta*** [*'~wow'*] *:/*
– *ez nagy habi votl ;) jókép olyan értelmes ezért olyan pontjó ;)*
– *na szal lááááájk;)*♥ [Facebook message wall, 2010]

Example 14.3: Expressions drawing and directing the attention of the listener; exotheses

Ömm uhh ['er uhm'] *ezt a mesét még non szerettem ...a maci utálta a kelbimbót és mindig vándoroltak a nőjével valamerre világgá xD :D* ***nah mind 1*** [*'well, whatever'*] *az a cime h ehagyott játékok és nemtom sehonnan letölteni ha vki megtalájja akk szójjon mááár Oo :D* [Facebook-üzenőfal, 2010]

Csőőő ['hi'] *martin mizu? :/* [Facebook message wall, 2010]

héé ['hey'] *gyerekek viselkeggyetek mindenkiszép sorba xD* [Facebook comment, 2010]

Mongyuk [*'~as a matter of fact'*] *az tél tulzásvlt mikor adri reggel köszönéshelkyett kért félkiló darált húst xD* ♥ [Facebook comment, 2010]

bazeg [*'~dammit'*] *ezeket szedmárle vagy vágjki vagy vágdki a félhajam vagy csinájj vmit mert besírokXD* [Facebook comment, 2010]

na mivan [*'~now, what's up?'*] *elkapott a HÉV pajti? xD* [Facebook comment, 2010]

Example 14.4: Functional units indicating conditions, consequences, cause, purpose etc.

Ha jobban nevetnél, ['if you laughed louder'] *akk ikszdéé is lehet akár^^:DD* [Facebook-komment, 2010]

Az biztos, ***ha egyszer mi betámadunk*** ['once we decide to attack'] *az kasza lesz!!!:D:DPusz* [Facebook comment, 2010]

Example 14.5: Forms of address

merci *[pet name for Merzedes] te csak nelopd a mesém :D* [Facebook comment, 2010]

hmm pedig a kelbimbótól rendkívüljóllehet fogyni ***viktor*** ['Victor'] *:)♥♥* [Facebook comment, 2010]

Fogd a kezem ***viktor*** ['Victor''] *;) XDXDXDXD hozd a mátét is ♥♥ xD* [Facebook comment, 2010]

héé ***gyerekek*** *['~boys'] viselkeggyetek mindenkiszép sorba xD* [Facebook comment, 2010]

annyira hihetetlenül aranyosvagy! meg az idézetes lap is haláliii! imdálak ***Rékaaa*** ['Rebecca']*!♥ :))* [Facebook comment, 2010]

neked is naaagyon boldog karácsonyt ***bébii*** ['baby']*!♥ ;)* [Facebook comment, 2010]

Dávittemeg *['And you, David'] ecerüencsak hülyevagy ... xD♥* [Facebook comment, 2010]

Example 14.6: Self-identification [This category depends on text type. No fully representative example could be found in the corpus for this.]

Énvagyok a ZELEFÁÁÁNT ['I'*m a neeelephant'*] *Ő.Ő :D köszi :)* [Facebook comment, 2010]

Example 14.7: Imperatives influencing the discourse process

nemááár *[~'you've gotta be kidding'] :D:D azt addmár kölcsön :D:D* [Facebook comment, 2010]

Example 14.8: "Discourse markers" (Augmente)

xD Flóraaa labellópajtááás vagy de ennél személyesebbet akartam ***érted*** ['you know what I mean'] *:I xD: D* [Facebook comment, 2010]

na ***tudod mit*** ['you know what'] *Vivicsi?....te meg leszophacc ....szívessen csajszi ;Đ♥ am meg mér vok hülye? :D* [Facebook comment, 2010]

Example 14.9: Modalizers

paintbe rajzojjak neki muffint ***vagy mi?****['or what?']* [instant messaging, 2010]

Example 14.10: Text segmentation markers

am ['by the way'] *meg nem vagyok kozmetikus és nem is leszek ^^* [Facebook comment, 2010]

xDXD mondjuk ***jahh*** ['yeah'] *:D:D:D de akk is jobb..mert rajtam van...♥*

nah *[~'well'] nem...mind2 fasza de ittmost tied jobban adja ..de csak ITT :D* [Facebook pair of comments, 2010]

Hát ['well'] *Hédi mingyá kitöröllek xDD* [Facebook pair of comments, 2010]

Example 14.11: Units forecasting the type of speech act

haszabadkérdeznem ['may I ask'] *itt mi a jóeget csináltok? :DDD* [Facebook comment, 2010]

Example 14.12: Functional units referring to the mental status of the next unit

asszem ['I think'] *majd hétfőn is lehet fotózkodni :)* [Facebook comment, 2010]

Example 14.13: Reference to reported speech

dávid nekünk úgy áll a kezünk ***mintha azt mondanánk*** ['as if we said'] *Hogy micsi?? :DDD* [Facebook comment, 2010]

én ***megmontam h*** ['*I've already told you that*'] *nemadja vissza :)* [Facebook comment, 2010]

Example 14.14: Topicalization and reference complex [no examples were found]

Scherer (1984: 148) highlights the following three features of spoken language (he notes that these are not necessarily the most typical ones but they are definitely the most widely discussed features in the relevant literature): 1. text segmentation[32] and interaction-related signs; 2. reduction, ellipsis; 3. speech errors, slips of the tongue (German: Defektiva). Text segmentation signs make it easier for the recipient to perceive and process the text. This category includes not only verbal signs (e. g. particles) but also suprasegmental features (e. g. prosody, pause). Reduction and ellipsis may be interpreted in two ways, according to Scherer (1984: 153–5): either as the ellipsis of content (where the explicit or implicit nature of the utterance partly depends on the context, and the relationship and background knowledge of the communication partners) or as formal reduction. Schank and Schoenthal (1976: 29) also distinguish between two forms of ellipsis: the first one relies on the listener's everyday knowledge (e. g. clauses without verb or subject), the second one on the specific situational background knowledge (e. g. deixis). While reduction is activated by the context and the situation, speech errors and slips of the tongue occur by sheer coincidence (Scherer 1984: 166).

On the basis of Ilona H. Molnár's work, Borbála Keszler (1983: 169) collected the following features of spoken language: more elliptical and interweaving sentences, more interruptions, more inserted clause fragments, less multiple-clause sentences, and less participle or infinitive structures than in written language. Based on a corpus analysis, Keszler (1983: 181–202) identifies the following syntactic properties of spoken language texts: interrupted/unfinished sentences, interjections and related modifying clauses, and "irregular ways of composition". Tamás Szende (1973) compared spontaneous speech to written language and found that the former contains more repetitions, unfinished words and sentences, and more interjections.

32 They can be both polysemantic and polifunctional. For instance the German *ja* 'yes' may function as answer or as an opening formula (Scherer 1984: 150; this is also true for the Hungarian particle *ja* 'yeah').

As regards digital communication, we have to mention a formal characteristic which marks syntactic relationships: language users write certain words together to indicate that they treat them as a single intonation unit. (This observation is supported by the remark sent to the questionnaire in Chapter 4.2 pointing out that "*writing words under the same intonation as one word*" is a playful way to save characters.)

Example 15: Words written together to indicate intonation unit

****neröhögj*** [*ne+röhögj* 'don't laugh'] *:@****nemvicces*** [*nem+vicces* 'not funny']

összement...:/**babalett*** [*baba+lett* 'it's cool'] [instant messaging, 2010]

Te* ***nagyonhülyevagy* [*nagyon+hülye+vagy* 'you are so silly'] *xd...Az lesz a neve hogy little G.. Varga little G xD* [Facebook comment, 2010]

azszép [*az+szép* '~that's something'] *:D* [Facebook comment, 2010]

nemrossz [*nem+ rossz* 'not bad'] *xD* [Facebook comment, 2010]

nemáááár [*ne+már* '~you've gotta be kidding'] *:D:D azt* ***addmár*** *kölcsön :D:D* [*add+már (kölcsön)* 'you've gotta lend it to me'] [Facebook comment, 2010]

xD =/ ***énisszeretem*** [*én+is+szeretem* 'I like it too'] *at sonkás csülkeiteket xD* [Facebook comment, 2010]

It is clear from the above that from a grammatical perspective digilect is closer to spoken language than to written language. For this reason, digilect texts are characterized by 1. reduced grammar, "telegram style"; 2. the redundancy of spoken language (e. g. topic-repeating pronoun), and 3. responsiveness to linguistic change (e. g. the grammaticalization and lexicalization of the expression *asszem*).

In the following, the description of the grammatical characteristics of digilect will be restricted to character-saving and reduction techniques (traditionally called abbreviations), digilect-specific neologisms and methods of creating words, and syntactic properties. Reduction trends will be discussed in detail as this topic can be considered a relatively well-established field. The various methods of word creation are also discussed in depth (the highly productive *-(V)l* and *-(V)z* verb-forming suffixes, for example, are presented as a case study). As regards syntactic characteristics, however, due to the fact that this field is least exposed to change, a comprehensive examination would have gone well beyond the scope of this book. For this reason, instead of trying to give a full description, I will present two case studies which mark the possible directions of further research in the field of digilect syntax. The first one deals with topic-repeating structures, and the second one with the grammaticalization process of the expression *asszem*.

3.3.1 Character-saving techniques (abbreviating words and structures)

In instant messaging abbreviations are used primarily to approximate typing speed to normal speech rate. This principle is often legitimized in the field of connected communication with the metaphor "time is money", i. e. with the requirement to be economical and efficient. This is why mistakes and typos are tolerated more in digital communication, without imposing any social sanction for "committing" them (Storrer 2000; Fix 2001; Kilian 2001: 64; Frehner 2008: 95). Lower case typing and the omission of punctuation marks also facilitate quick writing (this is much more typical to the synchronous instant messaging than to asynchronous forum posts or e-mails where there is more time for composition, see: Frehner 2008: 47–50; Kilian 2001).

The character-saving techniques used for shortening words and word combinations are referred to as *character reduction* here in order to avoid conflict with the term *abbreviation* which is a word formation technique.

Textisms, that is the distinctive abbreviated word forms used in texting, not only appear in text messages (for instance: *c – see, txt – text, jk – just kidding, 2day – today*) but also in other means of electronic communication (Crystal 2008: 187). As a point of interest, in his book on manners István Sille lists English abbreviations used in fax messages (unfortunately without providing any source), and notes that "the English language uses phonetics creatively for the purpose of abbreviations in fax messages" (Sille 2009: 195). His examples for abbreviations are as follows (not all of them use phonetic rules): *u – you*; *pls – please*; *cld u – could you*; *cfm – confirm*; *swb – single room with bath*; *bst rgrds – best regards*; *thnks in adv – thanks in advance*; *tks – thanks*. Regarding the German language, he only mentions a closing formula (*mfg – mit freundlichen Grüßen* 'with best regards').

Döring (2002) distinguishes between three functions of abbreviations. Their economical function is to save space so that more information can be sent at a time. The use of non-normative abbreviations is a way of expressing group identity, so this is the identity function. And the third one is the interpretation function which, on the basis of the type and presence/absence of abbreviations, helps establishing the framework of interpretation, so it facilitates comprehension.

Main types of abbreviation in digilect (cf. Veszelszki 2009):

- abbreviation with the initial letter of the word (*hogy* 'that' > *h*);
- Arabic style writing, that is the omission of vowels (*puszi* 'kiss' > *px*, *sorry* > *sry*, *ok* > *k*, *szóval* 'so' > *szvl*, *szeretlek* 'I love you' > *szrtlk*, *magam* 'myself' > *mgm*, *velem* 'with me' > *vlm*);
- capitalising letters whose letter name is pronounced (*emlékszel* 'you remember' > *Mléxel* [where the letter *M* is pronounced *em* and the letter *x* as *ksz*], *jó estét* 'good evening' > *jóST-t* [where the letters *ST* are pronounced as *es+té*], *ember* 'man' > *Mber* [where the letter *M* is pronounced as *em*]);

- replacing letters with numbers, other letters or symbols (*pusz* 'kiss' > *pux*, *szia* 'hi' > *cya*, *szívesen* 'you are welcome' > *cív*, *köszi* 'thanks' > *köcce*);
- contraction (*vagyok* 'I am' > *vok* and *wok*, *holnap* 'tomorrow' > *hnap* and *hn*, *szerintem* 'in my opinion' > *sztem*, *nagyon* 'very' > *naon* and *non*);
- (special type) swear words in euphemistic form (*b+* [=*bazd meg* 'fuck you'], *qrva* [=*kurva* 'bitch']).

Abbreviations are not only typical to text messages and instant messaging, but also to e-mails, message walls and internet forums. There are many lists of recommended abbreviations on the internet which help shorten words and expressions to only a few characters. This is particularly common in the English language (see *Annex 2*).

3.3.1.1 Abbreviation with the initial letter. Acronyms

The most commonly used form of abbreviation is replacing the word with its initial letter, which is called *initialism* by David Crystal (2008: 181). This technique is primarily applied to conjunctions: *hogy* 'that' (*h*), *mert* 'because' (*m* or *m.*), *mint* 'as' (*M* or *m*), *vagy* 'or' (*v*); as well as to the most popular word for saying goodbye: *puszi* 'kiss' (*p* or *P*).

Example 16: Initialisms

*Bocsi,de változott az idöpont,mert elöbb haza kell mennem.Nem lenne baj,ha fél 9-re jönnél ki?Bocs,**h** most írom,de tegnap este tudtammeg és még nem volt ma idöm* [*hogy* 'that' > *h*] [text message, 2006]

*Gratula a vizsgádhoz!Szeretnék menni Ftóra,már toborzok.De ez csak szomb.este döl el.Majd megírom,**h h** állok!Puszi* [*hogy hogyan* 'that how' > *h h*] [text message, 2005]

*ÖRÜLÜNK, HOGY JÖSSZ! MI LAGZIBAN LESZÜNK **P**:GY* [*Puszi* 'kiss' > *P*] [sms, 2001]

Acronyms are words created from the initials of a series of words. They are either pronounced letter-by-letter (the so-called *alphabetisms*, e. g.: *ftf* or *f2f* < *face to face*) or as whole words (e. g.: *lol* < *laughing out loud*) (Crystal 2008: 176).

Crystal uses the term abbreviation for the reduced form a word, and ellipsis for the shortening of whole sentences (Crystal 2008: 176). Following the pattern of English abbreviations, many acronyms are created in Hungarian with compressed meaning (which corresponds to Crystal's ellipsis category): *P:NY* [< puszi: anya 'kisses: mom'], *NHE* [< nem hiszem el 'I don't believe it'], *nkv* [< nem kell válaszolni 'no need to answer'], *VM* [< vigyázz magadra 'take care'], *hv* [< hogy vagy? 'how are you?']; *lsz* [< légy szíves 'please']. The following new words invoke George Orwell's newspeak from his famous book *1984* (Orwell 2000): *hv* [< hogy vagy? 'how are you?'], *lsz, lc, lecci* [< légy szíves 'please'].

Example 17: Acronyms with compressed meaning

Hozza kell szoknom a sok vizsgahoz,eddig max3 volt igy 7lenne! Menjünk köv. szemeszterben ***lsz!*** [*lsz* < *légy szíves* 'please'] *Z.* [text message, 2001]

Rebuses are "words and sentences made out of letters, pictures, or logograms, such as *c u l8r* 'see you later'" (Crystal 2008: 184). The rebus *C U*, which is also termed as *homophonic acronym* (Kilian 2001: 70) can be solved by pronouncing the letters: *see you* (Dürscheid 2002). Similar examples: *Gute N8* (German *gute Nacht* 'good night'); *4U* (*for you*); *2L8* (*too late*).

3.3.1.2 Arabic style writing

I call Arabic or Hebrew style writing the popular form of abbreviation that leaves only those consonants of a word which are necessary for recognition and comprehension. In Aramaic, Hebrew and Arabic script vowels are not (always) indicated (Crystal 2003: 257). This character reduction technique is also called *consonant spelling* by Frehner (2008: 54). In Hungarian this form of abbreviation is primarily used for highly frequent words, such as the reflexive pronouns *magam* 'myself', *magad* 'yourself' etc. (> *mgm*, *mgd*) or the dative personal pronouns *neked* 'to you', *nekem* 'to me' etc. (>*nkd* and *nkm*).

Example 18: Arabic style writing

MG SK LYN BLDG NVNPT KVNK! [< MÉG SOK ILYEN BOLDOG NÉVNAPOT KÍVÁNUNK! 'We wish you many happy returns of your name day.'] *:-P Péntek, kb. éjl/2-töl Spigiboyjal az E-klubban, +felel? Csak kibírjam addig...Puszi* [text message, 2003]

*Huhu!eltüntél!milyenVtA7vgiVacsi&kirand?****nlm****Brutal7, tréningetTartottam,drOra,mostMegyekEgerbeKonfra,arraKészültem.anyu****Nlm****Vt,ezJo! jösszCsüt/velnesz?cuppak* [nlm, Nlm < nálam, Nálam 'at my place'] [text message, 2008]

A special type of the Arabic style writing is the capitalization of consonants which are pronounced with the letter name (an English example: *beefeater* can be spelled as *Bfeater* as *bee* is the letter name of the letter *B*—Translator's note). One of the respondents of my 2006 questionnaire defined this abbreviation technique as follows: "*I usually abbreviate words by capitalizing the consonant-vowel or vowel-consonant parts of the word which are pronounced the same way as the consonant in itself (and I also use numbers this way) e. g.: LNtétbN* [< ellentétben 'as opposed to'; where the *L* is pronounced as *ell* and the *N* as *en*], *1etlNM* [< egyetlenem 'my only one'; where the number *1* is pronounced as *egy*, the *N* as *en* and the *M* as *em*]" (Veszelszki 2006). Similar examples to this form: *1es üzNetekbN* [< egyes üzenetekben 'in certain messages'], *1etMN* [< egyetemen 'at the university'], *fLLevNítNi* [< feleleveníteni 'refresh our memories'], *küldtM* [< küldtem 'I've sent (sg)'], *MLM* [< emelem 'I lift it'], *rMélM* and *RMLM* [< remélem 'I hope'], *szRetlek* [< szeretlek 'I love you'], *SzRLM* [< szerelem 'love'].

Example 19: Capitalizing consonants which are pronounced with the letter name

Jo ***reggLit*** [< *reggelit* 'breakfast'],*Bibkò!Hat te* ***mRre***[< *merre* 'where'] *jarsz,még jelt* ***sM*** [< *sem* 'not even'] *adsz!****MilyN*** [< *milyen* 'what kind of'] *v tegnap a buli?Tartalmas 7végét* ***innN*** [< *innen* 'from here'] *az agyikobol.Hull a ho es hozik.Tanulni fogok pentekre.P:NY* [text message, 2003]

KLlMS heidLbRgLést [< *kellemes heidelbergelést* 'enjoy your stay in Heidelberg'],*maratonian* ***mRész*** [< *merész* 'bold'] *Kislanyok!xezz&ek sok élményt,vigyazzatok1(S)masra.Varom ahiradot mnd* ***mennyisegbN*** [< *mennyiségben* 'quantity'].*P:NY* [text message, 2004]

Hosszu&kevés hibas üzit ***küldtM*** [< *küldtem* 'I sent'],*7fn majd olvasom valaszod.M nkm v fogas(de* ***nM*** [< *nem* 'not'] *ruha)gondom:mozog&faj az alsoB1es fogam.****nM*** *tudom,meddig(****nM****) huzom ki.jobulizast!* [text message, 2004]

Crystal (2008: 179) uses the term *contraction* for the type of abbreviation where letters are omitted from the middle of the word (e. g. *but* > *bt*). Schlobinski's (2009: 70–75) Spanish, Portugal and Swedish examples also fall into this category which clearly shows that omitting vowels is not a Hungarian-specific abbreviation technique, but is also used in text messages and instant messaging in other languages, so it is a near-universal feature of languages.

Example 20: Arabic style abbreviations in foreign languages (Source: Schlobinski 2009: 70–75)

abrazos > *brzs* 'hugs' (Spanish)

beijos > *bjs* 'kisses', *comigo* > *cmg* 'with me', *desde* > *dsd* 'since', *hoje* > *hj* 'today', *nada* > *nd* 'nothing' (Portugal)

någon > *ngn* 'somebody', *något* > *ngt* 'something' (Swedish)

3.3.1.3 Replacement of letters (with numbers, other letters or symbols) based on homophony

"In the English language the substantial difference between written and pronounced forms enables the creation of playful abbreviations" (Beck–Beck 1999: 461). Many abbreviations are based on homophony, that is on the similar pronunciation of words with different spelling and meaning (in English: *4* – *for*; in Spanish: *salu2* – *saludos* 'greetings'; Italian: *6* – *sei* 'you are'; in Chinese: *88* /baibai/ – derived from the English *bye-bye*; Schlobinski 2009: 25).

Number homophony is a method for replacing letters with numbers based on similar pronunciation (*1 7* [< egy hét 'one week']; *1SMÁS* [< egymás 'one another'];

még1szer [< még egyszer 'once more']; *mind1* [< mindegy 'nevermind']; *mind2* [< mindkettő 'both of them']; *6ás* [< hatás 'effect']; *7FŐ* [< hétfő "Monday']). This method is described as follows by a respondent of my questionnaire on texting (see: Veszelszki 2007): "numbers in words, quantifiers with figures".

The digit *1* (spelled in Hungarian as *egy*) can be used to create the following forms: *1es* [< egyes 'certain'] *üzNetekbN, 1etértek* [< egyetértek 'I agree'], *1etM* [< egyetem 'university'], *1etMN* [< egyetemen 'at the university'], *1szer* [< egyszer 'once'], *1szerű* and *1xü* [< egyszerű 'simple'], *1ütt* [< együtt 'together'], *m1, m1ek* [< megyek 'I'm going'], *még1szer* [< még egyszer 'once more'], *mind1* [< mindegy 'nevermind']. The digit *2* invites English abbreviations to Hungarian: *2 'to'* and *'two'*. The digit *5* (spelled as *öt*) can be found in the word *ötlet* 'idea': *5let*. The digit *6* (spelled as *hat*) may replace both the stem *hat* 'influence' and the verb-forming suffix *-hat* 'can (modal auxiliary verb)'. *66* [< hathat 'can influence'], *6ás* [< hatás 'effect'], *+lát6od* [< megláthatod 'you can see it'], *+mond6od* [< megmondhatod 'you can tell it'], *felhív6sz, hív6sz* [< felhívhatsz, hívhatsz 'you can call me'], *i6* [< ihat 'can drink'], *minden6ó* [< mindenható 'almighty'], *taliz6* [< talizhat 'can meet'], *tud6om* [< tudhatom 'I should know']. *6+* is the abbreviation of an abbreviation (hat. meg < határozd meg 'define it'). The digit *7* (spelled as *hét*) mainly appears in words starting with *hét-* (*7en* [< héten 'this week'], *7fő* [< hétfő 'Monday'], *7v, 7vége, 7vg* [< hétvége 'weekend'], *a 7en* [< a héten 'this week']) but it can also replace the verb-forming suffix *-het* 'can (modal auxiliary verb)', as in *le7* [< lehet 'it may be']. The abbreviation *jó8* [< jó éjt 'good night'] plays on the similar pronunciation of the Hungarian *éjt* 'night' and the English *eight*. German examples: *8ad "achte auf dich"* 'take care'; *8ung "Achtung!"* 'Watch out!' (Schlobinski 2009: 27).

Example 21: Number or character homophony as a character reduction technique

*Szia!kösziAzAggodast.2tölFél6igVtamOtt.1oratVartamInjUtanIs,kimentA**6asa**,uhMütétKözbenKaptamMég**1et**..kabaVok.*[name]*EljöttÉrtem.tejfogdarabVt,nemCiszta.KösziMég**1x**!p!* [6asa < hatása 'its effect'; még 1et < még egyet 'another one'; még 1x < még egyszer 'once again'] [text message, 2009]

*C!+**biz6atlansag** netovabbja a telefonnal kapcsolatban.nincs még meg a sim,csak holnapra. ha délig meglesz,akkor indulok a 14:30as vonattal fegyencvarosba.csok* [+biz6atlansag < megbízhatatlanság 'unreliability'] [text message, 2004]

*Haho!Sokatkellettmeloznia**7en**is,devolt**1szabad¤om**,dolitirtam.Karacsonyivacsimatézaban.Sokvizsidvan?Szilvrölnincshirem.T?Hangolodokazünnepekre,te?Jótancsit!Puszka* [7en < héten 'this week'; 1szabad¤om < egy szabad napom 'a free day off'] [sms, 2005]

Szia [nickname], *Igen,dolgozom a Kosztolanyi gimiben,jol érzem magam.**Jovo7en** raerek a keddet kiveve. Jo tanulast,puszi:*[nickname] [Jovo7en < jövő héten 'next week'] [text message, 2006]

*Hol jarsz?Rendben utaztal?Nem volt nagy zajongas a vonaton?Tudtal aludni?Sok csomagod van?**Leg1xübb** taxizni,nem?Majd csörömpölj ide,ha **beszél7ünk**!Csok* [Leg1xübb < Legegyszerűbb 'the simplest way'; beszél7ünk < beszélhetünk 'we can talk'] [text message, 2005]

*csak mostesett le a tegnap éjjeli smsed értelme: „rémálomból ébredve..." milyen álom az?:-)))))))) **jó8**!*[initial of a name] [jó8 < jó éjt 'good night'] [text message, 2001]

*Elküldöm még **1x**,h **ért7ö l1en**: Nagyon Kellemes Ka#onyi Ünnepeket&Boldogsagban,sikere kben gazdag Uj Évet kivanok!Nalam maradt az ajandekod.* [1x < egyszer 'once'; ért7ö l1en < érthető legyen 'to be clear':] [text message, 2004]

Replacing consecutive letters with only one letter based on similar sounding can be a good character saving solution as well. The grapheme æ is used as a ligature in the expression *rærsz?* [< ráérsz? 'are you free?']. The letter combination *ksz* is often replaced with *x* (e. g. *emléxem* [< emlékszem 'I remember'], *ezexerint* [< ezek szerint 'accordingly'], *igyexem* [< igyekszem 'I'll do my best'], *kex* [< keksz 'biscuit'], *lax* [< laksz 'you live'). The digraph *sz* is often replaced with the German *scharfes S* (*ß*), irrespective of the fact that a letter is not part of the Hungarian alphabet (e. g. *KÖß* [< kösz 'thanks']; *PUß* [< pusz 'kiss']). Sometimes the Hungarian digraph *ly* is written as *j* clearly disregarding Hungarian spelling rules (the two letters mark the same phoneme—Translator's note). Several examples indicate that the grapheme *x* has become a replace character for *sz*, and is even doubled to indicate gemination (e. g.: *xomorú* [< szomorú 'sad'], *köxönöm* [< köszönöm 'thank you'], *xábbuuxxx* [< szábusssz < szervusz 'hi'], *köxee xépen* [< köszee szépen < köszi szépen 'thank you very much']*roxxindulatból* [< rosszindulatból 'out of malevolence'], *xoríts* [< szoríts 'keep your fingers crossed for me']).

Example 22: Saving characters by replacing consecutive letters with a single letter based on homphony

*fantaziad **kisréßeis** Lég,h LképzLd azt a kupit!1példa:kamraban ebédLtünk.De holnapra mar csak nehany simitas marad.a **ßegélyragaßtas** jo5l&t:takaritas könnyü ezRxP* [kisréßeis < kis része is 'even a small part of'; ßegélyragaßtas < szegélyragasztás 'gluing the edges'] [text message, 2008]

SZASZ![NICKNAME]*NAK VAN TELOJA*:[11111111].*UNONÉNIVEL MÁR TEGNAP IS MAJDNEM +**LÁTOGAΠALAK**,DE +IGÉRTAM,TEHÁT BE IS TARTOM.BALCSIZHATUNK,BENNE VAGYOK. JO VONATOZÁST!**PUßIKA*** [+LÁTOGAΠALAK < meglátogatlak 'I'll visit you'; PUßIKA < puszika 'small kisses'] [text message, 2003]

Nagyon fölment a pumpám az [nickname] *miatt.ÉN még **ijen hüjét** nem láttam.Asszem megint tudom,kinek lesz muszaj bulittartani.:(dea*[nickname]*tnemhivoneg.hogy vagy édesem?* [ijen hüjét < ilyen hülyét 'such an idiot'] [text message, 2004]

There are a variety of symbols at our disposal to express frequent meanings economically. They are mainly logograms, such as +, @, £." (Crystal 2003: 258). Logograms are written symbols that "represent a word or a meaningful part of a word" (Crystal 2008: 182). A popular special abbreviation is the + sign (pronounced as *meg*) standing for the verb prefix *meg* (e. g. *+beszél* < megbeszél 'talk over') or for the *és* 'and' or conjunction *meg* 'plus': *+besz* [< megbeszéljük 'we'll talk it over'], *+cslom* [< megcsinálom 'I'll do it'], *+hívás* [< meghívás 'invitation'], *+kaptam* [< megkaptam 'I've got it'], *+lát6od* [< megláthatod 'you can see it'], *+mond6od* [< megmondhatod 'you can tell it'], *+mondom* [< megmondom 'I'll say it'], *+van* [< megvan 'got it'], *+vár6om* [< megvárhatom 'I can wait it']. The graphic ¤ character standing for *nap* 'day' [based on the homonymy between nap 'day' and nap 'sun'—Translator's note] is an iconic individual sign: *hol¤* [< holnap 'tomorrow']. The section sign (§) is used to indicate the letter *s* (pronounced in Hungarian as *es*) in the female name *Emese*: *M§E*. The Greek *Π (pi)* letter appears in the expression *ΠhN* (< pihen 'rests'). The *$* symbol usually means money.

Example 23: Using symbols to reduce characters

Hahóó!apunak hamarabb ***mutatod*** *+ a kalapod mint nekem?+* ***vagyok sértödve****.goethe sem lehet jobb mint az én integráljeleim,br =***re** *még komáljuk 1mást.otthon31ezel?p* [mutatod+ < mutatod meg 'show it'; + vagyok sértődve < meg vagyok sértődve 'I'm offended; =re < egyelőre 'for the time being'] [text message, 2002]

*Nyuszika!****.osan*** *most akartalak minderröl tájékoztatni.Nem megyek be holnap, de +****kérek*** *vkit.*[name]*1tündér,a másik bige számomra ismeretlen!Puszika* [.osan < pontosan 'exactly'; +kérek < megkérek 'I'll ask'] [text message, 2004]

A szombatom tuti volt,Á-nál aludtam(:- >),Øtanulás.igy ***vas¤*** *voks+tan.a tvben azért* ***követgeΠem*** *az eredmt.nemsemmiszoros****leΠ.szép¤ot!****bye* [vas¤ < vasnap < vasárnap 'Sunday'; követgeΠem < követgettem 'I loosely followed it'; leΠ < lett 'have become'; szép¤ot! < szép napot! 'have a nice day!'] [text message, 2006]

The principle of minimization is applied to messages where non-letter characters are used instead of word-forms or morphemes. The message "*MegØhir*" would be traditionally written as "még nincs hír" 'no news yet', but the abbreviating technique saves seven characters for the sender. The same sign, which is commonly used in university note-taking for negation, appears in the predicative structure "*Ø volt rázós a zh*" (< nem volt rázós a zh 'the written exam was not at all difficult'). *1@* comprises the figure *1* (spelled as *egy*) and the at-sign (spelled as *et*), so the solution of the rebus is *egyet* 'one thing'. The multiplicative suffix *-szer* is replaced with an asterisk in the expression *1** (< egyszer 'once'). The eagerness to be unique may have motivated the creation of this expression: *1/2elem* (fél 'half' + elem < félelem 'fear').

Example 24: Graphemes and symbols replacing complete morphemes

Szia [nickname], *Mar elkezdodott-nemcsak a tanulas,de a* ***1/2elem*** *is:< Meg***Ø***hir a nyvizsgarol. Persze h szanok Rad idot:)! Puszi,jo tanulast:*[nickname] [1/2elem < félelem 'fear'; Ø < nincs 'no'] [text message, 2004]

Szia Barátném! ***Ø*** *volt rázós a zh, ne aggódj. Amúgy fogkrémet felejtettem el tegnap...Na puszi:* [nickname] *E:*)* [Ø < nem 'not'] [text message, 2003]

Hali!nem íroma szdolit,éppen megyek haza,fagyiztam ***1@****,elötte booktáraztam,persze semit sem talaltam.konzulenst meglatogattam,ari volt.a méhekhelyetAZAGYAMZSONG* [fagyiztam 1@ < fagyiztam egyet 'had an icecream'] [text message, 2004]

*Nagyon-nagyonügyivagy!!Gratulálok!!!Majd****1**** *elmehetnénkmegünnepelni,pl.Süssfelnap,bár lehet,erre majd csak a vizsgaidöszakutántudunksortkeríteni,sajnos.pusz* [1* < egyszer 'once'] [text message, 2006]

3.3.2 Neologisms and word creation methods in digilect

Digilect influences both the lexicon and the grammar of a language. The lexical-grammatical characteristics of digilect were illustrated above by the trend to take over IT terminology (e. g. *szkennel* 'scans', *szerver* 'server', *domén* 'domain', *driver*), the influence of foreign languages (chiefly English; e. g. *CapsLock, pendrive, fájl* 'file', *cset* 'chat'), and the large-scale production of new words and collocations (neologisms). The present subchapter takes a closer look at newly coined words specifically in relation to digilect. To this end, it may be useful to first clarify the meaning of neologism.

István Szathmári defines *neologism* as follows: "Neologisms are the new words, expressions, shades of meaning and grammatical forms which continuously enrich the language in parallel with the development of social relationships and thought" (Szathmári 2004: 154–155). The German linguistic encyclopaedia (Bußman 2002: 463) defines neologism as "a newly introduced or innovatively used linguistic expression. As opposed to ad-hoc expressions, neologisms have, to a certain extent, already become part of the lexicon but are still perceived by speakers to be new which makes them stylistically marked. Normally, neologisms are created to name new objects or ideas [...] but they may have an expressive or persuasive function as well" (translated from a Hungarian translation). The pragmatic value of neologisms is sometimes more than their function to name new extra-linguistic phenomena. They can also serve to express group identity, indicate innovation or simply raise awareness.

The classification of neologisms is always problematic: "Actually, the term *neologism* has some inherent theoretical problems which are hard to resolve, including in particular the basic question how new [...a word should be] to be called a neologism" (Aprile 2005: 57; as cited in: Zs. Fábián 2009: 110). In fact, "except for some valuable experiments, there is still no scientifically sound system of criteria for classifying certain words as neologisms" (Scotti Morgana 1981: 2, as cited in: Zs. Fábián 2009: 110). The situation is further complicated by the rapid change of the lexicon. Zsuzsanna Fábián summarizes this as follows: "The lexicon of any language today is characterized by a rapid increase in the number of neologisms. This change is inevitable and unstoppable and comes with scientific-technological development and globalization. Still, only a part of the new words and meanings integrates permanently into the lexicon; many of them prove to be short-lived and fall into oblivion as quickly as they rose to prominence." (Zs. Fábián 2009: 105).

There is a notable related concept, called *netologism*, which refers to neologisms created and used in an online context (*net* × *neologism*). My examples are largely netologisms (e. g.: *kommentel* 'post a comment'; *egoszörfölés* and *egoszörcsölés* 'egosurfing/egosearching, searching for the own name'; *szkájpozás/szkájpizás* 'use the telecommunication software Skype').

In the following, we are going to review the main word formation methods which are used to create neologisms.

Word formation means operations "which result in new words (units of lexical meaning, i. e.: lexemes)" (Ladányi 2007: 57). The most common types of word formation in Hungarian are morphological derivation and composition (Ladányi 2007: 57). Mária Ladányi in her classification of word formation methods mentions a third category, called conversion or zero derivation where a word is created from an existing word without any change in form (e. g.: the adjective *órás* 'one who has watches' > the noun *órás* 'watchmaker') (Ladányi 2007: 57, 65). Less common word formation techniques include back-formation (e. g. *szakdolgozat* 'thesis' > *szakdolgozik* 'writes a thesis'), clipping (e. g. *telefon* 'telephone' > *fon* 'phone'), abbreviation (e. g. *például* 'for example' > *pl* 'e. g.'), acronyms (e. g. *personal computer* > *PC*), morphological blending (e. g. *csatorna* 'channel' × *alagút* 'tunnel' > *csalagút* 'chunnel'). The latter types of word formation use non-morphological (extra-grammatical) means (Ladányi 2007: 65).

In addition to the above mentioned categories, Keszler (2000a: 307–320) and Lengyel (2000: 321–345) mention the following infrequent types of word formation and examples in their studies in *Magyar grammatika* (Hungarian Grammar): reduplication (*ikerítés*, e. g. *bim-bam* 'ding-dong'), clipping and derivation of clipped words (*szórövidülés és továbbképzése*, e. g. *labor* < laboratórium 'laboratory'; *fagyi* < fagylalt 'ice-cream'; *fincsi* < finom 'delicious'; *tetkó* < tetoválás 'tattoo'), back-formation (*elvonás*, e. g. *kajál* 'eat' > *kaja* 'food'; *autóbusz* 'autobus' > *busz* 'bus'; *gépírás* 'typewriting' > *gépír* 'type (verb)'), acronyms [mozaikszó, including: initialisms (betűszók), e. g. *MÁV* < Magyar Államvasutak 'Hungarian

State Railways'; contraction of two words (szóösszevonás), e. g. *maszek* 'entrepreneur' < magánszektor 'private sector'; contamination (szóalakvegyülés), e. g. *zavar* 'disturb' × *kerget* 'chase' > *zargat* 'harass'; egyéb mozaikszók (other acronyms), e. g. *Bácstej* < Bács-Kiskun megyei tej 'milk from Bács-Kiskun county)], word cracking (szóhasadás, e. g. *fia* 'his son' – *fiúja* 'his boyfriend', both derived from *fiú* 'boy' with the same suffix), folk etymology (népetimológia, e. g. *nyugdíjas* 'pensioner' > *nyögdíjas* '~ one with income only enough for groaning', *e-mail* > *emil* 'e-mail', *file* > *fájl* 'file'), genericised proper names (tulajdonnevek köznevesülése, e. g. *Diesel* > *dízel* 'diesel engine/fuel'), solidification (elemszilárdulás, e. g. the independent greeting formulas *dicsértessék* 'praised be' < dicsértessék a Jézus Krisztus 'praise be Our Lord Jesus Christ' and *csókolom* 'hello' < csókolom a kezét 'I kiss your hand', or *mozoghatnékja van* 'showing a persistent desire for action').

I've been collecting newly coined words and expressions in relation to digital media since 2008 (this collection was published in 2012 in my book titled *Netszótár* [Net dictionary]). The corpus was compiled mainly from digital texts including (warez) blogs, forums, instant messaging, tweets, posts and comments. Another important source was my 2010 questionnaire which was completed by 637 respondents. In one of the questions respondents were prompted to enter newly coined words. Later, this list of words was filtered by removing proper nouns (including the name Facebook, but leaving its variants *fácse, face, faszbúk, facézik, fácsézik*) and phrases which were outside the scope of the examination (e. g. *dob egy sms-t* 'send a text message', *felmegy MSN-re* 'log in to MSN', *ráír MSN-en* 'send a message on MSN'). With this process I managed to identify 677 new words.

These neologisms were classified using a combination of the system of Ladányi (2007) and the system of Keszler (2000a) and Lengyel (2000).

3.3.2.1 Word formation

Derivation is a type of word formation which creates "a new word by adding a *képző*-type affix to the root word" (Keszler 2000a: 309).[33] Mária Ladányi notes that "root word + derivative affix" type morphological operations do not fully cover the phenomenon referred to as derivation (*szóképzés* in Hungarian) (Ladányi 2007: 63).

Classifying the 130 derived neologisms in the corpus into word categories we found 92 verbs (e. g. *kamozik* 'use webcamera', *fészel* 'be on Facebook', *csézik* 'play with the video game Counter Strike'), 36 nouns (e. g.: *prezi* 'presentation', *telcsi* 'telephone', *trollkodás* 'trolling') and only five adjectives (e. g.: *bugos* 'has bugs (defects)', *javás* 'runs the Java platform').

33 Pioneers in this field also include Sándor Károly and Jolán Berrár.

Noun neologisms created through clipping are often derived further: *telcsi* 'telefon' or *üzi* 'üzenet', *pöndi* 'pendrive', *láwcsi* 'love' (*láwcsi* is a variant of the English *love*, adapted to Hungarian phonetics, modified by replacing *v* with *w* and provided with the Hungarian diminutive suffix *-csi*). Clipped word are not only derived by the *-i* or *-csi* diminutive suffixes but also by the suffix *-ó*, e. g.: *pityó* < pisztoly 'handgun', *szeró* < server 'server'. The *-ás*/*-és* noun-forming suffix was also frequent in the corpus (e. g.: *zippelés* 'compression of files', *meghekkelés* 'hacking of a programme or website').

Example 25: The occurrence of the word *üzi*

Hejhaj!mizujs?olvastad[nickname]***Üzijét****?:(szegény.takaritas?ottAlszotokMar?komfortérzet?nalunkFestöKész,szép!lehetTakaritani+padlozni.sokMunkaMég.4vizsVanHatra,énKi.puc* [üzi < üzenet 'message'] [text message, 2008]

The noun *tumbász* 'user of the tumblr.com website' is an unusual example for morphological derivation (used in a sentence: *Milyen régóta vagy tumbász?*' For how long have you been a tumbász?'). Its peculiarity is that a basic morpheme was created by clipping the name of the website (*tumb-*) which was then provided with the *-ász* suffix to form a noun which denotes an occupation.

The concept of *meeting* creates a special thematic group among the derived words. The basic word for one of the key functions of texting (i. e. the arrangement of meetings) in Hungarian is *tali* aka *találk* < találkozunk 'let's meet', találkozó 'meeting'. The noun *tali* is formed by the verb-forming suffix *-z* to create the verb *talizik* 'meet sy'. (An analogous example: the verbs *találxunk* and *taláxunk* are the reduced forms of the standard *találkozunk* 'we meet' designed to save characters.)

Example 26: Form variants of the finite verb *találkozunk*

Szia!az index nincs nalam,a hivatalban van.nkm [nickname] *intézte,m épp bent volt.de kérdezd meg a dr hivatalban,h kell.ma van félfogadas,ha jol tudom.holn* ***tal****!puszi* [tal < találkozunk 'we'll meet'] [text message, 2009]

hal!semmi gond,csak az utsó napok itthon.ma grilleztünk,sajna csak családi körben.holnap 5kor fotómutogatós bicaj?iparinál ***tali****?aztán sóstó?jó8pusz* [tali < találkozó 'meeting'] [text message, 2009]

szia [nickname]*m!siófokolsz még?kivel?sokat dolgozol?melyik cukiban?az utsó napot töltöm itthon,holn visszaköltözöm.sze-pé du* ***taliz6nánk****!szólj,ha van kedved&idöd!puszó* [taliz6nánk < talizhatnánk 'we could meet'] [text message, 2006]

De a holnapi sörözésre elgyühettél volna! Na majd 10én ***találxunk****! Sintaxis elöadást meg ilyesmit 1ütt kéne felvenni!*[initial of a name] [találxunk < találkozunk 'we'll meet'] [text message, 2002]

*képzeld,elfelejtettem,hogynem****talizunk****,kerestelek:)* [*hogynem talizunk* < *hogy nem találkozunk* 'that we'*ll not meet*'] [text message, 2006]

duvoltunkmajsán,erlebnisbadoltunk.hol¤valószinümegyünk[name]*valfehértózni. hope,**taláxunk**!puszkó* [találxunk < találkozunk 'we'll meet'] [text message, 2007]

Szia,[nickname]*!boldogNévnapot!mostSzerdanTudszJönni?**tali**f7korAKondiElött!7tölHas LabPopsi!jelezz!még1xBoldog*[given name]-*napot!puszik!* [tali < találkozó 'meeting'] [text message, 2009]

Haho!bocsiACsöndért.[nickname]+*lepi7vgreVittGyörbe.maRegMarVizsgaztam,tegnTanulas. kösziAKöszöntöt!mikor**Tali**?csütEste?*[nickname]*val1eztetnemKell+angoltLemondani.. hogyVagy?p!* [tali < találkozó 'meeting'] [text message, 2009]

Verbs formed with the verb-forming suffixes *-(V)z* and *-(V)l* are remarkably highly represented in the corpus. The following case study was conducted to find out why.

3.3.2.1.1 Case study: the verb-forming suffixes *-(V)z* and *-(V)l* in digilectic neologisms

The denominal verb-forming suffixes *-(V)z* and *-(V)l* are highly productive as "they can create new words under given semantic and syntactic conditions" (Keszler 2000b). The verb-forming suffixes *-(V)z* and *-(V)l* are competitive variants of each other, both meaning in general: 'to deal with something'. Mária Ladányi found that *-(V)z* is the general and *-(V)l* is the specific variant, and the distribution of the two suffixes is subject to special productivity rules (Ladányi 2007: 105–113). The set of rules is provided in *Table 8*. Ladányi (2007) examines the frequency of the suffix from the perspective of productivity mainly within the theoretical framework of natural language theory (1999). These concepts are not discussed here, as Ladányi deals with the whole phenomenon providing detailed theoretical explanations and practical examples.

The present case study aims to verify Ladányi's productivity system through own examples.

Of the 92 verbs in the corpus 25 were verbs ending in *-ik* in the third person singular (this special category of Hungarian verbs is hereinafter referred to as "-ik verbs") and formed by the suffix *-(V)z* (e. g. *emesenezik* 'uses the MSN software', *facézik* 'surfs on Facebook', *guglizik* 'uses the Google search engine'). There were only six verbs in the corpus which were provided with the -(V)z suffix but which not ended in *-ik* (hereinafter "non-ik verbs"), e. g.: *regiz* 'register, log in', *kommentez* 'comment on a post'). A much larger number, altogether 58 verbs had the *-(V)l* suffix (e. g. *tegel* 'tag a person in a photo', *szkennel* 'scan a document', *szkájpol* 'use Skype, the telecommunication software'). The following three verbs did not fit into either category (*shiftelődik* 'move/slide away' e. g. on a bench, *trollkodik* 'intentionally disturb internet communication; act as a troll', *kockul* 'become addicted to video games/internet').

Table 8: Distribution of scope when attached to (foreign) nouns—phonological rules (based on Ladányi 2007)

-(V)z
noun stems ending in a vowel
multisyllabic noun stems ending in a consonant
monosyllabic noun stems ending in *l* or *r*
-(V)l
monosyllabic noun stems ending in a consonant (except: with *l* or *r* ending)
monosyllabic foreign verb stems
multisyllabic foreign verb stems with front vowels or both front and back vowels (suffix with back vowel: *-ál*)
a specific group of nouns ending in *-ing*

Applying Ladányi's productivity system to the verbs formed with the *-(V)z* or *-(V)l* suffix in the corpus results in the following classification:

The following *-(V)z* suffixed verbs fit perfectly into the given morphophonological system: Noun stems ending in a vowel: *csézik/CS-zik/cs-zik* 'play with the video game Counter Strike', *csipázik* 'fights with telescopic rifle in shooter video games', *csörizik* 'ring sy's mobile phone', *facézik* 'use the Facebook website', *guglizik* 'use the Google search engine', *ikszdézik* 'smiles in a way that resembles the emoticon XD',[34] *PvP-zik* 'plays in a "player versus player" mode in video games'; *becécéz* 'sends an e-mail to a secondary recipient', *legugliz* and *meggugliz* 'use the Google search engine to obtain information', *regiz* 'register'; *CTRL + C – CTRL + V-zik* 'use the copy-and-paste function'. Multisyllabic noun stems ending in a consonant: *emesenezik/MSN-ezik* 'uses the MSN software', *internetezik* 'uses the internet', *jútyúbozik* 'uses the YouTube video sharing website', *twitterezik* 'uses the Twitter microblog'; *kommentez* 'comment on a post'. No monosyllabic noun stems ending in *l* or *r* appeared in the corpus. Although Ladányi's productivity rules do not apply to them, the so-called cross-semantic rules (Ladányi 2007: 108–109) can describe sports and game names, as in *kodozik/COD-ozik* 'plays with the video game Call of Duty', *WoW-ozik* 'plays with the video game World of Warcraft); as well as the usage of instruments, as in *kamozik/camozik* 'use web camera', *ircezik* 'uses the IRC, Internet Relay Chat service', *netezik* 'uses/surfs on the internet'). Verbs with non-foreign stem include *gépezik* 'uses/plays on a

34 *XD* (spelled as ikszdé in Hungarian) is an emoticon, i. e. an icon expressing emotion or mood. As an icon it is based on similarity: if rotated 90 degrees, it depicts a human face where *X* symbolizes the eyes and *D* the mouth. It means laughing heartily with tightly closed eyes. It may also mean that the information preceding the icon should not be taken seriously.

computer', *homokórázik* 'waits for computer response' (*homokóra* = hourglass), *kettőspontdézik* 'laugh in a way that resembles the emoticon :D'; *becsizmáz* 'boot in' (playful transliteration, where *be* = in, *csizma* = boot).

The suffix *-(V)l* is attached to monosyllabic noun stems ending in a consonant (except for the consonants *l* and *r*) in the case of the following 17 verbs: *belinkel* 'hyperlink sg', *bezippel* 'compress files', *bippel* 'ring sy's mobile phone once', *fészel* 'use Facebook', *floodol* 'send a lot of texts in a short time frame during instant messaging', *gémel* 'plays a video game', *gengel* 'look and walk around in a video game', *hekkel* 'hack a website', *klikkel* 'click with the mouse button', *linkel* 'provide URL', *mapol* 'create a map for a video game', *pingel* 'measures the time delay of data traffic between server and computer, typically indicated by a moving table tennis ball like icon', *posztol* 'post a message online', *regel* 'register', *szkájpol/skypol* 'use Skype, the telecommunication software', *szörföl* 'surfs the internet', *tagel/taggel/teggel* 'tag a person in a photo'. The *-(V)l* deverbal verb-forming suffix were also attached to monosyllabic foreign verb stems, accounting for one-third of all neologic verbs in the corpus: *addol* 'add sg or sy to a list', *becsekkol* 'check in', *beloggol* 'log in', *bootol* 'load in/boot', *cheatel/csítel* 'cheat in a video game', *csekkol* 'check', *csetel* 'communicate through instant messages', *hájdol/hide-ol* 'hide in a game', *hílel/healel* 'restores the life of (heals) a character in a video game by means of magical spells', *killel* 'kills/shoots down the opponent in a video game', *laggol* 'the game runs at reduced frame rate due to slow internet connection', *lájkol/like-ol* and *meglájkol* 'click on the Like button in Facebook to express liking', *lávol* 'love sy/sg', *nédel* 'use grenade to liquidate the enemy in a video game', *plantel* ~ *plantol* 'activates the timer on explosives in a video game', *pókol/poke-ol* 'pokes at sy on a social networking website to draw his/her attention', *printel* 'print a document', *share-el* 'share content', *szkennel* 'scan a document', *szpemel/spamel* 'send unsolicited emails (spams)', *walkol* 'walk in a video game'. Multisyllabic foreign verb stems with front vowels or both front and back vowels also fall into this category: *copy-paste-el* 'use the copy-and-paste function', *diszlájkol* 'dislike', *kommentel* 'comment on a post', *update-el* 'update'. The exception made to the rule for verbs with back vowels (which were formed by the *-ál* suffix) was only illustrated by one example: *formattál* 'format the computer hard disk'. The corpus did not contain any verbs derived from English gerunds. Although it has a Hungarian stem, the verb *tetszikel* 'recommend a content with the Like ('tetszik') button' is still special as the *-(V)l* suffix follows the *-ik* closing morpheme in it.

Only the multisyllabic complex foreign stems could not be placed in the system: *downloadol* 'download', *forwardol* 'forward', *fotosoppol* 'edit/modify with the software Photoshop', *multitaskol* 'multitask, perform multiple tasks at a time'.

The origin of the verb *bippel* is an etymological curiosity. The word *bippel* was borrowed from an ironical Transylvanian song (Open Stage Székelylend: I love you, Aranka) and it means 'ring sy's phone once'. The word was probably influenced by Romanian and the variety of Hungarian used in Transylvania.

Example 27: Origin and use of the verb *bippel*

Küldtem neked SMS-t / de te arra sem válaszoltál, egy csomószor ***bippeltelek****, / utána meg kikapcsoltad.* [*egy csomószor bippeltelek* 'I gave you a one-ring call many times'] [lyrics]

1&lenM!eddigTornaztunk+dumcsiztunk [nickname] *val.joVt.ugyeRendbenOdaértél, lécci1****bipp****!szépNapunkVt+int,köszönömATulit,helyretett(él). holnNagyonDrukkolok. nyugisKészülést!* [lécci 1bipp 'please ring my phone once'] [text message, 2009]

Bippelj *meg 1x,ha ideérsz 1710re :-) puszko!* [Bippelj meg 1x 'ring my phone once'] [text message, 2009]

17:30 tali ecserin az autónál? 1 ***bipp*** *ha ok* [1 bipp 'ring my phone once'] [text message, 2008]

To sum up: the case study presented the morphophonological and, to a smaller extent, semantic productivity of neologic verb-formation through verbs formed by the suffixes *-(V)l* and *-(V)z*. The corpus-based examination showed that most of the examples containing *-(V)l* or *-(V)z* suffix can be classified in Ladányi's phonological (and partly semantic) system with only four exceptions.

3.3.2.2 Formation of new words by composition

The 90 compounds in the corpus were dominated by nouns (e. g. *fotómegosztó* 'photo sharing website or software', *hírfolyam* 'news feed', *bétaváltozat* 'beta version') and included only a few adjectives (e. g. *Facebook-függő* 'Facebook-addict; one who uses Facebook excessively', *iwiw-gyanús* 'successfully designed [e. g. photo] which is worth sharing on the social networking website iwiw').

The compounds in the corpus typically combine two nouns (e. g. *billentyűzár* 'keypad lock', *forráskód* 'source code', *profilkép* 'profile photo', *videókonferencia* 'video conference'). Compound nouns can also be created by combining participles and nouns (e. g. *böngészőprogram* 'web browser, such as Mozilla Firefox, Internet Explorer', *keresőportál* 'search engine, such as Google', *kezdőlap* 'startup page of a browser', *nyitóoldal* 'startup page of a website'), or nouns and participles (e. g. *e-mail szolgáltató* 'mailbox provider', *adathordozó* 'data storage device', *helyesírás-ellenőrző* 'spell check function', *képernyőkímélő* 'screen saver', *médialejátszó* 'media player', *spamszűrő* 'spam filter'). Adjective + noun combinations are less common (e. g. *kékhalál* 'blue screen (error screen displayed after fatal system error)', *miniblog* 'mini blog, e. g. Twitter', *okostelefon* 'smart phone').

In the following, instead of analyzing all compounds, we will only discuss examples which are interesting on the basis of their meaning, formation or components.

In a novel type of compounds a metric prefix (typically used in IT) is added to the English word *like* written with Hungarian phonetics: *megalájk* 'mega like', *gigalájk* 'giga like', *teralájk* 'tera like' (all meaning: 'like very much'). The same logic is behind the expression *megalol* 'absolutely funny' where the intensifier prefix *mega-* is joined with the internet-specific acronym *LOL* (*laugh out loud*).

New compounds often contain acronyms in the first place (e. g. *PIN kód* 'PIN code', *UTP-kábel* 'UTP cable', IP-cím 'IP address', *RSS-csatorna* 'RSS channel', *SD-kártya* 'SD card', *SIM-kártya* 'SIM card'). The first component is always an acronym borrowed from English (see: *RSS*: *Really Simple Syndication*).

The word *oprendszer* 'operating system' is a compound of the abbreviated form of *operációs* 'operating' and *rendszer* 'system'. The now obsolete word *oplogó* < *oplogó* 'operator logo (a logo which appears on the status screen of mobile phones)' was formed the same way.

The word *tapipad* is a playful transliteration of the English *touchpad* (a pointing device used on laptops as an alternative to a mouse). The *tapi* element of the compound refers and resembles to the English *touching* or *tapping*, that is the movement of the finger operating the device (*tapogat* 'touch' > *tapizik* 'fondle' > *tapi*).

The analogy of the word *körlevél* 'multi-recipient letter; circular' influenced the formation of two other compounds as well: *kör-e-mail* 'multi-recipient e-mail' and *kör-sms* 'multi-recipient text message'. Both word forms raise questions in terms of spelling: *kör-sms* has to be hyphenated as the second component is an acronym (the form *SMS* is increasingly replaced by the lower case form: *sms*), while *kör e-mail* is written in two words to avoid double hyphenation, similarly to *e-mail-cím* 'e-mail address'.

A foreign (chiefly English) component is combined with a Hungarian one for example in *domainnév* 'domain name', *spamszűrő* 'spam filter', *e-mail-cím* 'e-mail address', *e-mail-szolgáltató* 'mailbox provider', *nicknév* 'nickname', *websüti* 'web cookie', *geekduma* 'geek speak (jargon used by technology-obsessed people)', *linkgyűjtemény* 'collection of URLs', and *webalbum* 'web album'. In certain compounds the first component is an originally English technical term written with Hungarian phonetics: *fájlcserélés* 'file sharing (fájl < file)', *fájlformátum* 'file format', *ímélcím* 'e-mail address (ímél < e-mail)', *lézernyomtató* 'laser printer (lézer < laser)'. The opposite order (Hungarian component + foreign component) was represented only by one example in the corpus: *ingyenfrag* 'literally: free frag; a player in video games who is 1. idle or away from the computer or 2. an easy target for others to shoot down'.

The first component of *überface* is the German *über-* 'super, over', which is also used in Hungarian slang as an intensifier (see: *überel* 'outperform'), and the last component is the English *face*. The compound means: 'great guy, popular person'.

3.3.2.3 New words formed through clipping, back-formation and blending

Although clipping (szócsonkítás), back-formation (szóelvonás) and blending (szóösszerántás) are types of abbreviation in the sense that they shorten word-forms, they are discussed among the types of word formation in line with the traditional grammatical classification (new words created through clipping can be considered abbreviations in the short run, but word formations in the long run).

Clipping is used here in the meaning of any operation which creates a new word-form by deleting the beginning or the end of the original word-form.

As clipping is closely related to back-formation, it may be useful to briefly summarize literature on the latter. Back-formation is the reverse operation of morphological derivation (Bencédy–Fábián–Rácz–Velcsovné 1974: 166–167). Its definition emphasizes the role of language users' linguistic instinct: "sometimes language users feel simple words, which cannot be analysed morphologically, as derived or compound words due to false analogy, and they omit the alleged word ending to give new life to the remaining part of the word [...]" Back-formation is considered to be a less common type of word formation (compared to composition and derivation) together with word cracking (szóhasadás), acronyms (mozaikszó-képzés), morphological blending (szóvegyülés), folk etymology (népetimológia), oronyms (szóhatár-eltolódás) and word-form imitation (szóalakutánzás). The term is defined as follows: "Back-formation is a type of word formation by which a non-existent word is deduced (abstracted) from an apparently derived or a wrongly segmented word" (A. Jászó 1991: 295). According to the authors of *Magyar grammatika* (Hungarian Grammar), back-formation is the phenomenon "where an existing or alleged morpheme (usually a képző-type affix or component of a compound) is clipped from a simple word and the remaining morpheme or group of morphemes are used as a full word" (Keszler 2000: 340–1). There are two types of back-formation: either a morpheme is clipped from the word (*sétál* 'walk (verb)' > *séta* (walk 'noun'), *zabál* 'consume excessively' > *zaba* 'food (slang)') or an element is deduced from an alleged compound (*zűrzavar* 'confusion' > *zűr* 'confusion (slang)', *autóbusz* 'autobus' > *busz* 'bus'). The latter type is classified by other authors as a separate category (semantic shortening). The book also mentions the term *clipping* (szócsonkítás), a method preferred by neologists during the Hungarian language reform in the 19th century. They often abstracted existing or alleged suffixes from words to make them shorter:

sávos 'striped' > *sáv* 'stripe, lane', *gépely* 'pulley' > *gép* 'machine', *gyökér* 'root (of a plant)' > *gyök* 'root (as in mathematics and linguistics)'.

Károly (Minya 2003: 69–75), who studies Hungarian neologisms coined in the 20th(-21st) century, distinguishes between four basic types of back-formation in modern Hungarian relying on the relevant literature:

1. "abstraction of the (alleged) root (gyökelvonás)" "creating new root words by clipping off word endings thought to be képző-type suffixes and giving new meaning to the remaining part of the word" (P. Fábián 1998: 46–47); for example: *kapál* 'hoe (verb)' > *kapa* 'hoe (noun)', *ficamodik* 'become sprained' > *ficam* 'sprain';
2. abstraction/animation of a suffix (képzőelvonás, szóvég-megelevenedés): the word ending is treated as a suffix to create new words; for example: *-nc* (*újonc* 'raw recruit'); *-da, -de* (< *csárda* 'country tavern', *kaloda* 'stocks'; > *cukrászda* 'patisserie', *sütöde* 'bakery');[35]
3. abstraction from a compound (Kovalovszky 1977: 166–168): a word is taken as a compound and is shortened to one of its components but the meaning of the compound is preserved, for example: *zűrzavar* 'confusion' > *zűr* 'confusion (slang)'; *rövidzárlatos* 'short-circuited' > *zárlatos* 'short-circuited', *takarékpénztár* 'savings bank' > *takarék* 'savings bank (colloquial)';
4. compound verbs coined through back-formation: compound nouns are shortened to unusual compound verbs; this is the most common type of back-formation today; e. g.: *nagytakarítás* 'thorough cleaning of the house' > *nagytakarít* 'thoroughly clean the house', *gépírás* 'typwriting' > *gépír* 'type', *végkielégítés* 'severance pay' > *végkielégít* 'offer severance pay'.

Back-formation is also used in English: *mobile phone* > *mobile* (UK), *cell phone* > *cell* (US). According to the *Magyar grammatika*, these two examples are inferred forms as the second element is omitted from the original expression.

As a special subtype, clipping (Crystal 2008: 178) is used to create new words by leaving out the last letter or letters (e. g. *comin* < *coming*). A Hungarian example to this is *csörr* which is clipped from *csörrents* or *csörögj* 'call me on the phone'.

The word *mobiltelefon* 'mobile phone' is shortened to *mob* by keeping only the beginning of the word, as well as to *fon* by keeping the last three letters (possibly on the analogy of the English *phone*) or to *tel* by keeping the first three letters of the component *telefon*. This form of clipping is termed as *csonkítás* by Ladányi (2007) and as *szórövidülés* by *Magyar grammatika* (Lengyel 2000: 339–340). For

35 New example: *Gombócda* (< gombóc 'ice cream ball'; the name of an ice cream parlour in Budapest).

the sake of terminological consistency, these two terms are merged here under the name *szócsonkítás* (clipping). Further examples to clipping include *holn* [< holnap 'tomorrow'], *tegn* [< tegnap 'yesterday'], *im* [< imádlak 'I love you very much'], *regg* [< reggel 'morning'], *rem* [< remélem 'I hope'].

Example 28: Clipping as a type of word formation

Hi!Végeztem:a mai 4lett.Boldog vagyok!NYÁR!Holnap reggel megyek haza.Te mikor+hogy mégy [nickname]*hoz?Otthonról,anyu* ***fonjáról*** *majd felhívlak.Edinának jó szórakozást!p* [name initials] [fonjáról < telefonjáról 'from his phone'] [text message, 2005]

ha ok, akkor csengj egyet a ***mobomra!*** [nickname] [mobomra < mobilomra < mobiltelefonomra 'to my mobile phone'] [text message, 2002]

I use the term contraction (szóösszerántás) as a subtype of clipping where the middle of the word is omitted. Some examples to this: *sztem* [< szerintem 'I think'], *tok* [< tudok 'I know/can'], *mek* [< megyek 'I go'], *tnap* [< tegnap 'yesterday'], *vok* and *vaok* [< vagyok 'I am']. These shorter forms are linked to pronunciation variants resulting from increased speech tempo, as explained by Mária Laczkó: "Regarding the Hungarian language, many studies have shown that speech tempo is becoming faster" (M. Laczkó 2009: 447; cf. Gósy 1988; Kassai 1993; Vértes 1987, 1989). Mária Laczkó demonstrated with experiments that increased speech tempo has a (negative) impact on speech quality (M. Laczkó 2009: 463). The same word can be pronounced differently resulting in different pronunciation variants (see *Table 9*).

Certain pronunciation variants (e. g. *szerintem* 'I think' > *szentem*; *szóval* 'so' > *szal*) may be related to the written form used in digilect, though it is not obvious whether less articulated pronunciation influenced the shortened written form or vice versa.

Table 9: Pronunciation variants (based on M. Laczkó 2009: 463)

tehát **'so'**	***például*** **'for example'**	***szerintem*** **'I think'**	***mostanában*** **'these days'**	***szóval*** **'so'**	***általában*** **'generally'**
tát	példal	szentem	mostaába	szova	általába
tet	pélál	szintem	mostába	szoal	általáb
	péla	szeintem	mossába	szal	áltlába
	péda			sza	általba
	példa				áltába
	példá				
	pédál				
	péau				
	pédáu				

The following words are probably taken from orality: *egs* [< egészségedre 'bless you'], *mek* [< megyek 'I'm going'], *má* [< már 'yet, already'], *tok* [< tudok 'I know/ can'], *vót* [< volt 'was/were']. The verb vagyok 'I am' is shortened in a rather peculiar way to *vaok* or even more frequently to *vok*. The word *valószínűleg* 'probably' has several variants: *valszeg, valszi, vsz, vszínű, vszleg*. We can see several mutations in the case of the verb *találkozunk* 'let's/we'll meet': *találkozunk* > *találkszunk* > *találxunk* / *taláxunk*, or the conjunction *mert* 'because': *mert* > *mer* > *mR*. Further examples to contraction: *hnap* [< holnap 'tomorrow'], *tnap* [< tegnap 'yesterday'], *sztem* [< szerintem 'in my opinion'], *naon* [< nagyon 'very (much)', *ien* [< ilyen 'such'].

Example 29: Contraction as a type of word formation

hajh!bpztem!TztamTegn+ma.[name] ***VszínüNincsPént.mindenkiJön,de"késikPicit":) a"mitHozzak"rádBízva.*** [Vszínü < valószínűleg 'probably'] [text message, 2004]

*Szia! Még nem tudom a menüt,de **vszinü** nem bogracsos pöri lesz. Pezsgö tuti,de falavatassal ovatosan,m5centi vastag:) Varlak titeket! Kiss* [vszinü < valószínűleg 'probably'] [text message, 2008]

*Sokatgondoltamrada7en.katasztrofa7emvolt,megmindignincs1asztalomse.marjaratatobbikollegais.tegnszarvasonvoltunkanyuvalkertipartin.tutivolt.holngondrad!jovo7**vokP*** [vok < vagyok 'I am (available)'] [text message, 2005]

duvoltunkmajsán, erlebnisbadoltunk. hol¤valószinümegyünk [name] *valfehértózni. hope, **taláxunk**! puszkó* [találxunk < találkozunk 'we'll meet'] [text message, 2007]

*Szia!**vtam** az irodaban,leadtam a **szlat**. A németes könyv mikor jön? Az is nyomtatando?* [name]*röl még mindig semmi?ha*[name] *hazajön,rajuk kellene kérdezni.köszi mindnt* [vtam < voltam 'I was'; szlat < szlát < számlát 'the invoice'] [text message, 2009]

The following variants were created using a technique which does not fit into any category of word formation but is similar to contraction: *sörtünk* [< söröztünk 'we drank beer'], *elgyünk* [< elmegyünk 'we'll go'], *dulunk* [< indulunk 'we depart'], *csörgi* [< csörögni 'to call on the phone']. In texting users sometimes only use root verbs without rag- or jel-type suffixes. In my 2006 survey a respondent defined this technique as follows: "*verb root without morphological suffixes + dot and similar transformations: csörgi = csörögni* 'to ring sy's *phone*'" (Veszelszki 2006). Commonly used words remain recognizable even if their middle (or beginning or ending) is clipped.

These forms resemble German inflectives which were transferred from the language of comics into chatspeak, then SMS language. Certain motions, actions are expressed by root verbs, e. g.: *grins* < *grinsen* 'grin (verb)', *heul* < *heulen* 'yell (verb)' *dichknuddel* < *ich knuddele dich* 'I embrace you repeatedly with love'. The creation of inflectives in German is commonly attributed to Erika Fuchs who tried to adapt

English sound words to German when translating Micky Mouse comics in the 1950s (Schlobinski 2009: 20–21). Crystal (2008: 184) links these onomatopoeic words with the phatic function of communication: "the social function of language, used to show rapport between people or to establish a pleasant atmosphere. Many textisms involve a phatic function, such as *g* ('grin')". Metalevel action-describing expressions are distinguished from non-metalevel units by asterisks (*) (cf. gesture-describing expressions in Chapter 3.5.1.3: *facepalm*, *arcpálma*, *headdesk*).

Example 30: Action-describing expression, inflective

*ich gehe jetzt schlafen *wink** [I'm going to bed *hand-wave*; source: Fix 2001: 60]
**sigh* well I could still catch the 4-hour train* [Facebook comment, 2011]

3.3.2.4 Other types of word formation

Furthermore, special lexical elements include word creation methods which do not fit into the word formation categories of traditional grammar: the consonant combination *ksz* replaced by the letter *x*, combinations of letters replaced by numbers (*5let*), or the prefix *meg-* and the conjunction *meg* replaced by the + sign. This is not only a question of spelling. It may also modify the word itself, as demonstrated by the expression *müxik* (< *működik* 'operates, works'). *Müxik* is a shorter variant of the word *működik* but playfully formed as if it had both d- and sz-forms [like the verb *verek**sz**ik/vereke**d**ik* 'fight', but *műkö**d**ik* has only d-form, the sz-form (**mű**k**sz**ik*) is ungrammatical – translator's note]. An example for its usage in a text message: *Már most közlöm veled h nagyon fogok neked szoritani!!menni fog&a jelzölámpa is **müxik** már.kell atis-poló nem?bár nekem nem küldenek.mindent bele schaffner(in)!*

In the language of electronic communication (e-mail, text message, multimedia message, internet forum, message wall etc.) a characteristic type of word formation is to spell words phonetically (as pronounced) even if they are normally spelled according to the principle of word analysis (e. g. *lécci* 'please', *eccer* 'once' instead of *légyszi* and *egyszer*). The expression *légy szíves* 'please' > *lécci* went through a similar process of change as the word *asszem* (main clause like expression > particle). Phonetic spelling creates a one-to-one relationship between phonemes and graphemes (English example: *what* – *wot*, see: Crystal 2008: 184). The approximation of spelling to pronunciation is called *pseudo-phonetic spelling* by Frehner (2008: 49).

Acronyms and initialisms are considered to be an independent category (e. g. *ppt* 'Powerpoint presentation', *FB* 'Facebook', *rszü* < rövid szöveges üzenet 'short text message'). Ladányi and Keszler categorize blending differently: Ladányi uses the term *morphologic blend* for the creation of new words by merging two others

(e. g. the already mentioned *csalagút (=Chunnel)*). By contrast, Lengyel (2000: 342–343) differentiates between *blending* (*szóösszerántás*, where the beginning of one word is added to the end of another, e. g. *citrom* 'lemon' × *narancs* 'orange' > *citrancs* 'grapefruit') and *contamination* (*kontamináció*, where words are combined in any other way, e. g.: *zavar* 'disturb' × *kerget* 'chase' > *zargat* 'harass'). The type called morphologic blend by Ladányi and blending by Lengyel is represented by only one example in my data base: *szpemetel* (*spam* × *szemetel* 'litter').

Though it is not a direct word formation technique, but a more broader method for expanding the lexis, *borrowing* and the *adaptation of words* (*szóátvétel* and *szókölcsönzés*) must be mentioned here due to their high frequency. The donor language of foreign words is usually the English. Words can be borrowed in unchanged form (e. g.: *task manager*, *browser*, *inbox*); or with phonetic adaptation (e. g.: *divájsz* < *device*, *fícsör* < *feature*, *kapcsa* < *captcha* 'verification code'). Calques are less frequent examples (e. g.: *kékfog* < *Bluetooth*, *adatbázis* < *data base*, *képbolt* < *Photoshop*).

Sometimes already existing words gain new meaning (expansion of meaning), for instance: *mappa* 'folder' + 'folder on the computer', *becsomagolás* 'wrap up' + 'compress files', *ráír* 'write on it' + 'initiate conversation with somebody through instant messaging'.

Non-categorized examples of word formation include *emil* [< *e-mail* 'electronic mail'] and *láma* [< *lame* 'ineffectual']. Similar example is the Hungarian word *gamma* which derives from the English *game*.

3.3.3 Impersonal forms, depersonification

(Deverbal) noun + the verb "to be" is a special impersonal structure. Most probably, the avoidance of marking the agent in a grammatically explicit way is a face saving act by the speaker (in the sense used by Goffman): *Szercsi van. Lávcsi/ Láwcsi van.* 'I love you'; *Rajság van.* 'we are so cool'. (*Lávcsi and Láwcsi* are the variants of the English *love* written with Hungarian phonetics and with the Hungarian diminutive suffix *-csi.*)

Further examples in my collection: *kurvajóság lesz* '~it's going to be great'; *csetség van* '~there is chatting online'; *nagy szeretés van* '~there's great love here' (*Image 9, 10*). Further examples from spontaneous speech: *boldogság van* '~it's happy here'; *harag van* '~it's angry here'; *királyság van* '~it's cool here'; *élet van* '~it's full of action here'; *kaja van* '~it's eating time here'; *munka van* '~it's working time here'; *unalom van* '~it's boring here'; *szerelem van* '~it's love what we have here'; *mezőgazdaság van* '~it's agriculture what we have here'.

Image 9: Examples for depersonification *nagy szeretés van* '~there's great love here' (Facebook image comment, 2010)

Image 10: Examples for depersonification *kurvajóság lesz* '~it's going to be great', *csetség van* '~I'm chatting online' [Facebook message wall, 2010]

3.3.4 Syntactic characteristics. Spoken language features in writing

The use of spoken language elements in writing is often referred to as "telegram style". Telegrams were systematically compared to text messages by Frehner (2008: 191–201).

Digilect is closer to spoken language than to written language from a syntactic aspect as well.[36] The syntactic characteristics of digilect will be illustrated

36 The differences between written and spoken language are discussed in Chapter 1.4 and the introductory part of Chapter 3.3.

through two case studies below. This limitation was necessary as describing all the characteristics—demonstrating the impact of spoken language on digilect syntax—would have gone beyond the scope of this book due to their continuous expansion, volatility and large number. For this reason, I selected two markedly different phenomena which can provide a pattern for further case studies.

Both the presence of topic-repeating structures and the grammaticalization process of the meta-operator *asszem* prove that digilect are primarily affected by spoken language rather than the norms of written language.

3.3.5 Case study: topic-repeating structure as a spoken language characteristic

The corpus for the analysis of topic-repeating structures were compiled from instant messaging texts taken from the software DirectConnection which is primarily used for chatting, and secondarily for file sharing. The software enabled me to follow conversations without taking part in them which made it possible to avoid the observer's paradox. These conversation texts were used to build a corpus which finally consisted of 21,300 words. Using the word processor's search function all demonstrative pronouns in a topic-repeating function were highlighted to create a collection of examples—together with context—consisting of more than 70 pronouns (topic-repeating pronouns account for 0.34 % of the total word count in this collection). This was the first corpus-based analysis on topic-repeating pronouns. In DirectConnection discourses usually focus on topics such as computers, the use of different software, troubleshooting, and less commonly on personal or public topics. Normally, participants do not know each other personally only by nickname (e. g. [HUN]Einherjarblut or [HUN]Gaben86), though there are some examples for real personal relationships as well (which are noticeable in the conversations). Verbal aggression is not rare in the texts: instances range from simple bantering[37] to really offensive remarks.[38]

It is not only formal written characteristics but also grammatical structures that move these texts closer to spoken language. The usage of demonstrative

37 For example: < [HUN]Barbi > balika333, you are so silly:)"; „< [HUN]slider_keny > face palm Gabat /O\; < [HUN]GaBaT > yeah, palm your face; < [HUN]GaBaT > lest it falls to the ground:)); < [HUN]!!-+=Pistu > your teeth are so big they're pulling it off"

38 For example: „< [HUN]gabkaa33 > what the fuck should we wait for?; < [HUN]Escarioth > kiss my ass [HUN]petike007 :DDD; < [HVIP]Mr.Vain > Thomas??what would you kick it with I would beat your head before you could even move your leg!"

pronouns in particle function was identified as a characteristic feature of spoken language "standing close to expletives" by Borbála Keszler (1983: 175) who analysed informal conversations.

3.3.5.1 Pronouns deixis and coreference

According to *Magyar grammatika* (Hungarian Grammar, Keszler 2000: 68–69), the university course book on Hungarian grammar, syntactically, pronouns are independent parts of a sentence, morphologically, they can be affixed (by rag-type and occasionally képző-type affixes), and semantically, they have indirect denotative meaning (see: Károly 1970: 71) that is they have only partial meaning compared to basic parts of speech. "Pronouns do not refer directly to the concrete world by denominating the referent but in a general, indirect way by means of the context or speech situation" (Laczkó 2003: 314, 2004: 470). "Naturally, this does not mean that pronouns lack any constant general meaning. Still, it is evident that such meanings are not homogeneous but represent a different level of abstraction for each pronoun type, thus, pronoun subclasses are linked to different textological and/or grammatical, and logical functions" (Laczkó 2004: 470).

Pronouns play an important syntactic role; their textological capacity results from their pragmatic and referential nature (see: Tolcsvai Nagy 2001: 169–171). Pronouns (or proforms) are grammatical forms of textual meaning at micro-level; they contribute to the text-level relationship between elemental units (Tolcsvai Nagy 2001: 169).

Referential relationships may be established through one of the two basic categories of textological forms, namely *deixis* and *coreference*.

According to the functional approach, deixis as a textological operation brings the representation of an entity of the extra-textual world into the intra-textual world (Tolcsvai Nagy 2001: 175), thus building a bridge between them. In other words, deixis connects the personal, spatial and temporal relationships within the text with the speech situation and the context by simply referring to them (Tolcsvai Nagy 1999: 161). This reference is primarily achieved through demonstrative and personal pronouns. Szilárd Tátrai (2000: 228) writes that "deictic linguistic elements are those demonstrative and personal pronouns, adverbial pronouns, adverbs, and different lexical and grammatical means which enable us to refer directly to the context of language use, and to certain components thereof (see: Lyons 1977/1989: 637; Levinson 1983/1992: 54–55; and Bencze 1992: 41; Tolcsvai Nagy 1999: 161–162)".

Deictic reference can be exophoric (referring to something outside the text) and endophoric (referring to something within the text; see: Levinson 1994: 856). (For further details on exophoric and endophoric deixis see: Laczkó 2004.)

As opposed to deixis, coreference is a structurally symmetric textological operation as it is made up by the relationship between two linguistic items which refer to the same thing within the text (Tolcsvai Nagy 2001: 180). Earlier and later items in the text can be referred to by anaphors (which refer to antecedents) and cataphors (which refer to postcedents), respectively. Antecedents and postcedents are typically expressions with conceptual meaning, while anaphors and cataphors are typically pronouns. Therefore, coreference marks an asymmetric relationship from a functional point of view: "because one of the two elements (the phoric pronoun) can only be understood if the other (the expression with conceptual meaning) is known" (Laczkó 2005: 78). The meaning of the pronoun is made complete by the context: the coreferential linguistic element referred to by the pronoun is there in the context.

3.3.5.2 Information structure: theme–rheme, topic–comment, topic–focus

In linguistics, it has long been evident that the information structure in communication has to be divided into given and new information. It is, however, disputed whether this distinction means a dichotomy of opposing categories or a continuum (cf. the considerations of Gyuris 2003). Information structure is traditionally described by the following concept pairs: theme–rheme, topic–comment, topic–focus, focus–presupposition (Vallduví 1990). These concepts are defined in several ways (though the definitions overlap to some extent). In Firbas's view (1964: 272), for example, themes convey the least and rhemes the most amount of information in a sentence (Vallduví considers this definition analogous with topic–focus definitions). By contrast, Halliday (1976) defines theme as "what is being talked about" and what is placed first in a sentence. The latter definition makes it easier to identify the theme but it also implies that all sentences have a theme (even wh-questions where this function is allocated to the question word). The theme–rheme progression in the text was termed thematic progression by Daneš (1970, 1974).

Topic–comment definitions attributed to Mathesius (1915) hold that topic is what the utterance is about and comment is what is said about it. The topic is usually placed in a sentence-initial position (though not necessarily, see in more detail: Vallduví 1990).

According to the dichotomy termed topic–focus articulation by the Prague school, topic is old and focus is new information. This can be considered a cognition-oriented approach as it centers around when the information is disclosed.

The focus–presupposition dichotomy was established by Halliday (1967). He holds that focus has the same meaning as rheme, so it marks the most informative part of the sentence, while presupposition (or background) is the information in the discourse which is already known to the speaker and the listener.

This subchapter presented the dichotomy of information structure within a sentence from different views. The common element in the definitions is that one part conveys the new, unknown information (which is also marked by intonation) and the other part refers to old, known information which the sentence talks about (see: Gyuris 2003: 9–13).

The term *topic* is not only used to describe the information structure of sentences (utterances). Structural syntax, for example, defines topic as an element which occupies a specific syntactic position and has characteristic syntactic, semantic and prosodic features.

3.3.5.3 The topic-repeating structure

Topic-repeating structures are discussed by diachronic, traditional, functional, generative and structuralist grammars from different aspects. In the following I will describe these approaches and try to find what they have in common.

The term *topic-repeating structure* is not at all widely accepted: depending on the linguistic approach, it is called topic-repeating pronoun (témaismétlő névmás), topic-repeating pronoun structure (témaismétlő névmásos szerkezet) topic-repeating structure (topikismétléses szerkezet), relative statement (relatív állítás) or contrastive topic (kontrasztív topik). The term topic-repeating structure is used throughout this case study, except for the presentation of the different approaches where the term corresponding to the specific approach is used.

The most frequently quoted definition was provided by Edit Szalamin: "a phrase in topic function or a verbal complement at the front of the sentence is followed by a focusing pronoun (in an agreed form), mostly without stress or pause when pronounced" (Szalamin 1988: 91–92).[39]

3.3.5.3.1 Topic-repeating pronoun from a historical perspective

Looking through the *Collection of Transylvanian Hungarian Etymology* (Erdélyi magyar szótörténeti tár) in the framework of a diachronic examination of spoken language, Piroska B. Gergely (1997: 65, 72–74) found a variety of topic-repeating structures (under the following headwords: *a, az, e, ez* és az *aztán*), but these do not seem to be the most dominant types in the history of the language.

There are only a few examples for pronouns which shift focus to the topic to mark contrast: for example "1766: »Székelly Jánost Domokos Ferentz... kezeb(en) levö Fiatal Faval Föbe akara ütni nálam léven ekkor... Lovász Mihálly, *a'* kapá meg a botott hogy Föbe nem ütötte« ['Ferentz Domokos wanted to hit János Székelly on the head with a club, but there was with me Mihálly Lovász *who* withheld the club

39 As my corpus was based on written texts, it was not suitable for examining pause and intonation (more on this matter: Kádár 2002; Károly 1958; Tompa 1964).

preventing the blow'] [Szilágycseh; *a* mn al.]" (B. Gergely 1997: 72). This structure is more common in identifying clauses, such as: "1756: »a' mely Hidat... ő Nga tsináltatott, *aztat* minél hamarább őszve vagdaltattná« ['the bridges he made himself, *those* he wanted to be cut into pieces as soon as possible'] [Koronka MT; *az* mn II. al.]" (ibid.).

"The most frequent structures in the earlier history of the language—and as such the potential starting points of the evolution of topic-repeating structures—were the ones where the topic and the comment were separated from each other by inserted clauses or complements. Here pronouns stabilized the relationship between the topic and the comment and thus promoted comprehension" (B. Gergely 1997: 72–73). For example: "1606: »azok a zemeljök kiket kj hagjot Daniel Peter uram *azok* nem attak kezest« ['the persons whom were left out by Mr. Peter Daniel, *those* had not provided surety'] [Vargyas U]. 1754: »ã Deszkát... a' meljekből mük dolgoztunk, *azakat* külön tétette« ['the planks, which we used for work, *those* were ordered to be separated from the rest'] [Gernyeszeg MT]. 1817: »Szilágyi Ádám... a házi Mobiliáit mi édes annyárol maradott *azokot* adogattya el és él« ['Ádám Szilágyi ... the tangible assets in the house, which we inherited from our dear mother, *it is those* he sells to make a living'] [M. igen AF; Benkő Sámuel lev.]" (B. Gergely 1997: 73).

Gergely B. found an early, 16-century instance for the reverse topic-repeating structure which is quite rare even today (1997: 73). "1592: »*Azt* penigh en az *levelet* feierwari josa kezebe attom« [~'And *that*, I say *that letter* I gave to the hand of Josa Feierwari'] [Kv]". In the diachronic corpus the main tool for topicalization is the expression *azután* ~ *aztán/osztán* which is used as a particle.

3.3.5.3.2 Topic-repeating pronoun as function-less sign of hesitation

Topic-repeating structures have been subject to debates in Hungarian linguistic prescriptivism since the early 20th century (cf. Kádár 2002).

Prescriptivists have usually discouraged the use of these structures on the grounds that they lack any function. János Horváth went so far as to call them "un-Hungarian" structures borrowed from a foreign language (Horváth 1911: 70). Similarly to Horváth, József Tompa (1964: 359–363) also considered such pronouns as expletives or hesitation signals. According to Tompa these structures stylistically belong to the unsophisticated, substandard register, though he noted that they may be characteristic to spoken language. J. Béla Nagy argues that these structures are used primarily for rhythmic purposes (Nagy J. 1965: 352–354; 1963).

Edit Kádár, who made Edit Szalamin's definition more specific (Kádár 2002), believes that this language use issue is not nearly as central and stigmatized as, for example, the *amely/ami* distinction. Kádár is sure that the theory of prescriptivists cannot (and should not) be applied to the usage of topic-repeating pronouns in informal written texts (e. g. instant messaging), as normative language use is clearly far from such texts. The syntax of these texts also shows that they share many characteristic with spoken language, which may explain the relatively large number of topic-repeating pronouns in the corpus.

3.3.5.3.3 Topic-repeating pronouns as part of relative statements

Edit Kádár (2002)—using an analytic model built on the concepts of topic, comment and focus (É. Kiss 1983)—examined a unique form of topicalization: a primarily spoken language structure in which a (topic-repeating) demonstrative or adverbial pronoun signifies that the topic is marked.

She examines sentences containing such structures as relative statements which are created on the basis of a special system of conditions.

3.3.5.3.4 Topic-repeating pronouns marking contrastive topics

In 2003 Beáta Gyuris studied topic-repeating structures as markers of contrastive topics, i. e. as the sign of units which are placed on the periphery of the sentence, pronounced with ascending intonation, followed by a significant pause in speech, and used to mark semantic contrast. Gyuris refutes that the contrastive topic is the second focus of the sentence or a subtype of the topic, and argues that sentences containing a contrastive topic (i. e. topic made emphatic by a topic-repeating pronoun) are structures with special syntactic and prosodic characteristics.

3.3.5.3.5 Topic-repeating pronouns as part of an atypical anaphoric relationship

Krisztina Laczkó (2003: 321, 2005: 83) regards topic-repeating structures as atypical anaphoric relationships because the anaphoric pronoun and its antecedent are in the same clause and agree in case and number. The pronoun does not have an identifying function in the topic-repeating structure but its role is similar to that of a discourse particle. Topic-repeating structures can be regarded as a special case of discourse deixis in which “the sole function of the pronoun is to place some kind of an emphasis” (Laczkó 2001: 105, e. g.: *A sakk* ***az*** *egy jó játék.* [‘Chess **that** is a good game.’]).

Laczkó demonstrates that “despite the functional similarity, topic-repeating pronouns cannot be considered as particles in terms of their form. Particles take no prefix or suffix, they do not form morphological or syntactic relationship with other words, and they may not function as a part of a sentence (cf. Kugler 2000a: 275). Topic-repeating pronouns, however, do not meet these criteria because they agree with the antecedent in number and case” (2003: 322, 2005: 83).

In her 2001 study Krisztina Laczkó suggests that topic-repeating structures could be considered as a special type of interpretative structures, however, later in her 2003 and 2005 studies—referring to Sándor Károly (1958: 32, 68)—she adds that “they are interpretative only in form but not in fact” (Laczkó 2003: 322 and 2005: 83). Sándor Károly argues that this structure emphasises the preparatory part of the sentence and that it came to existence due to pragmatic reasons: the pronoun has a rhythmic function and marks hesitation.

3.3.5.3.6 Topic-repeating structure and fronted proposition

The so-called fronted proposition (előrevetett propozitum) is an anaphora-like structure which resembles topic-repeating structures. For example: “*Búza, rozs, kukorica, krumpli, répa, disznó, liba, tyúk, apró szemű sárgabarack, vörösszilva – ezt adhatják a kúnhalmiak...*” [‘Wheat, rye, corn, carrot, swine, goose, hen, apricot and red plum—this is what Kúnhalom can offer...] (Keszler 2000: 468).

3.3.5.3.7 A common element in relevant literature

This relatively detailed review of the relevant literature aimed to demonstrate that different linguistic schools describe and classify topic-repeating structures differently. However, they have something in common: they all describe topic-repeating structures as a characteristic of spoken language.

3.3.5.4 The function of topic-repeating structures

I applied the grammatical theory to the corpus and in the following I will use the results to classify the functions of topic-repeating structures. Although there are many references to the various uses of topic-repeating structures, no detailed, systematic summary or statistical data have been provided about the linguistic role of topic-repeating pronouns.

Edit Szalamin reckons topic-repeating demonstrative pronouns primarily as instances of spoken language redundancy but she also presents many additional functions of these pronouns (Szalamin 1988: 92). Krisztina Laczkó takes a holistic functional linguistic approach and grades the functions of topic-repeating structures on a scale: “The range of emphatic topic is naturally not homogeneous; there are different grades ranging from the strong expression of contrast to the usage as fillers” (K. Laczkó 2003: 321, 2005: 83).

Topic-repeating pronouns may take the following functions in order of frequency: highlighting, semantic contrast, identification, fronted proposition, anaphoric connection. The corpus contained no examples for topic-repeating structures in emphatic function (i. e. used as fillers) as it was compiled from written texts.

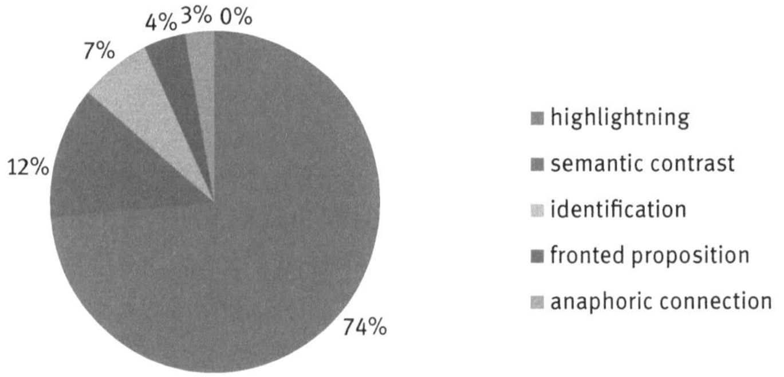

Figure 4: The function of topic-repeating pronouns in instant messaging

The primary function of topic-repeating pronouns is emphasis, as documented by Szalamin (1988) and Laczkó (2003: 321 and 2005: 83). In modern spoken Hungarian the speaker may use the following main tools to emphasize his/her purpose: *ráérni ráérek* type root-repeating infinitive structures aimed at emphasizing the verb (cf. Nádas 2007); structures involving topic-repeating pronouns structures aimed at nominal items, and certain words used as particles (e. g. *aztán*) (B. Gergely 1997: 72–74).

The corpus analysis justified the common observation in relevant literature that the main function of topic-repeating structures is to highlight the preceding nominal word or expression to better reflect the purpose of the speaker. Two-thirds of topic-repeating pronouns appearing in the examined instant messaging texts serve to emphasize something (see: *Figure 4*). Particle-like pronouns place the topic into the range of emphatic topics (Szalamin 1988: 95).

Example 31: Emphasis as the main function of topic-repeating pronouns

< [*HUN*]*Gyöngyi* > *Az ogm*[40] ***az*** *milyen formátum? Nemvagyok szellemifogyi, csak kezdö* ['What kind of format is OGM? I'm not a retard just a beginner.'] [chat, 2007]

< [*HUN*]*gomeZZZ* > *a TTH*[41] ***az*** *hogy muxik pontosan?* [*'How does TTH work precisely?'*] [chat, 2007]

< [*HUN*]*barkoba* > *feltöltési sebesség* ***az*** *melyik?* [*'which is the upload speed?'*] [chat, 2007]

In this respect, Edit Kádár points out that topic-repeating pronouns can be considered as a special form of spoken language redundancy: "its function

40 File format.
41 TTH is the abbreviation of "Tiger tree hash", a commonly used form of hash tree.

is to make explicit a topic which is already emphasized by stress and intonation" (Kádár 2002: 310). Redundancy in this case does not mean being superfluous as the presence and absence of the pronoun influences the meaning of the sentence.

The corpus, which abounds with IT-related technical terms, abbreviations and proper nouns (software and video game names), shows that the topic-repeating pronoun mainly appears after these unusual and less known terms and thus shifts the focus to them (*Example 32*). In my corpus numbers are also highlighted, emphasized this way (*Example 33*).

Example 32: Topic-repeating pronoun after foreign words or technical terms

< *[HUN]Foox®* > *strong dc* ***az*** *jó* ['strong DC is a good thing'] [chat, 2007]

< *[HUN] [ADSL]Tyrael* > *ndc* ***az*** *egy 306os verzioju magyaritott zdc* ['ndc is an adapted version 306 zdc'] [chat, 2007]

< *[HUN]gidacska* > *strong* ***az*** *többszálasd* ['strong is a multiple-tread ...'] [chat, 2007]

< *[HUN]128kb* > *rol 325i* ***AZ*** *VIRUS* [rol 325i is a virus] [chat, 2007]

Example 33: Topic-repeating pronoun after numbers

< *[HUN]Ateec* > *7* ***az*** *neked gyorsss???* [*'7 is fast for you?'*] [chat, 2007]

< *[HUN]-prot-* > *256* ***az*** *kicsit kevés lesz* ['256 is going to be a bit short] [chat, 2007]

In addition to shifting focus, topic-repeating demonstrative pronouns also serve to harmonize the topic–comment relationship when there is a distance between them, and to refer back to the topic as real anaphors (Laczkó 2003: 321 and 2005: 83). In the corpus 2.86 % of the pronouns play this role. This function is often allocated to the so-called focus-repeating pronouns (fókuszismétlő névmások) which are similar in structure to topic-repeating pronouns. Focus-repeating pronouns effectively facilitate coherence in complex sentences containing inserted elements by placing the focus into a pronominal position (in this case the topic and the pronoun do not stand next to each other).

Example 34: Topic-repeating pronouns in sentences with inserted elements

< *[HUN]Suwayyah* > *nem tudja valaki, hogy az Aliens – Special Edition (ez ugyebár a második rész)* ***az*** *mitől special edition?:))* [*Anyone knows why Aliens – Special Edition (which is the second part, isn't it?) is called special edition?:))*] [chat, 2007]

Topic-repeating structures are also suitable for expressing semantic contrast. More than 10 % of the examined structures serve this function. These structures usually contrast two phenomena; and in some cases they even place a pronoun after both contrasted nominal items.

Example 35: Topic-repeating pronouns marking semantic contrast

< *[HUN]ZzZeno* > *www.atw.hu* ***az*** *van. de ftp.atw.hu* ***az*** *nincs. hol a hiba?* [*www.atw.hu is available but ftp.atw.hu is not, where is the problem?*] [chat, 2007]

< *[HUN]Csiribiri* > *az oraszám* ***az*** *megvan de viszont az oradij* ***az*** *karcsi:sss* [I've got a lot of classes but don't get much money for them:sss] [chat, 2007]

< *[HUN]HEGYI* > *nem kék külömböző kocsiról beszélsz? Dodge Charger meg a Dodge Ram* ***az*** *két külön tious* [you're talking about two different cars, aren't you? Dodge Charger and Dodge Ram are two different types.] [chat, 2007]

In indentifying sentences the pronoun may be used for rhythmic purposes. The subject and predicate of such sentence is the same lexical element; and the pronoun typically follows the subject and precedes the predicate. This structure is present in 7 % of the corpus.

Example 36: Topic-repeating pronouns in identifying sentences

< *[HUN]AngelinaJoli* > *Bocs, de a tény* ***az*** *tény* ['Sorry but facts are facts.'] [chat, 2007]

< *[HUN Brainstorming* > *hiába földi* ***az*** *földi :D* ['well, earthly is earthly'] [chat, 2007]

< *[HUN]WIKTOR* > *a szály*[42] ***az*** *szabály :) nem tudsz mit csinálni.. megmondta Polyák anyja.* [*a rule is a rule :) you can't do anything with it.. Polyák's mom has already told us.*] [chat, 2007]

The present case study on topic-repeating structures not only aimed to provide statistical information about and summarize the functions of such structures but also to prove that this spoken language characteristic may be found in written texts, more specifically in secondary literacy as well.

Secondary literacy and orality share many grammatical characteristics. This observation, however, needs to be supported by further examinations on the frequency of topic-repeating pronouns in spoken language.

This type of structure is not mentioned in the Rules of Hungarian Orthography (i. e. it has no standard form of spelling), this is why only spaces are used for segmentation. The punctuation issues relating to this structure may provide further topic for research (but are not discussed in detail here).

42 This word is mistyped. Correctly: "szabály".

3.3.6 Case study: grammaticalization of the Hungarian expression *asszem*

The influence of spoken language on digilect is illustrated here in another case study which deals with the grammaticalization process of the discourse marker *asszem*. Although grammaticalization as a process dates back earlier than digital communication, certain phases of it can be witnessed in texts marked by digilect.

Grammaticalization has a dual meaning: on the one hand, it refers to research, or a theoretical framework dealing with the origin of grammatical forms, on the other hand, it refers to the phenomenon of language change during which certain linguistic items become more grammatical (Hopper–Traugott 1993: 1–2; Dér 2013a, 2013b). In this book, I will use the term *grammaticalization* with the latter meaning.

The case study below tries to support, using a Hungarian example, Traugott when she argued that the process of grammaticalization may have resulted in discourse markers (which are still subject to change). Before this, however, we will briefly review the various definitions of the process and look at its main characteristics and concomitants, as well as the main factors of canonical grammaticalization and Lehman's parameters. We will also clarify the notions of meta-operators and discourse markers, together with their main features and functions. Finally, I will describe the expression *asszem*, a currently evolving discourse marker of a written-spoken language, according to the main aspects of grammaticalization.[43]

Example 37: The use of *asszem* on the social networking website Facebook (messages posted on Facebook wall and related comments, 2010)

*Ma délután elindultam a belvárosba, hogy bevigyek egy könyvet és hogy majd 5–6 felé hazaérek és tanulok a ma reggeli vizsgámra. Ehhez képest most értem haza, miután 2,5 társasággal 3 különböző helyen (**asszem**) 7 sört megittam. Azt hiszem, itt vagy nagyon stimmel valami, vagy nagyon nem. :-)*

[Comment No. 1] *Sztem minden a lehető legnagyobb rendben történt. :X :D*

[Comment No. 2] *Ha átmegyek, akkor tuti! De **asszem** amúgy is.*

43 Although the examples are taken from a corpus analysis conducted in 2009–2010 and the examination of a social networking website in 2006, the phenomena illustrated through them and the conclusions drawn from them seem to have remained valid. Iwiw, the Hungarian social networking website was permanently closed in 2014.

3.3.6.1 Grammaticalization

3.3.6.1.1 Various definitions

The two interpretations of grammaticalization,[44] the various theoretical frameworks, and the difference between diachronic and synchronic perspectives have resulted in many diverse definitions of grammaticalization. The expression itself was first used by Antoine Meillet (1912: 131, as cited in: Diewald 1997: 6) to refer to the process where an autonomous word takes the role of a grammatical marker. This general definition was later modified by Christian Lehmann, who argued that during the grammaticalization process a linguistic item becomes or is made more grammatical (Lehmann 1991: 493, 1995: 9).

The broader and narrower definitions of grammaticalization may be reconciled as follows: a lexical item takes on grammatical functions or a grammatical item becomes even more grammatical (Hopper–Traugott 1993: 2; also: Heine–Claudi–Hünnemeyer 1991: 2; Heine–Kuteva 2002: 1). Lehmann's differentiation between lexicon and grammar (2002: 1–3) is too binary: he looks at grammar as a system of regularly formulated signs treated analytically, and lexicon as a system of irregularly formulated signs treated holistically. As opposed to this binary distinction, Martin Haspelmath views grammar and lexicon on a scale: "grammaticalization pushes the linguistic expression towards the functional end of the lexical-functional continuum" (Haspelmath 1999: 1045, as cited in: Dér 2008: 14; *Figure 5*). Heine describes the process as a series of changes, though in his *World Lexicon*, which deals with grammaticalization through examples from 500 languages (Heine–Kuteva 2002), he only focuses on source and target forms.
To sum up, I accept the sufficiently complex and general definition of Csilla Ilona Dér who points out that grammaticalization is a type of linguistic change during which "in a highly specific pragmatic and morphosyntactic context lexical or less

Figure 5: Grammatical and lexical continuum (Figure by the author on the basis of Haspelmath's description)

44 In fact, the word has one more meaning: it refers to an early stage of language development where children abandon pre-grammatical, one-word expressions and engage in utterances which observe grammatical norms (compare: Feilke–Kappest–Knobloch 2001: VII). However, this will be disregarded here.

grammatical/pragmatic items transform into more grammatical/pragmatic items resulting in greater integrity between the items" (Dér 2008: 117).

3.3.6.1.2 (Concurrent) phenomena of grammaticalization: formal (phonological) change, structural modification and semantic-pragmatic change

In their definition of grammaticalization, Heine and Reh do not examine the source item, the process or the target item, but they rather list the stages of the linguistic change: "In essence, we understand the term 'grammaticalization' as a development process in which linguistic items lose their semantic complexity, pragmatic significance, syntactic freedom and phonetic substance" (Heine–Reh 1984: 15, as cited in: Dér 2008: 9).[45]

Heide Wegener (1998: 37) distinguishes between three processes which occur during grammaticalization (and uses them to analyze the evolution of German modal particles). These processes are as follows: 1. reduction of phonological substance (the linguistic item loses some of its phonological integrity, i. e. it becomes less emphatic; full vowels become schwa vowel sounds and even become clitics), 2. loss of semantic substance (the linguistic item loses some of its specific/referential/denotative meaning, and at the same time it gains an abstract/grammatical meaning), 3. loss of syntactic freedom (the item may not take any position in the sentence, and it may not appear in all types of sentence). Wegener mentions pragmatic significance, as used by Heine and Reh, only later, in the detailed elaboration of the topic under, terming it pragmatic/illocutionary/meta-communicative strengthening (Wegener 1998: 43).

In the introduction to their lexicon of grammaticalization, Heine and Kuteva mention four mechanisms inherent in grammaticalization: desemanticization (semantic bleaching); extension (the scope of use extends to new contexts); decategorization (loss/modification of morphosyntactic characteristics); erosion (phonetic reduction, loss of phonetic substance).

Following Dér (2008), I will highlight two out of the several aspects: the formal-structural and the semantic-pragmatic changes. As function and structure always change together—or as Lehman (1995: 126) put it: they go "hand in hand"—they can only be separated for analytical purposes: they are integral part of each other both chronologically and in terms of the chain of causation.

It is reasonable to accept the distinction made by Csilla Dér (2008: 31): formal changes shall be understood as changes appearing in the surface structure, and structural changes as changes affecting the deep structure. For example, the change of word category is a structural change, while phonological

45 Although the hypothesis that grammaticalization is a unidirectional process and the phenomenon of degrammaticalization are mentioned by all studies in this field, I will not discuss these topics as they are not really relevant here.

reduction and phonemic merger are formal changes. Dér in her paper from 2013 takes grammaticalization as “a combination of three specific mechanisms that are invariably present in any instance of grammaticalization: grammaticalization = specific semantic change + category shift (+ destressing)” (Dér 2013a: 176). This view excludes the interpretation of grammaticalization as purely formal or purely semantic change, but includes pragmaticalization among grammaticalization processes. Agreeing with Hopper–Traugott (1993: 4), structural changes are taken here to include not only individual words but also full constructions and linguistic changes affecting even syntagmas.

Grammaticalization may take place without formal erosion, but not without structural change (structural reanalysis, change of word category) (Heine–Kuteva 2002: 3; Dér 2008: 32). Thus, grammaticalization is—if not always, but quite often—accompanied by phonological (surface) changes, either in the form of phonological reduction (for example: *találom* ‘I find’ > *talám* > *talán* > *tán* ‘perhaps’), or as the fusion of linguistic items (for example: *az* ‘that’ + *ki* ‘who’ > *azki* > *akki* > *a’ki* > *aki* ‘who’ [Haader 2003: 507–508]).

In relation to grammaticalization, the following five semantic aspects are typically highlighted: reanalysis, analogy (see: Meillet 1912), metaphor, metonymy and iconization (Di Meola 2000: 12–19; Hopper–Traugott 1993: 32–62; Diewald 1997: 42–63; see also: Dér 2008; Lehmann 1995; Heine–Claudi–Hünnemeyer 1991). This subchapter only gives a general description of semantic change, but it may be worth examining grammaticalization from these aspects as well (especially regarding Hungarian discourse markers).

Grammaticalization needs specific context, therefore it might as well be described as context-induced reinterpretation (Heine–Kuteva 2002: 2). Joan Bybee explains semantic change occurring during grammaticalization with habituation, that is certain linguistic items gain more general meaning as a result of frequent usage (Bybee 2003: 618). Semantic change is often described as the loss or bleaching of semantic characteristics (e. g.: Heine–Kuteva 2002: 2), still this is not emptying or depletion (desemanticization, compare: Lehmann 1995) but change. From a semantic-pragmatic point of view it also involves enrichment or strengthening (Traugott 2003). According to the distinction made by Hopper and Traugott (1993: 69), semantics primarily deals with meanings which are relatively stable irrespective of the context, typically arbitrary, and may be analyzed if certain logical conditions are met; while pragmatics focuses on the expectations, conclusions, assumptions and communicative purposes of the participants in communication.

The semantic change makes the linguistic item lose some of its specific meaning constituents and gain several abstract constituents (compare: cognitive metaphor theory).

Inference is the process of deriving logical conclusions. In a communication situation it helps the receiver interpret the sender's utterances and understand the conversational implications, that is the messages suggested but not expressly said by the speaker (Molnár 2002: 13). Grammaticalization is the conventionalization of these implications (Ladányi 1999: 411; Dér 2008: 27). Pragmatic strengthening means the stabilization of inferences, the integration of pragmatic meanings into the structure's meaning. This phenomenon might as well be termed as epistemification (Epistemifizierung, compare Ágel 1999: 181), as it refers to the evolution of linguistic means which reflect the speaker's judgement on the truth-content of propositions (Molnár 2002: 13).

3.3.6.1.3 Lehmann's parameters

Lehmann's grammaticalization parameters are central to any examination in this field (Lehmann 1982: 120–179, 1985: 306). This is also reflected by the fact that all studies dealing with grammaticalization at least mentions these factors, and many give detailed analysis of them (inter alia: Di Meola 2000: 6–12; Diewald 1997: 22–29; Eckardt 2006: 24–26; Dér 2008: 32–34).

Lehmann argues that the degree of grammaticalization of a linguistic sign is primarily reflected by its autonomy (independence): the more grammaticalized the item, the less autonomous it is, i. e. their relationship is marked by inverse proportionality (Diewald 1997: 22). The concept of autonomy is difficult to access, and is made more accessible by Lehmann through three categories. These categories (weight, cohesion and variability) mark the rows of the 2 × 3 matrix, while paradigmatic and syntagmatic dimensions give the columns (Table 10).

Weight (Gewicht) relates to the distinctness of a linguistic item. Paradigmatic weight or integrity indicates the semantic and phonological size of a sign. Due to phonological attrition or desemanticization, the integrity of the linguistic sign

Table 10: Lehmann's parameters, supplemented with arrows indicating the degree of grammaticalization[46]
(Source: Lehmann 1995: 123; supplemented by Ágnes Veszelszki)

dimension / **parameter**	*paradigmatic*	*syntagmatic*
weight	integrity ↓	structural scope ↓
cohesion	paradigmaticity ↑	bondedness ↑
variability	paradigmatic variability ↓	syntagmatic variability ↓

46 The ↑ arrow indicates that if the degree of grammaticalization is stronger, the parameter is also stronger, the ↓ arrow indicates that the parameter is weaker even if the degree of grammaticalization is stronger.

taking part in the grammaticalization process becomes weaker than that of less grammaticalized signs.

Structural scope (Skopus) indicates the linguistic item's scope of action within the construction. It narrows down with the condensation process of grammaticalization.

Cohesion marks the relationship of the linguistic sign with other signs: the more a linguistic sign is linked to other signs, the less autonomous (and the more grammaticalized) it is. Paradigmatic cohesion or paradigmaticity refers to the cohesion of a sign with other signs in a paradigm, i. e. the degree to which the sign is liable to enter into a paradigm. This parameter is enhanced by paradigmaticization (paradigmatic integration).

Syntagmatic cohesion or bondedness (Fügungsenge) marks the intimacy with which a linguistic sign is connected with another sign to which it bears a syntagmatic relation, and its degree may even extend to merger/fusion.

Variability (Variabilität) is the mobility compared to other linguistic signs. In the paradigmatic dimension it signifies the extent to which the linguistic sign may be used (up until becoming obligatory), while in the syntagmatic dimension it refers to the linguistic sign's mobility within the syntagma. Grammaticalization reduces both paradigmatic and syntagmatic variability.

These grammaticalization parameters do not have absolute values; it is more reasonable to place them on a scale. The degree of grammaticalization may be expressed by summarizing these six parameters and comparing with other linguistic items.

These six parameters are commonly referred to as the canonical parameters of grammaticalization which characterize a typical grammaticalization process but which may not or only with restrictions be applied to cases different from prototypical cases (e. g.: the evolution of discourse markers) (Eckardt 2006: 26; Traugott 1997). Csilla Ilona Dér tries to resolve this contradiction by calling the evolution of discourse markers *pragmaticalization* on the basis of the most characteristic stage of the process, in agreement with Waltereit (2002) (Dér 2008: 110–117).

3.3.6.2 Meta-operators ~ discourse markers

Several theories set out to explain the evolution of discourse markers. Traugott (1988) believes that they came to existence through grammaticalization other than the canonical one (i. e. not determined by Lehmann's parameters), while Waltereit (2002) argues that they developed through pragmaticalization. In the following I will focus on this subfield of grammaticalization theory.

Csilla Ilona Dér (2009) compiled a detailed list of the various (more than 70 Hungarian and foreign) names and definitions of discourse markers, indicating their similarities and differences. The list will not be reproduced here; only the three most influential theories are highlighted.

The elements holding this function are called modifying sentence part by Borbála Keszler (1986), following Ilona H. Molnár. "In these sentences [...] the most important part of the utterance is not contained in the main clause but in the subordinate clause; the main clause only provides secondary information about the speaker's relationship to the message contained in the subclause" (H. Molnár 1968: 45).

Janusz Banczerowski (2000: 132–143) attributes similar function to the so-called meta-operators which are involved in the organization of conveying messages and contain statements on the parts/elements of the text.

The term discourse marker also refers to function rather than word category (compare: particle). The most frequent definitions are the following: elements connecting discourse segments (Bruce Fraser), metapragmatic elements (Traugott 1997: 3). The functions of discourse markers: they connect discourse segments, contribute to the segmentation of the text, serve as epistemic signs, regulate the relationship between the participants in communication, and promote turn-taking (Auer–Günthner 2003: 1).

Using the findings of Dér (2008) and Wegener (1998), I collected the most important properties of discourse markers. Meta-operators do not influence the truth-content of the sentence, so they are not propositional items. They have procedural rather than conceptual meaning, since they serve as the speaker's instructions and give metatextual work to the recipient. They are sometimes deemed optional and omittable, though this is indeed doubtful as psycholinguistic examinations show that meta-operators largely contribute the recipient's comprehension. Their presence is often linked to orality. (On Hungarian discourse markers: Dömötör 2009; Schirm 2009.)

3.3.6.3 The grammaticalization of the discourse marker *asszem*

Alexandra Markó took note of the semantic and formal change of *asszem* as early as in 2003. In her paper titled *Asszem – is it possible that words become modifiers/particles in synchrony?* (Markó 2003) assumes, on the basis of Old Hungarian analogies, that *asszem* is on the way of becoming a modifier or particle. Ilona H. Molnár came to a similar conclusion (1961: 356): "the function of this main clause, which has largely lost its original meaning, [...] is similar to that of modifiers". Alexandra Markó examined private emails because, compared to other written forms of communication, they "contain much more phrases typical of spontaneous speech and much more abbreviated forms imitating and uncovering the loosening techniques of spoken language" (Markó 2003). The following forms of internet literacy/digiliect probably originate from medial orality (compare: Koch–Oesterreicher 1994) and came to existence through looser pronunciation: *nem tudom* 'I don't know' > *nemtom* 'dunno', *mit tudom én* 'I don't have the faintest' > *mittomén, azt hiszem* 'I believe' > *asszem, megyek* 'I'm going' > *mek* and *mék, miért* 'why?' > *mér, mert* 'because' > *mer, egészségedre* 'bless you!' > *egs, valószínűleg* 'probably' > *valszeg*. The primary motivation behind these reduced forms includes simplification, faster typing and humour/invention.

Comparing the innovative expressions of the digilect, we will find that *müxik* (the contracted form of *működik* 'it works'), *aszonta* (the contracted form of *azt mondta* '(s)he said') and *nemtom* (the contracted form of *nem tudom* 'I don't know') are only the formal variants of the full expressions, but *asszem* has undergone some pragmatic change as well. The forms *nemtom* or *nem tom* are not newly created expressions, they "have been present in written tradition for centuries: it was first recorded by the Hungarian linguist and philosopher Albert Szenczi Molnár in his Hungarian Grammar from 1610 as an example to elision in the form *Nem t'om*" (Markó 2003). *Nemtom* is the merger of *nem* 'not' and *tudom* 'I know', imitating the pronunciation of the expression. *Nemtom* may appear in a main clause position (*nem tudom, hogy*... 'I don't know whether'), as the main clause of the object clause introduced by the conjunction *hogy*, though this conjunction may be missing from the beginning of the subordinate clause (Haader 2000: 487–490). No other (epistemic or pragmatic) meaning is attributed to *nemtom*; it is the abbreviated and merged form of *nem tudom*, a logical-semantic relationship between a negative particle and a verb (Kugler 2000b: 289).

Example 38: The use of *nemtom* in various forms of texts typical of the digilect

Szú!emlegeПem én is az átállítós buli,m [name]*volt-tömeg-,tök depis a 7végém de* ***nemtom*** *még.csüt du úgy mint múltkor be tudok ugrani hozzád egy röpkét,majd besz.p*
[text message, 2006]

Kedves most végzett bölcsészek,
nemtom *feletek milyen az infóáramlás, de én véletlenül tudok meg mindent, írásban a világért se értesítene a btk, pláne olyan apróságokról, mint a diplomaosztó..*
[iwiw message wall, 2006]

Hát [given name], én még mindig úgy érzem magam, mint akit agyonvertek...
szal[47] ***nemtom****, van-e értelme a folytatásnak!* *De azért mindezek ellenére megérte, egész jó volt, főleg amíg az alkohol csillapította a fájdalmat.* 
[iwiw message wall, 2006]

47 *szal* = *szóval*

The change of *asszem*, however, is not restricted to mere formal modification. I will present this process through a corpus complied from internet forum comments.

For the examination I selected an internet forum which had operated since November 2004. Although communication is not simultaneous here, the language of messages posted to internet forums share the characteristics of spoken language (compare: secondary literacy). The website primarily deals with linguistics and scientific life, but there are a lot of personal remarks on the pages as well. I chose this website because it stored all older comments, it had a sophisticated search function (which made it possible to find *asszem* forms in comments made five years earlier), and forum users often engaged in debates which brought their comments closer to spontaneous language use—though such forum comments may not be compared to the spontaneity and quickness of online chat communication. Between November 2004 and February 2009 15,000 messages were posted to the forum. I used the search function to filter only those messages which contained the expression *asszem*. After manually removing repetitions (i. e. inserted quotations of previous comments) I found 228 *asszem* word forms in the corpus. These words were copied together with their context. User nick names were deleted and replaced with a code (A1–A43). 41 forum users[48] wrote down the word *asszem* more or less frequently (user A10 took the lead with 37 occurrences), the average being 6.34 (*Figure 6*).

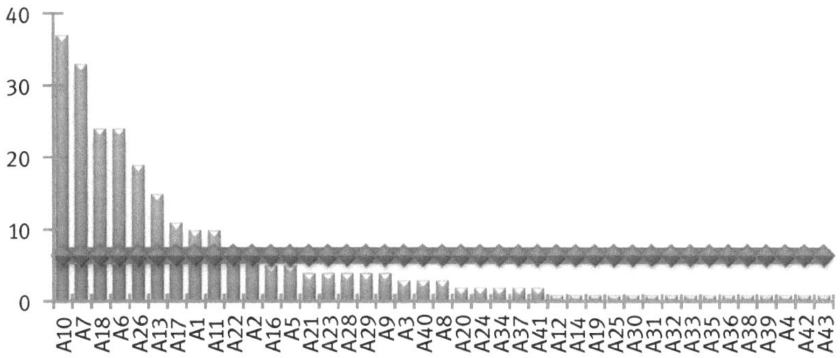

Figure 6: Frequency of *asszem* by commenter

48 The number of codes and users writing down *asszem* differs as I removed repetitions after allocating the codes to users.

Since the collected forum messages were copied together with the time of their posting, I examined the messages for potential correlations between the lapsing of time and the frequency of the use of *asszem*. I found that though usage is not evenly distributed on the timeline, the use of *asszem* is increasingly common (showing a peak between January and June 2007; *Figure 7*).

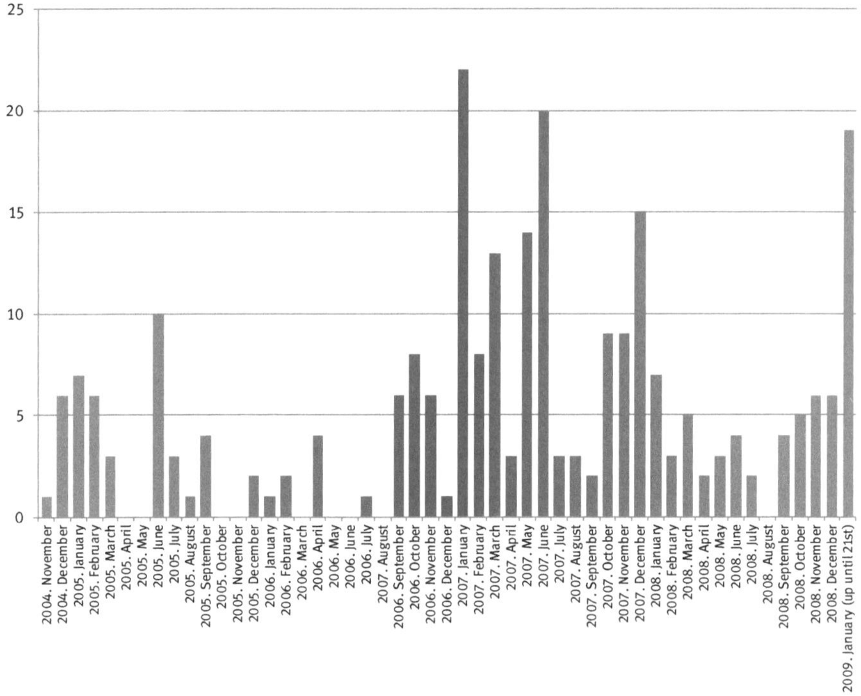

Figure 7: Frequency of *asszem* shown on the timeline

The database of Hungarian National Corpus (Magyar Nemzeti Szövegtár, MNSz.) was used as control corpus. In the corpus, which consisted of 187,644,886 words in June 2009, the word *asszem* occurred 1480 times (7.89 word/million words). Looking at the sub-corpuses, I found that the word occurred 1466 times in the sub-corpus of "personal texts" (accounting for 99.05 % of the total occurrence), and appeared only a few times in scientific texts, the language of the press and fiction, and there was no occurrence in official texts (*Table 11*).

I followed a holistic approach during the examination, the aspects were merely separated for the sake of presenting my findings. In the following the corpus will be examined from morphological, syntactic and semantic-pragmatic aspects.

Table 11: The word *asszem* in the Hungarian National Corpus (2009)

	word (pcs)	word / million words (pcs)	total occurrence (%)
personal	**1466**	**78.73**	**99.05 %**
scientific	8	0.31	0.54 %
press	5	0.06	0.34 %
literature	1	0.03	0.07 %
official documents	0	0	0.00 %

From the perspective of formal reduction, a similar process unfolds in the case of *asszem* than in the cases of *lám*, *talán* and *tán* which became modifiers/particles in the Old Hungarian period: the loosening techniques of spontaneous speech resulted in formal reduction (*Látom* 'I can see', ... > *lám* [...]; *Találom* 'I find', ... > *talám* > *talán* 'perhaps'; D. Juhász 1991: 509). From a phonetic point of view, the formal reduction may be explained by the tendency of two open syllables: "It clearly meets the requirement of the middle phonetic position, as the middle syllable is elided. The tendency of two open syllables gives the final form of *asszem* from the form *asziszem* created from *azt hiszem* through lenition and the elision of consonants" (Markó 2003).

Having examined the forum corpus, I found that *azt hiszem* > *asszem* fails to establish a complete paradigm: I only found two present and two past tense indicative forms (*asszem, asziszi, aszittem, aszitték*). There were much more form variations in the Hungarian National Corpus. It should be stressed here, however, that these word forms, which seemingly establish a paradigm, are merely in the state of formal reduction, semantically they fully correspond to the appropriately conjugated form of *azt* 'that' + the verb *hisz* 'believe', and they contain the "believing" meaning component of *hisz* 'believe'. Thus, from a semantic point of view, these forms may not be identified with *asszem* which is in the process of grammaticalization and bears epistemic meaning as well. This can be illustrated through some examples from the national corpus:

Example 39: Incomplete paradigm in the Hungarian National Corpus

Eddig ***aszittem*** *ez a RaSek fiu csak hulyeskedik... Szegenykem, ugy tunik komoly a baj...* [MNSz.]

násznép nézett nagyokat. mimeg ***aszittük*** *h betépett a pap* [MNSz.]

[...] *szépen és jól játszott... 90 percig. 4-0-nál talán* ***asszitték****, hogy simán megvan a meccs.* [MNSz.]

nézd b+ van 58 perced jáccatok asztán tünés, a közönség teljesen sux, szerintem ***aszitték*** *a vhk-n vannak. na mind1.* [MNSz.]

Now, let us examine the corpus from a syntactic point of view in the spirit of Givón's witty remark: "Today's morphology is yesterday's syntax" (1971: 413). In synchrony, the word *asszem*, it seems, parts with *azt hiszem* and gives up its main clause function to become a function word. The words *talán, lám, vajon, hadd, akár* became modifiers from independent clauses and are now capable of reflecting the speaker's subjective attitude (Haader 2008: 78). "They were mainly modal main clauses but due to their meaning they could transform to modifiers relatively easily: [úgy] *találom* ['vélem'], ... → *talán*; [úgy] *látom*, ... → *lám*; [úgy] *valjon* ['legyen'], ...? → *vajon* (valyon); while the change of *hagyd* → *hadd* could have started in the Early Old Hungarian period" (D. Juhász 1991: 505).

According to Lea Haader's examinations (2008: 78) the main clause of subordinated clauses introduced by the conjunction *hogy* 'that/whether' has usually lost the force of its meaning, as the proposition is given by the subordinated clause (so in these types of sentences the subordinate clause has an explanatory function). The connection between the clauses is weak; the subordinate clause is relatively independent. If a so-called modifying main clause is shrunk, it is not the subordinate clause which fades into the main clause but vice-versa: the main clause fades into the subordinate clause (H. Molnár 1961: 147–152; Keszler 1977: 131 and 1986). This type of subordinate clause is usually affected by a grammaticalization process during which a clause disappears, i. e. the main clause turned into a modifier, a particle or a conjunction in the course of previous language change processes (compare: "the function of this main clause, which has largely lost its original meaning, [...] is similar to that of modifiers" [Tompa 1961–62 II: 356]). Haader (2008: 80) collected the basic criteria of such clause-deleting grammaticalization processes: main clause–subclause order of clauses; subclauses with explanatory function; attitude marker; main clause expressing sensation (1SG), will (2SG) or confirmation (3SG); and communicative-pragmatic factors. Sentences beginning with *azt hiszem, hogy...* 'I believe that' seem to fulfil all four of these criteria.

I conducted various tests to find out whether the main clause *azt hiszem* has faded (or has started to fade) into its own subclause; *Figure 8*). Ilona H. Molnár (1968: 43) suggests the following syntactic tests for distinguishing between modifying sentence parts in fixed syntactic relationships and main clauses in free syntactic relationships: "1. Does the conjunct *hogy* appear in the sentence? If not, would it fit into the sentence? 2. Is there any stressed demonstrative adjective in the main clause referring to the subclause? If not, would it fit into the sentence without disturbing the equilibrium between the main clause and the subclause. 3. Can the predicate in the main clause extended any further? 4. Can the predicate in the main clause freely altered in terms of person, number, tense and mood? 5. Can the predicate in the main clause intensified? I added some more criteria to this list. In 18 % of the sentences the conjunction *hogy* may not follow the expression *asszem*. It indicates the fusion of the two components, if the expression cannot be segmented to its constituents in the given sentence. In half of the forum sentences the order of the two words cannot be reversed like this: *hiszem azt*. The distribution test shows that the syntagma *azt hiszem* cannot be replaced with *úgy hiszem* 'I believe that' (11 %), or with *azt gondolom* 'I think that' (25 %). In 18 % of the sentences the expression does not qualify as an answer to the question *azt hiszed, hogy...?* 'Do you think that...?'

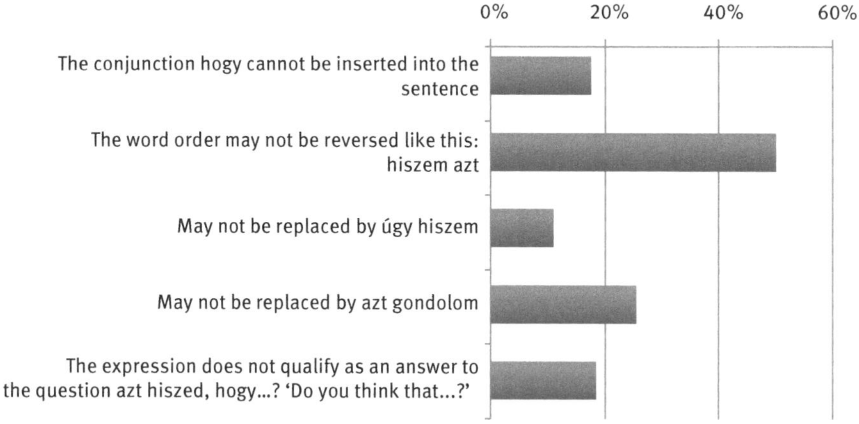

Figure 8: Syntactic tests on the word *asszem*

Example 40: Syntactic tests on the word *asszem*

Example 40.1: The conjunction *hogy* cannot be inserted into the sentence

Ez speciel tényleg anglicizmus, van rá szép magyar szó vagyonbiztosítás. Persze [...] *a tradicionális élet- és vagyonbiztosítás mellett van/lehet érteleme a nem életbiztosításnak is, de azt meg nem úgy írjuk (**asszem**).* [Forum, 2008]

*(Viszont a webfordítás pont hu tényleg azt adja ki, h his liver loop... Hát akkor meg is van a megoldás, **asszem**.* [Forum, 2008]

[...] *Ebből a szempontból tanulságos az örmény önkormányzat körüli háborúskodás. Egyik oldalon vannak az Erdélybe a 17. században (**asszem**) érkezett örmények leszármazottai, akik őrzik az őseik emlékezetét, talán hagyományait is, magukat örménynek (is) tartják* [Forum, 2008]

Example 40.2: The word order may not be reversed like this: *hiszem azt*

***Asszem**, feladom, megadom magam, lemondok, bekussolok.* [Forum, 2008]

*A megy **asszem** urái, de minimum finnugor. A jön ugor, de van egy bizonytalan finnségi egyeztetése is, a potenciális megfelelőjének ott* 'marad' *a jelentése.* [Forum, 2008]

Example 40.3: May not be replaced by *úgy hiszem*

*ezt ismeri valaki – **asszem** épp vmi iesmire gondoltam...* [Forum, 2008]

*Én **asszem** azt értem rajta, hogy az ember nemcsak maga él a saját meggyőződése szerint, hanem úgy gondolja, hogy másoknak is a saját meggyőződése szerint kellene élniük, és ezt hajlamos rájuk is erőltetni.* [Forum, 2008]

Example 40.4: May not be replaced by *azt gondolom*

*Nem ugyanolyan értelemben mondjuk az LFG-re meg a Hamletre, hogy„nekem tetszik". (Talán... **Asszem**...)* [Forum, 2008]

*Most már annyira nem értelek, hogy **asszem** feladom az egész vitát, elmeorvoshoz fordulok, és csak kisérővel megyek nyilvános helyre.* [Forum, 2008]

Example 40.5: The expression does not qualify as an answer to the question *azt hiszed, hogy...?* 'Do you think that...?'

*Bár **asszem** – és ezt a jelek szerint kevesek mondhatják el itt magukról – nekem még sosem volt kezemben a nagy mű.;)* [Forum, 2008]

*Legutóbb általános másodikba hallottam valami ilyesmit, csak már **asszem** kiment a fejemből.* [Forum, 2008]

Although the punctuation of its written form does not directly relate to the linguistic change (particularly not in the case of forum texts which tend to disregard spelling and punctuation conventions), still it may be interesting to examine the punctuation issues of *asszem*. Borbála Keszler (1986) notes that, by principle, "modifying sentence parts which have preserved their subordinate function from a formal perspective" (e. g.: *azt hiszem* 'I believe', *úgy lehet* 'it may be', *mint említettem* 'as I have mentioned', *képzeld, ahogy mondani szokás* 'imagine, as the saying goes' etc.) are separated from the rest of the sentence with a comma or a pair of commas. In 68 % of corpus occurrences *asszem* was not preceded or followed by any punctuation mark. 17 % was followed, 8 % was preceded and 4 % was both preceded and followed by a comma. As far as other punctuation marks are concerned, only brackets (2.6 %) and dashes (0.8 %) could be found in the collected sentences. Examples for the above-mentioned types:

Example 41: Punctuation marks and *asszem*

*Mindenesetre nem ide való, úgyhogy töltögessétek, aztán megkérjük az adminokat, hogy gyomlálják ki. És **asszem** kellenek moderátorok is majd.* [Forum, 2008]

*Nos, **asszem**, az vitathatatlan tény, hogy az őszinteség, nyíltság sallangokkal nem valósulhat meg.* [Forum, 2008]

*Ugyanis ha katolikus házasodik össze protestánssal protestáns szertartás szerint, azt a katolikus egyház nem tekinti érvényesnek (**asszem**).* [Forum, 2008]

*Én nem ismerem az illetőt, nem szóltam bele az előző vitába – **asszem** -, stb.* [Forum 2008]

In the following I will discuss *asszem*, and its functions in particular, from a semantic and pragmatic aspect.

It is a clear sign that a linguistic item has lost its semantic substance, if it can appear together with its antonym or heterosemic form (Wegener 1998: 40). This latter would be exemplified by the sentence introduced by *Asszem azt hiszem* "I think I believe". The growing metacommunicative power (and increasing semantic emptiness) of *asszem* is shown by the fact that it relates to another cognitive process.

Example 42: *Asszem* does not refer to the cognitive process "I believe"

*Legutóbb általános másodikba hallottam valami ilyesmit, csak már **asszem** kiment a fejemből.* [Forum, 2008]

Alexandra Markó (2003) found that "in certain cases the form *asszem* does not convey the semantic meaning of *azt* + *hiszem* 'that + I think' but it rather plays some kind of pragmatic function", so it may indicate the speaker's attitude. Markó identifies the following functions of *asszem* as a discourse marker: offering apology, requesting permission in some way, reducing own responsibility; form of politeness, making reference to an assumed common knowledge; "dampening the edge of the utterance"; filler, discourse organizer.

It may be considered as the pragmatic strengthening (see: Traugott 1988) of *asszem* that the "believing" meaning component of *hisz* 'believe' is hardly present in the corpus, i. e. it is better to say that *asszem* has a pragmatic function. In more than half of the examined corpus, *asszem* indicated (epistemic) uncertainty.

Example 43: *Asszem* as the indicator of epistemic uncertainty

*Régebben olvastam, hogy egy párizsi nőt is szekáltak, azt **asszem** éppen Milkának hívták és varrodája volt. Meg levédett doménneve.* [Forum, 2008]

***asszem** a jogi terminus nem a fajtalankodás, hanem a fajtalanság.* [Forum, 2008]

Én 4 órát aludtam, hulla másnapos vagyok és nagyon fáj a bal vállam!  ***Asszem** a következő bulit kihagyom.* [iwiw message wall, 2006][49]

***Asszem**, hogy holnap lesz valami szalagavató, vagy ilyesmi*  [iwiw message wall, 2006]

49 The examined collection of examples were complemented with the examples taken from the wall message corpus to demonstrate that *asszem* may not only appear in a pragmatic function in internet forums. The sources of examples may vary, so they are indicated in square brackets.

A peculiar type of the examples is the one which indicates the source of the proposition (the source of information). In this case *asszem* can be replaced with *szerintem / X. Y. szerint* 'in my/somebody's opinion'. This type, which is used as a face saving strategy, occurs in 39 % of the examples.

Example 44: *Asszem* as the marker of origin of the proposition

A bűncselekmények áldozatainak segítség, kímélet és jogorvoslás jár, nem pedig az, hogy a törvény romlottnak vagy fertőzöttnek nevezze őket. ***Asszem*** *ezzel egyet lehet érteni.* [Forum, 2008]

Hogy még az expertek is csinálnak helyesírási hibát? ***Asszem*** *nem sikerült kellően hangsúlyossá tenni az öniróniát, vagy nemtom... Mongyuk annyi szmájlit tettünk ki, hogy szerintem elég egyértelmű volt.* [Forum, 2008]

If these utterances are examined in terms of the subject of consciousness, the subject, who is responsible for the information conveyed, "may be the speaker itself or the addressee of the utterance or even a participant of the utterance" (Tátrai 2005). The speaker may indicate or make explicit that (s)he assumes responsibility for the information provided, for instance through the autonomous sentence part *azt hiszem*.

The melioration type is meant to save the partner's and not the speaker's face, *asszem* mitigates the corrective, criticising, or—less commonly—insulting or reprimanding edge of the utterance.

Example 45: *Asszem* in a meliorating function

Asszem *nem csak véleménykülönbség. A hunok ugye még az első ezredforduló előtt eltűntek (mint hunok; mert a génjeik nyilván benned még virgonckodnak)?* [Forum, 2008]

Asszem *inkább te értetted félre, amit idéztél.* [Forum, 2008]

Asszem has a discourse organizing role if it introduces a new topic or draws the attention to a piece of information (highlights something).

Example 46: *Asszem* as a discourse organizer (highlighter)

ezt ismeri valaki – ***asszem*** *épp vmi iesmire gondoltam.... http://www.szintan.hu/nevO.htm* [Forum, 2008]

In spontaneous speech fillers are used to fill up hesitation pauses or to save time for the vocalization of the message. *Asszem* may have similar function as fillers in spontaneous speech (which is particularly surprising as I have examined written texts—though conceptually they were closer to spoken language texts).

Example 47: *Asszem* as a filler

mer ***asszem*** *hiába tanultam szótagolva olvasni, azér az egész alakot ismerem fel,*
[Forum, 2008]

Én ***asszem*** *azt értem rajta, hogy az ember nemcsak maga él a saját meggyőződése szerint, hanem úgy gondolja, hogy másoknak is a saját meggyőződése szerint kellene élniük, és ezt hajlamos rájuk is erőltetni.*
[Forum, 2008]

Although my examples can be classified into various categories, the specific types show some overlapping. However, it is certain that when *asszem* occurs, "through the meaning of epistemic uncertainty the expression may take on mitigation/reduction function as an illocutionary force".[50] Functions are not seen here as a continuum of successive groups, but rather as overlapping or complementary fields.

3.3.6.4 Summary

To conclude this case study, we may point out the following: Written orality has created a special grammaticalization phenomenon. While in a "traditional" grammaticalization process, as observed in orality, structural change precedes phonetic-phonological change, in the case of the discourse marker *asszem*, formal reduction seems to have preceded grammatical change. Initially, *azt hiszem* in its original meaning turned into the contracted form *asszem* only because it was shorter that way (the same method produced the forms *asszed*, *aszisszük* etc.). The phonologically modified form does not always mean the grammaticalized variant in the forum texts. Grammaticalization influences the already reduced form, therefore the ungrammaticalized and the grammaticalized forms (as well as the forms currently undergoing the process of grammaticalization) may coexist with special overlappings.

3.4 The formal characteristics of digilect

The easy-to-identify formal, more specifically typing-related characteristics of digilect are the following: abbreviations, irregular spelling, lower case typing, omission of diacritics and punctuation marks, and the use of emoticons. Abbreviations have already been detailed in the previous subchapter, among the

50 Note by Nóra Kugler to Alexandra Markó's paper.

grammatical characteristics of digilect. This subchapter deals with variations in spelling and the use (or lack) of punctuation marks.

3.4.1 Phonetic spelling

Crystal (2008: 184) defines phonetic spelling as "a spelling system which represents speech sounds in a regular, one-to-one way" (such as: *wot* – *what*). Lakoff (1982: 252–253) illustrates phonetic spelling with examples taken from comics. This spelling type does not mark variation in pronunciation but nonstandard spoken communication (e. g.: *yuh* and *ya* – *you, ta* – *to*; *Omigosh* – *Oh, my gosh*; *wuz* – *was*).

Phonetic spelling, however, does not only appear in English but in Hungarian as well: *nemtom* [< nem tudom 'I don't know'], *evvan* [< ez van 'that's the way it goes'], *ijen* [< ilyen 'such'], *pész* [< pénz 'money']). This phenomenon seems to overrule word analysis, one of the four principles of Hungarian spelling (e. g.: *kézzed* [< képzeld 'imagine!'], *mongyuk* [< mondjuk '(let's) say'], *jáccani* [< játszani 'to play'], *aggyál* [< adjál 'give me'], *csinájjuk meg* [< csináljuk meg 'let's do it']).

Example 48: Phonetic spelling

móni, ***nemtom****, ki az a marypen, de mondja, ha mondja, mer akkor én mondom, ha mondom* [nemtom < nem+tudom 'I don't know'] [iwiw message wall, 2006]

tom' *hogy gáz, ezért van kb 5 éve hotmailem... nagyon bosszantó a freeamilban a rengeteg nem várt fostos üzenet+az hogy időnként lefagy és h sz@r az egész. hehehehehe.* [tom' < tudom 'I know'] [iwiw message wall, 2006]

Spelling foreign, primarily English and German expressions with Hungarian phonetics may be motivated by users' ambition to be unique and funny. The adaptation of foreign words and expressions to Hungarian phonetics may be regarded as an idiolectic characteristic. When communicating in small groups, it is essential for participants to use a common code. Quite frequently, the members of the group develop a unique linguistic code which is hardly—if at all—comprehensible to the members of other groups. By this, people often intend to separate themselves from other members of society. Intimacy in communication means that the message is encoded in a way that it may only be decoded by a small group of receivers (Fülöp 1996). Perhaps it is specifically this desire for intimacy—and for the expression of group identity—that appears in texts which at first seem to be indecipherable.

Example 49: Foreign language expressions adapted to Hungarian phonetics

wäj nat wi sztart szpiking ingliss end vräjt vit hangerian fonetiksz *[=why not we start speaking English and write with Hungarian phonetics]* ***Ju gonna LÁJK it*** *:). [= you gonna like it]* [Forum post on the Facebook page "A lájkol egy magyar ige" 'Lájkol is a Hungarian verb', 2010]

áj vana bíí an enzsinier*!!!!!!!!!!!!!!!!*[= I wanna be an engineer] ***plíííz gad help míí****..... lool* [= please God help me] [message wall, 2006]

*jaja,****dajcslanddajcslandüberallesz****!*[= *Deutschland, Deutschland über alles*] [message wall, 2006]

ja és unom már hogy a médiamunkásoknak ilyenkor csak Micimackó meg hózik jut eszébe, mint esőkor az ***iccarénidéj*** *[=It's a rainy day]*, *halgasson inkább mindenki zoboraljáról Ghymestől Tánc a hóbant!!!* [message wall, 2006]

kislányom miről beszélsz? A te anyád a ***mánsztörofdövúúúd*** *[=monster of the wood].* ***Stimmt****!? [=Stimmt]* [message wall, 2006]

cimbi egy pöcs...:9 Mi?? Just do the cabaret belefeküdve a hóba **next to drinktrészi** [= next to drink Tracy ~ Dick Tracy] [message wall, 2006]

partee tunájt!! yeahh *[=party tonight! yeah!] Mónika meg all that jazzz lesz,* ***ofkorsz****!* *[=of course]* [message wall, 2006]

A similar motivation may be the knowledge of a foreign language which makes it possible to mix foreign and Hungarian words. Such hybrid words are examples of the so-called macaronic language. This is, however, not a novel phenomenon: it has long been called "macaronism" when foreign language morphemes (prefixes, suffixes), words or sentences get mixed up with a native language text, and thus become a source of humour. Presently, Hungarian is dominantly influenced by English in this regard, and the resulting macaronic language is called "Hunglish" (Hungarian + English) or "Magygol" (magyar 'Hungarian' + angol 'English'). Examples: *Van egy* ***mcszidzsed****.* 'You have a message.'; ***Soppingolunk*** *délután?* 'How about going shopping in the afternoon?'; ***Kóllod*** *van a kettesen.* 'You've got an incoming call on channel two.'; *Mikor lesz a* ***míting****?* 'When does the meeting start?' (Balázs 2005c: 13).

The verb *lájkol* (*like* + -[*o*]*l*) has spread rapidly among Internet users due to the popularity of Facebook, and is still widely used by them. The "Like" button is a Facebook feature which enables users to express their liking for a post, comment or content. Pressing this button is called *lájkol* in Hungarian, which is a hybrid word of the English *like* and the Hungarian *-l* verb-forming suffix (*Image 11*). Prefixed versions of the word include: *meglájkol* and *belájkol*. The English verb *dislike* used as the antonym of *like* has also entered the Hungarian in the form *diszlájkol*.

Example 50: Maccaroni expressions

RE-Klám: http://www.halasrapraja.ini.hu < — ***Checkold*** *meg te ish!!! (és ha teccett írj a fórumra ish) THX for ALL!* [*Checkold* 'Check it out' < *check (English verb) + -old (Hungarian morpheme)*] [message wall, 2006]

lájkol: *Like, love, enjoy (something). It originates from the "Like" button of the social networking website Facebook where it is used to express positive attitude toward a shared content. The Facebook "Like" had an impact on the entire internet communication and even on the real-life speech of internet users.* [Slang Blog, 2010]

Lájkol *is a Hungarian verb. The* ***counter-page*** *is ready. The* ***likeol*** *vs* ***lájkol*** *match has started. Live social media experiment.*[51] [post on a Hungarian Facebook page, 2010]

like= ***lájk lájkolni*** *és csoki ha nemtetszik* ***diszlájkoljad*** *vazze[like=lájk, lájkolni and if you don't like it, dislike it]* [post on a Hungarian Facebook page, 2010]

lájkolom *hogy valaki megvédi a* ***lájkolást****, és hogy ezt többen is* ***lájkolják****. és* ***diszlájkolom*** *azokat akikt* ***diszlájkolják*** *a* ***lájkolást****, és azokat akik* ***lájkolják*** *a* ***lájkot****.* [*I like those who stand up for liking and that this is liked by many people but I dislike those who dislike liking and those who like liking.*] [post on a Hungarian Facebook page, 2010]

Image 11: *Lájkol* as a macaronism (Facebook page, 2010)

The verb *lávol* has evolved similarly to *lájkol. Lávol* is the English word *love* spelled with Hungarian phonetics and formed with the Hungarian verb-forming

51 *Lájkol* is a Hungarian verb. Facebook page. http://www.facebook.com/pages/A-lajkol-egy-magyar-ige/ [27. 01. 2010; not available, 2016]

suffix *-(o)l*. It means: to like, to love. Variants: *lâwol, lúvol, lúwol*, luwol (*Image 12*). (For more examples to macaronic language see Chapter 3.3.2 on word formation and neologisms.)

Image 12: *Láwol* as a macaroni expression (Facebook post, 2010)

3.4.2 Upper and lower case

In internet and mobile communication it is a essential to type and send messages quickly. This eagerness to be quick is particularly apparent in spelling, for example in starting sentences with lower case letters.

Example 51: Starting sentences with lower case letters

sziasztok! újra itthon vagyunk, túléltünk mindent, és jó volt nagyon, sok sajtot meg croissantot meg bagettet, és rengeteget kirándultunk! itthon mizu? pusza [iwiw message wall, 2006]

mert a nők sokkal okosabbak, de ha ezt a férfiak által mérni nem kívánt tulajdonságot nem vesszük figyelembe, azért a BIBÓban akkor is több a lány. ez tény.  [iwiw message wall, 2006]

Capital and small letters can have defining, highlighting and attention-directing functions as well: they may highlight information which is considered important or may express louder speak (Crystal 2003: 229). All-caps, that is typing all letters capitalized, is interpreted as shouting in certain contexts and thus is disallowed by netiquette.[52]

Example 52: Defining role of letter case

ÉVFOLYAMTÁRSAK!!!!! nem tudja véletlenül valaki, h meddig lehet az indexeket leadni, és hogy hány órakor van to??SŰRGŐS!!!! köszi!! PUSZI!! [message wall, 2006]

remélem,én vagyok ma az elsö,aki BOLDOG SZÜLINAPOT KIVÁN.az égiek áldása legyen rajtad!:-) puszi a hazából [text message, 2002]

52 Behaviour in internet communication is regulated by the so-called netiquette, a set of social conventions developed and recorded by users. The word is a portmanteau of *network* and *etiquette*.

Gyarapitottam a szogyujtemenyunket:VAJLING, PAPUNDEKLI, HABISZLI.Ismered oket?Ha nem, majd kedden meselek.(ez a feszultsegkeltes:-))CUPP [text message, 2004]

Hi there! CONGRÄTULATIONS!!:-)) you did it! Hope I can shake your hand in person too, maybe in the car! Kiss Äron [text message, 2002]

Case alternation can be a source of humour as well.

Example 53: Case alternation within the same word

SeEhOwAwFuLiTiStOgEtMeSsAgEsWhIcHaReFuLlOfUpPeRaNdLoWeRcAsElE tTeRsBuTnOsPaCeSaSaNdArEcOmPlEtEnOnSeNse? [text message, 2002]

Siska! HaPpY BiRtHdAy!!!! [message wall, 2006]

3.4.3 Punctuation. Overused characters

"The omission of punctuation marks conveys a message in itself, though it also creates ambiguity and thus makes comprehension more difficult. This practice should be avoided in everyday written communication." (Keszler 2004: 35–36). The practice, however, is certainly present in digital communication. To catch up with normal speech tempo, in instant messaging sentences are started with small letters; punctuation marks are limited to exclamation and question marks in an expressive function (almost exclusively at the end of a message); and full stops are almost always omitted. Where in traditional written texts a comma would be required after a clause or a list element, in instant messaging this is replaced by hitting the Enter key (i. e. starting a new line). According to Zoltán Bódi, quick responses reduce the chance to lose information due to formal redundancy, so such written texts—which share many characteristics with spoken language—are formed only to the extent necessary to stay comprehensible. The use of sentence-initial capital letters, clause-separating commas and sentence-ending punctuation marks may still be enough to preserve the clarity of the message (Bódi 1997: 166 and 2000: 206). The aim is to make typing as fast as possible, approximating the tempo of live speech.

Example 54: Texts without punctuation marks

TELJESEN ELVAGYOK HANYAGOLVA SEMMIRÖL NINCS ÉRTESÍTÉSEM [I'M TOTALLY NEGLECTED I DON'T GET NOTICE ABOUT ANYTHING] [text message, 2001]

barcsak minden vagyam igy teljesulne [if only all my dreams came true like this] [text message, 2001] *jah és nem és nem értem h miért kell szavaznom a városunkra ha már felvették az iwiw-be mint várost szal nem tudom mit akartok* [oh and I really don't understand why I should vote on our city if it is already on < image > as a city so I don't see what you want < image >] [iwiw message wall, 2006]

Nem tudsz majd eljönni értem most varok még csak a mütétre kicsit nem vagyok jol. Puszi [Won't you be able to come for me I'm still waiting for the operation I'm a bit unwell. Kiss] [text message, 2009]

The use of punctuation marks is often restricted to ending sentences: users do segment their text, only separate sentences from each other. Units of thought, clauses and even items in a list are written without commas. This is, however, not a new phenomenon: As early as in 1982, Lakoff noted that in comics, for instance, sentences seldom end in full stop but rather in question or exclamation mark—probably to attract attention, indicate importance or express emotion (Lakoff 1982: 252).

Example 55: Punctuation marks used only at the end of sentences

Szia a melot visszamondtam. Most az [name] *bank van kilatasban. A horg jo volt most megyek hajduszoboszlo haverok buli fanta aztan jövö hét horvat...* [text message, 2004]

A TELCSIMMEL SEMMI BAJ CSAK KIKAPCSOLTAM.ITT MINDEN RENDBEN VAN NAGYOKAT FÜRDÜNK PASIZUNK éS KIRáNDULUNK.PUß:[nickname] [text message, 2002]

Nagyon szivesen lanyok maskor is-orulok h jol sikerult a buli. [nickname] [text message, 2002]

Kakukk!akkor jössz ugye?elötte mindenki cong,van kedved ugye neked is?majd onnan kimegyünk.szólok h hány tájt.pucc [text message 2004]

Punctuation marks may also be used for graphostylistic purposes (Érsok 2004: 296); this is suggested by the overuse of characters to indicate passion or intensity. This practice aims to visualize talk or to create mock oral writing (see: Frehner 2008: 59).

Example 56: Overused characters

Forza Italiaaaaaaaaaaaaaaaaaaaaaaaaaaaaaa!!!!!!!!!!!!!!!!!!!!!!!!!!! [iwiw message wall, 2006]

*Ne küldjetek továbbított leveleket!!!! Nem akarok interneten keresztül meggazdagodni, sőt azt is lesz*rom ha az iwiw fizetős lesz, vagy csődbe megy, vagy hupililára festik az egészet... AAAAAAAAAAAAAAAAAARGH).* [iwiw message wall, 2006]

BOLDOG NÖNAPOT!!! [text message, 2002]

Nem ugy gondoltam, ne sértödj meg!!!!!!!!!!!! *csak hüjjéskedem!* [text message, 2001]

Ha bekövetkezik az, hogy vagy a svájciak,vagy az ukránok, vagy az ausztrálok juthatnak be az elődöntőbe, soha többet nem nézek focivébét!!!!!!!!! [message wall, 2006]

SOS kellene anyukad!!! [text message, 2010]

Boldog névnapot kívánok!!!!!!!!!!!!!!! [text message, 2009]

Double punctuation marks are stronger than single ones. A double full stop, for example, does not have the same function as a sentence-ending (single) full stop, but is certainly not as strong as three full stops in a row. The same is true for double exclamation and question marks. The traditional row of three full stops, which marks ellipsis, therefore is often shortened to two full stops as to save characters (and typing time) without losing function.

Example 57: Doubled punctuation marks

A labamon alig-alig allok,mégis szeretnek a sracok, Angyalom,galambom,szeretem a bort..%-)
RemekVoltAzEgriSzépasszonyvölgyiBorkostolas:7félétKortyintottunk:-)!Cp [text message, 2004]

Persze,persze,az fontos.ablak megvar.+hatha nyitva lesz 6ig.milyen munka?hu,izgi!!-kacaraszunk [name]*val,de igy kivarom a végét.tarts ki,rad gondolok. < 3* [text message, 2008]

Fiokvezetö??? Aa csucs!! [text message, 2010]

Mutasd meg a dokinak az arcodat!! [text message, 2010]

Irregular spaces before punctuation marks go against the trend of shortening texts: users sometimes include spaces before punctuation marks though this is not permitted by Hungarian spelling rules (Keszler 2004: 21–28).

Example 58: Irregular space before punctuation marks

Na akkor nem is zavarlak téged a tanulásban . Ügyeskedjél és ha végeztünk BULIHEGYEK ! Puszancs neked is jó sokszor !!! PÁ [text message, 2002]

Jó reggelt kívánok! Sikerült elérni a fél 1-est ? ui: a maci sálját kösd meg ha majd a fütést éppen lekapcsolod nehogy megfázzon !!! pusi [text message, 2003]

SZIOKA AGI ! BOCSI A KÉSÖI JELENTKEZÉSÉRT!! JELENTEM OTT LESZEK HOLNAP !:)) ADDIG IS SOK SOK PUSZI [text message, 2004]

Boldog új évet kivánok !!! [given name] [text message, 2010]

Parts of the text are often highlighted by putting underscore characters before and after the expression. Asterisks can also be used for this purpose: they indicate winking (cf. Crystal 2001: 35). In German instant messaging texts inflectives are marked this way.

Example 59: Highlighting with special characters

*_**figyelj**_ nem mondom 2x* [instant messaging, 2010]

ez annyira gáz ***facepalm*** [instant messaging, 2010]

*Daily WUT: a buszra biturbó sebességgel felszálló idős néni _véletlenül_ beleül egy nyugdíjas bácsi ölébe, mondván _**másfelé nézett**_, és nem vette észre – azért ez már stílus, nem? A következő mondatban már be is mutatkozott. Így élünk.* [Facebook message wall, 2011]

3.5 New literacy and image-based communication. Visuality in digilect

Using visual information together with texts is an often-mentioned feature of digital change (cf. Schlobinski 2009: 6). The role of visuality is much more significant today than at any time in the history of communication: this is clearly shown by the graphic interface of computer programmes, the visual design of websites, the use of smileys in texts and the flood of images shared on social networking websites.

This return to archaic visual communication after the Gutenberg Galaxy is termed "iconic turn" by Gottfried Boehm (1994) and "pictorial turn" by W. J. T. Mitchell [...]. With the development of visual communication techniques, visual communication is becoming part of our everyday life ((Nyíri 2001, 2003: 273), people are exposed to an increasing flood of images (Giuliani 2006: 185, cf. Chapter 1.3), and the limitless production of images is becoming the activity of a global network which is impossible to hide away from, in fact some go as far as to call this process the omnipresence of images (Omnipräsenz der Bilder; Maar 2006). According to cultural anthropologists, psychologists and media scholars, the flood of images may ultimately force the monopoly of uttered and written words to images (*Homo Loquens et Audiens* becomes *Homo Videns*; see: Giuliani 2006: 185).

"Communication through multiple media affecting multiple sensory organs at a time" is a natural element of human life, as "humans primarily think in images and only secondarily in abstract words" (Nyíri 2000, cited in Benczik 2001: 243). This enables multimedial communication to dominate over monomediality. Kristóf Nyíri argues that "the epistemological examination of the problems of multimediality today primarily means the discovery or re-discovery of the logic of images and the analysis of possible ways to integrate texts and images. [...] Sometimes an image says more than a thousand words: texts record, *describe* how facts are connected but images *present* how to deal with things" (Nyíri 2000, italics included in the original text). Images can be understood without knowledge of the cultural context or the language, and they can depict even complex processes in a simple form (Maar 2006: 11–14).

Vilém Flusser says that in order to classify technical images according to their meaning "we have to set up other classification criteria which correspond to the character of technical images. We could classify them by the degree of information content: more or less informative, surprising, predictable images" (Flusser 2004). This distinction between redundant and informative images does not tell us about *what* images mean but about *how much* they mean. "Customarily, technical images are put into two categories (chemical and electronic images) on the basis of their technique rather than their meaning. Chemical images can be further classified as silent and static images (photos) and as loud and moving images (motion picture), while electronic images can be classified into categories ranging from videos to computer images" (Flusser 2004).

According to certain culturally pessimistic views, splitting texts into chunks of information (chunking) and filling these chunks up with images/icons will ultimately deprive web-based texts of their textual character, changing what was formerly a text to mere letter-icons and reading to mere viewing (a summary on culturally pessimistic views: Runkehl 2005: 206–207).

As opposed to verbal language with its conventionalized symbols permitting only linear decoding, visual language can be decoded quickly and simply. Images fall outside the control of conscience; they get round our mind and influence our thoughts and emotions through their suggestive power (Giuliani 2006: 185). Psychological studies (Runkehl 2005: 208) showed that texts and images are perceived by alternating saccades (rapid eye movements) and fixation (maintaining the visual gaze on a point). One saccade is enough to perceive 10–12 letters, while all the main elements of an image can be perceived by a few saccades (almost at a glance).

Hybrid "image+text" combinations, which lie on the verges of orality and literacy, are classified by Géza Balázs (2006) into binary oppositions ("visual–written" and "static–dynamic") as follows (*Figure 9*):

Emoticons, for example, fall within the visual graphic (+ textual) category and they can be both static and dynamic.

		static		
	graphic	graphic + textual	textual	
visual				*written*
	graphic	graphic + textual	textual	
		dynamic		

Figure 9: Typology of signs (Source: Balázs 2006)

3.5.1 Emoticons. Semiotic considerations

Emoticons are graphic signs expressing emotions; combinations of characters which can be displayed on computers and mobile phones (Laczházi 2004). Hungarian variants of the term used in the relevant literature include *vigyorkód* 'smile code', *mosolygó* 'smiler', *arcocska* 'little face', *kedvjel* 'mood sign', *ábraarc* 'figure face', *érzkép* 'emo-picture', *képecske* 'little picture' (Schirm 2002: 35), *smiley*, *szmájli* (smiley spelled with Hungarian phonetics), *hangulatjel* 'mood sign' and *emotikon* 'emoticon' (Bódi–Veszelszki 2006).

The interaction of text and image in a multimedial setting changes language into text–image conglomerates which clarify the context and meaning of each other (Schmitz 2003: 244). This is how digilect texts and emoticons are connected: emoticons are incomprehensible without context (or have only very general meaning), they help to avoid misunderstandings in written texts (more on emoticons: Amaghlobeli 2012; Bódi–Veszelszki 2006; Keszler 2004; Runkehl 2005; Veszelszki 2005b).

(Written) language, which is based on words, is expanded with the new dimensions of non-verbal communication. This function is provided by linguistic aids, such as emoticons—though in a very primitive way according to Kristóf Nyíri (2000). In internet communication smileys try to compensate for the lack of facial expressions and gestures when the message would be ambiguous without them. Emoticons lack any specific meaning as symbols do not qualify as codes (see: Sperber–Wilson 1996: 26). They only have a general meaning which becomes specific in the given context, so it is exclusively the context which concretizes the meaning of emoticons. This argument is supported by Kristóf Nyíri's observation that smileys are pictures as they have the most important property of pictures from a usage theory point of view: they have no meaning in themselves, they only gain it when used (Nyíri 2003: 267). "Sometimes an image says more than a thousand words—but what isolated static pictures say to us can only be interpreted in the context and explained by words" (Nyíri 2000).

Borbála Keszler calls emoticons "the new graphemes of internet language" which mostly signify the suprasegmental and non-verbal characteristics of speech (Keszler 2004: 85, 2006: 66; Bódi 1999); they satisfy the need to have different kinds of graphemes (Keszler 2004: 151).

The main function of emoticons is to express emotions and to substitute suprasegmental and extra-linguistic means. The importance of their expressive function has been evidenced by examinations in the field of non-verbal communication. Participants in personal (face-to-face) communication receive information through multiple channels at a time. According to Albert Mehrabian's examinations, in human communication words account for only 7 %, tone of voice accounts for 38 %, and body language accounts for 55 %. Birdwhistell came to a similar conclusion: he found that in interpersonal communication non-verbal channels account for 65 % and verbal channels for only 35 % in delivering the message (Pease 1989: 8). In addition, emoticons can make up for visual messages (i. e. information conveyed through webcams), speed up communication, break the overwhelming dominance of text, make the text more appealing to the eye, and even become a source of humour (Bódi–Veszelszki 2006). However, the excessive use of smileys can make it difficult to read the text; this practice is referred to as "smiley-flooding".

Example 60: Emoticons in various digital texts

Example 60.1: Facebook quiz and responses (comments) on a message wall

Mi vagy ki hiányzik az életedből? [name A] *tudd meg!: sex* ***;)***
[name B] *Ez ennyi:***)))***Én szóltam...***:DDD**
[name A] *tudtam h lesz hozzáfűzniwalód.***:D** *hááát...ezwan.***:DD**
[name B] *tehetünk ellene***:P***Anett túlélte az utat?***:DDD***Nem mondott semmit?***:)**

Example 60.2: Facebook wall post and response

[name C] *úú, az eru sajtból van új íz: a "különleges puszta íze". biztos pusztít.* **:D**
[name D] *Az olyan lehet mint a tejföl amit vettünk. Apu szerint tehén íze van. Megkérdeztem h az milyen, aszondta olyan mint a szaga. :D*

Example 60.3: Excerpt from an instant messaging text

A: *majd szólj, mit vigyek – a spar útba esik* ***:)***
A: *megmutassam a személyimet megint?* ***;)***
B: *oké, én meg intézem a kaját.*
B: *és a szülinapi szörprájzt.* ***:):):)***

Example 60.4: Excerpt from a blog post

nacső ***:D*** [nickname] *vagyok* ***xD*** *annyira hülyeség ez a bemutatkozásos dolog* ***:D***
ha valaki olvassa a blogom, úgyis kialakít magában egy képet rólam ***:D***
az lehethogy helytálló lesz, de szerintem nem, mert engem meg kell ismerni, és csak utána lehet elítélni ***:P xD***

3.5.1.1 Types of emoticons

There are two types of emoticons: one type is made up of ASCII characters (mainly of punctuation marks), the other type are graphic symbols. The latter also has two subtypes: static (non-animated) and dynamic (animated) emoticons. Static and dynamic emoticons are meant to convey the extralinguistic and paralinguistic information of spoken language (Bódi–Veszelszki 2006).

According to Kristóf Nyíri, any item can become part of the iconic language supplementing mobile communication if it is easy-to-create, easy-to-recognize, resembles the elements of the real world, can be combined with and connected to other items, has symbolic meaning denoting abstract notions, can complete written texts, is independent of culture and historical age, can be animated (Nyíri 2001: 77).

3.5.1.1.1 The original smiley

The original yellow smiley depicting a smiling face had nothing to do with electronic communication (*Image 13*).

Image 13: Smiley

While politically loaded symbols (such as ribbons in different colours, the dove of peace, the Mercedes logo like peace sign or the rainbow as a peace symbol) have not been quite profitable as commercial products, the smiley icon, which depicts a smiling face, have been a success story in the second half of the past century. By 1971 more than 50 million smileys had been sold on buttons only in the United States. The original icon was designed in 1963 by Harvey Ball, a commercial artist who was paid $45 for his work. The customer, State Mutual Life Assurance Company, which had just purchased another insurance company, wanted to use the icon on buttons, cards and posters to raise employee morale which declined as a consequence of the merger. Later on the smiley gained worldwide popularity: it appeared on posters, T-shirts and mugs. In the early 1990s it became the symbol of the acid house and underground music and subculture, and now even appears on Ecstasy pills.

3.5.1.1.2 Smiley made up of ASCII characters

The electronic smiley, which is a combination of colon, dash and parenthesis, is also more than 30 years old. The icon originates from Scott E. Fahlman who first used it in the following message in 1982: "I propose that the following character sequence for joke markers: :-) Read it sideways. Actually, it is probably more economical to mark things that are NOT jokes, given current trends. For this, use :-(19-Sep-82 11:44; Scott E. Fahlman" (W6; Frehner 2008: 77). Later, Fahlman made the following note on inventing this character combination: "Yes, I am the inventor of the sideways »smiley face« (sometimes called an »emoticon«) that is commonly used in E-mail, chat, and newsgroup posts. Or at least I'm *one* of the inventors" (Fahlman s. a.). According to Scott E. Fahlman, in the early 1980s students and teachers at Carnegie Mellon University used a bulletin board system (BBS) to share messages in various topics, including professional matters, political issues and jokes. Sometimes simple misunderstandings led to serious debates as the sender did not indicate nonverbal signs in his or her message and it was impossible to decide whether the message was a serious one or only a joke. It was in such a debate when Fahlman proposed, allegedly only half seriously, the use of the emoticon mentioned above. Although the symbol marking seriousness or jokes has carved a highly successful career since its invention, it was originally used merely for functional purposes: to make texts shorter and more accurate with visual signs. As an essential requirement, the emoticon was made up of ASCII characters which enabled users to type it quickly and have it displayed on all computers (cf. Schirm 2002: 34). Fahlman says that his :-) symbol was preceded by earlier attempts to mark humorous posts with similar visual signs. The most interesting one of these was the _/ symbol which was meant to depict a smile (W7).

Emoticon lists are collections of character combinations which depict a simple fact, an attitude, mood or personal trait relating to the communication situation on the basis of resemblance (Laczházi 2004; *Example 61, Annex 4*).

Example 61: Emoticons

:-)	I'm glad
:-#	I can keep secrets. My lips are sealed. I won't talk.
:-t	I'm sulking.
\o/	Hands up!
=:)	Bunny

A website[53] claims that the following emoticons are the most important one (*Table 12*).

53 "Emoticons typically represent a facial expression and are used chiefly to mark the tone of the preceding sentence or to indicate the writer's feelings" (W8; webpage content has changed).

Table 12: A list of emoticons

<table>
<tr><td>:-)</td><td>happy (smiley)</td><td>:-(</td><td>sad</td></tr>
<tr><td>:-D</td><td>laughter</td><td>:-c</td><td>very sad</td></tr>
<tr><td>;-)</td><td>wink, “only joking”</td><td>:’-(</td><td>crying</td></tr>
<tr><td>:-*</td><td>kiss</td><td>:-Y</td><td>comment (off-topic)</td></tr>
<tr><td>:-o</td><td>surprised</td><td>:-||</td><td>angry</td></tr>
<tr><td>:-P</td><td>tongue sticking out</td><td>:-V</td><td>shouting</td></tr>
<tr><td>:-|</td><td>bored, indifferent</td><td>|-O</td><td>yawn</td></tr>
<tr><td>:-/</td><td>sceptic</td><td>:-&</td><td>“tongue-tied”</td></tr>
<tr><td>:-Q</td><td>don’t understand</td><td>:-X</td><td>sealed lips</td></tr>
</table>

Table 13: Highly complex, thus rarely used emoticons

D	the user wears eclipse glasses
=`:´,o(	“I’m so depressed”
@-`-,—	“better use the language of flowers”
=`:o)	punk with mohawk hairstyle
:^)	the user a broken nose!
C=:-)	the user is a chef
C=} >;*{)	drunk, demonic chef with ugly wig, moustache and broad smile

The imagination of emoticon creators is only limited by the finite number of characters. There are also highly complex character combinations which are only understandable with some explanation (*Table 13*).

Users often note that there is no emoticon to express their mood or state of mind.

Example 62: Explicit reference to lack of emoticon

it’s morning :D but no emoticon can describe how my head looks like :D
[Facebook post, 2010]

3.5.1.1.3 Japanese emoticons

While “Western” style smileys depict human face rotated 90 degrees to one side, Japanese style emoticons, the so-called *kamojis* ‘facial signs’, can be understood without tilting one’s head. Japanese emoticons were invented in 1985 and have gained wide popularity even in Europe ever since. A smiling face, for example, is expressed by the icon (^^) or ^^ (depicting only the smiling eyes); wide smile by ^___^; and blushed faced by (*^^**). Quite often, these symbols do not only include ASCII characters but also Japanese typographic symbols, primarily katakanas as well. Japanese mobile phones can now display pictorial or even animated emojis in messages (Schlobinski 2009: 95; *Table 14*).

Table 14: Japanese style emoticons

(^_^)	happy, smile
(;_;) or (T_T) or (TT_TT) or (T^T)	sad, cry
(^_~) or (^_-)	wink
(-.-)Zzz	sleep
\(^_^)/	hurray!
d(^_^)b	1. headphones 2. thumbs up
(?_?)	don't understand/know
(o.O) or (o_O)	scared, confused, raised eyebrow
(*_*)	surprised

3.5.1.1.4 Dynamic emoticons

Animated smileys (or dynamic emoticons; cf. Veszelszki 2005b, Bódi–Veszelszki 2006) are primarily used in computer-mediated communication, such as instant messaging, e-mails or internet forums. Such emoticons are no longer limited to a colon–dash–parenthesis combination, but are linked to an animated image file (*Image 14, 15*).

Image 14: Japanese dynamic emoticons

Image 15: Dynamic emoticons (originally moving)[54]

Moving, animated smileys are categorized as dynamic emoticons. Animation is the technique of creating the illusion of motion by means of rapidly displaying a sequence of static images.

According to their function, dynamic emoticons can stand for facial expressions or gestures: 1. "Facial expression emoticons" supplement facial expressions which accompany interpersonal communication. They usually show one facial element in a simplified or exaggerated way (such as a mouth occupying the whole

54 See more examples in *Annex 12*.

face). Emoticons available in MSN mainly emphasized the eyes and the mouth, sometimes supplemented with raised eyebrows for expressing surprise, or long eyelashes for flattery. The showing of teeth indicates laughter but also has the connotation attached to the saying "showing one's teeth". Tongue-out smileys have several meanings: longing, mocking or cheekiness. Smileys seldom have hair as hair makes the face specific, while emoticons want to be general so that they can be applied to anyone. 2. "Gesture emoticons" include hands showing specific gestures: thumbs down for example means negative attitude or disapproval; thumbs up means positive attitude or approval; a lifted index finger moving left and right means warning; but we can find emoticons with greeting, waving or embracing hands too. As a peculiarity, gesture emoticons do not have a body: hands and legs are attached directly to the head (without making the icon weird or amorphous).

Further types of dynamic emoticons according to their design: occupation emoticons, style emoticons, logo emoticons, historic emoticons, grapheme emoticons, figures and emoticons holding signs. Their function is to express emotions, refer to daily routine activities, express aggression, present erotic/sexual content, maintain relationship (phatic function), and joking (for more details on these groups see: Bódi–Veszelszki 2006; examples in *Annex 12*).

Textual emoticons constitute a unique group as they function as the metalanguage of smileys: what cannot be expressed with static or dynamic images are expressed through a combination of image and text. This is clearly a return to verbality. There are two types of textual emoticons: speaking (or speech bubble) smileys and smileys holding a sign.

Another special type is *animoticon* (a portmanteau word coined by blending *animation* and *emoticon*) which combines image and sounds. As opposed to emoticons, they do not integrated into but appeared above the text during MSN messaging, unavoidably occupying almost the entire screen. Animoticons were generally used to draw attention (e. g. knocking on the screen, sticking out the tongue, sending a kiss or show a burst of anger).

Although the classic definition holds that in primary literacy writing can be considered as "frozen acoustics", in secondary literacy it is the other way round: with emoticons writing have become "animated acoustics".

3.5.1.2 Gesture-describing expressions

Although gesture-describing expressions are not emoticons, they are definitely signs expressing emotions which can make up for the missing aspect of nonverbal communication in writing. Such expressions, which are often used in online communication (e. g. in instant messaging), include for example *facepalm* which refers to the gesture of placing one's hand flat across one's face. In internet communication this word means shame, frustration or any form of disapproval. The expression is used both in writing—usually in the form *facepalm* or as the emoticon *m-(*—and in oral conversation. The word even exists as an intentionally twisted calque in Hungarian: *arcpálma* (*arc* 'face' + *pálma* 'palm tree') which is used as a reaction to stupid or ridiculous statements: *"Geee, you must be joking! Facepalm"*.

Another gesture-describing expression is *headdesk* which refers to hitting one's head to the desk as an expression of frustration.

The expression *a-a* also stands for a non-verbal gesture: head shake (referring to the sound made when shaking the head; *Example 63*, *Image 16*).

Example 63: Verbalization of the a non-verbal sign *a-a* (Facebook comments, 2010)

A: *Nah mivan?tanár behúzta a telod?:-o*
B: *Jaja komojan.mindenki ezt nyomta azt tölem is eveték:d tesim lesz. Nemleszek xd*
A: *:-)nekem most volt tesim... Qpert futsz?háhá [Are you running the Cooper test?]*
B: ***A-a** az majd köv hét [a-a 'no-no' it will be next week]*
A: *Gya angol lesz,figyelek.úgy néz ki.*
B: *Oké.én nem.aszok inkab:)*
B: -.-'
A: *Olyasmi:-)4es doga jóh:-):-P*

Image 16: Verbalization of the a non-verbal sign *a-a* (Facebook-komment, 2011)

3.5.1.3 Concluding remarks on emoticons

Emoticons have gone through major changes in terms of function and visual outlook, and have become one of the most characteristic features of internet culture. Although typing had already been a wide-spread form of writing before the age of the internet, the need for emoticons only arose with the emergence of digital communication. The main reason for this is that internet language is almost as fast as spontaneous speech so it became highly important to mark prosody, paralinguistic features and emotions simply and quickly, that is to broaden the narrow communication channels of writing (Crystal 2001: 38).

3.5.2 Combinations of images and texts: ASCII art

ASCII characters can be used to create drawings and special visual signs. They can be understood by rotating the display of the device (or the paper on which they are drawn) 90 degrees. For example, the character combination longing @- >— depicts a rose, and *< < < – or *ΔΔΔ* a Christmas tree. The character combination [-------] depicts a barbell referring to weightlifting or training in general (this sign does not require rotation). Further icon-type signs include @-&— ('flower'), < ~ > ('female sexual desire'), < 3 ('love', 'heart'). These signs appear(ed), among others, in instant messaging, online chatting and message wall communication.

Example 64: ASCII drawings

@- >—*s kedv& kivanok,v* ***@- >—****illat kLtsN fL szombat hajnalan,Cica.Piacra készülünk,még sül a krumpli.Nagymosás lß+fözés.Szép7végezést.P:NY* [the characters in bold depict roses] [text message, 2008]

< < <** *-Nagyon boldog ka#onyt kivanok Neked es szeretteidnek!*[name]ΔΔΔ*** [the characters in bold depict three Chrismas trees] [text message, 2004]

*mit külgyünk neki?:****--------------))*** *Pinocchio* [the characters in bold depict Pinocchio's long nose] [text message, 2002]

Special characters may be used as symbols or signs with predefined meaning. Some of the respondents in my text message surveys stated that they use the number sign (#) to indicate new topic, the Greek psi (Ψ) to indicate that the user is waiting for a response, and the Greek gamma (Γ) to say "write no more". ASCII drawings are often inserted to texts for decoration purposes.

ASCII characters can be used to create multiple-line drawings as well (*Image 17–22*). These are prohibited on public chat sites as they flood the message wall but they can be found in e-mails and social networking websites. Awareness-raising campaigns spreading on Facebook also utilise the potential of ASCII drawings, like the following one which fights against breast cancer.

Example 65: Text of breast cancer campaign using ASCII characters

Silicone breasts (+)(+). *Perfect breasts* (o)(o). *Cold breasts* (^)(^). *Grandma's breasts* \./\./ *Big breasts* (o Y o). *Small breasts* (.)(.) ... *let's take care of each type! post this to your message wall, show* ┌Π┐ (◉_◉) ┌Π┐ *-to breast cancer.....*

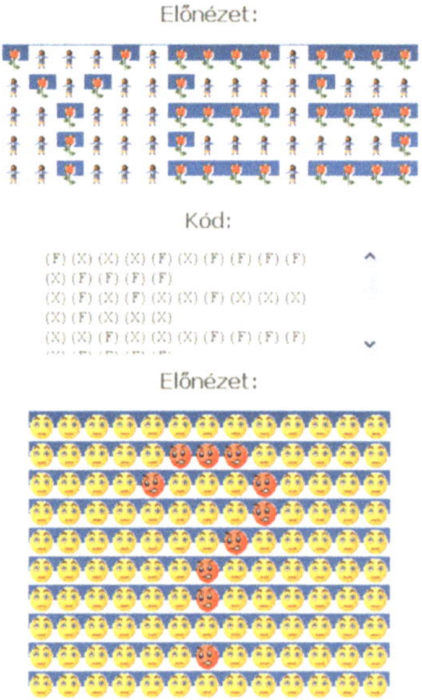

Image 17: Building complete images from emoticons—using MSN shortcuts

```
__________*/|_________
_________*/\_________
________*(_)_________
________.|‾|_________
________.||_________
________.||_________
________.||_________
________.||_________
_____,__.||_________
_____/#|_.||__/\_____
___(##(_,||,_)))____
___\###/ ||/&&/_____
____\##)&___&&(______
______)#/&&___&&\_____
_____/#|&&&___.&\\____
___(##\__&&&_'._)|___
_____\ ######### //____
______"+,______,+"______
```

Image 18: ASCII art #1

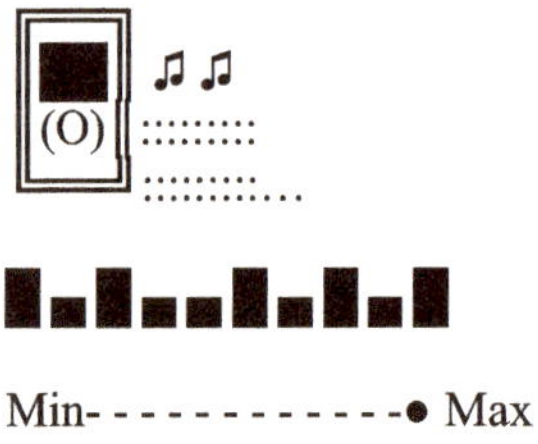

Image 19: ASCII art #2

Image 20: ASCII art #3

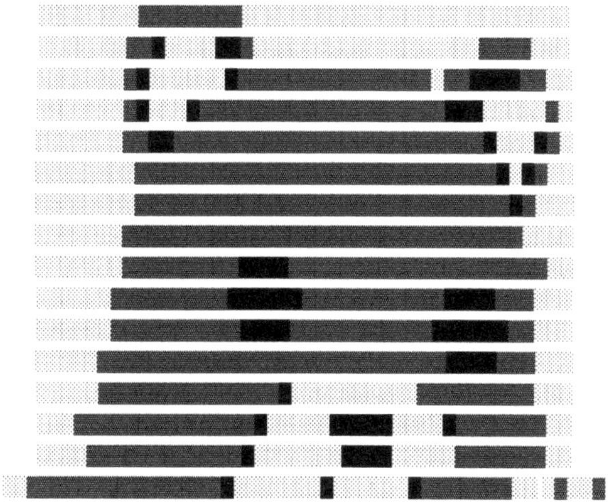

Image 21: ASCII art #4

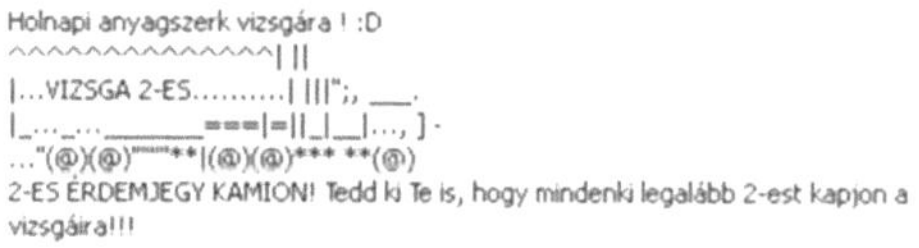

Image 22: Graphic element on Facebook „exam truck" (Facebook message wall, 2011)

3.5.3 Reaction-gifs

Whereas emoticons are built up either by punctuation marks or graphic symbols attached to them, and can be static or dynamic, reaction-gifs are moving pictures representing a series of short movements. Reaction-gifs can be sent as response in different conversational situations in digital communication to express emotions, but in a more intensive way than emoticons; they also present scenes extracted from mostly well-known movies, and one of their main components is humour.

Vis-à-vis the statics of emoticons (including dynamic emoticons), gifs, being dynamic and motion picture-like, involve the time-dimension in the range of tools to express emotions. Continuous repetitions (when the animation ends, it restarts again) exclude the sense of perfection, termination, or caducity from the act of communication. Compared to both static and dynamic emoticons, we can state that reaction-gifs, in most but not all cases, include verbal components, or textual elements to be quoted later on. While emoticons are graphic, thus most probably universal, figures, the characters of reaction-gifs are usually well-known or unknown faces (there can be exceptions, of course, e. g. animals, cartoon characters). The majority of emoticons can be easily interpreted without any further background knowledge, whereas in the case of reaction-gifs, the knowledge of pop-culture (films, series, music, features, typical gestures and reactions of politicians), or that of certain subcultures is necessary. Reaction-gifs, just like memes, spread and change quickly as trends require (Veszelszki 2015b, 2015c *Image 23*).

Image 23: Reaction-gif (without moving)

3.5.4 Memes

Richard Dawkins' original concept of meme, which was rooted in the Darwinian theory of evolution, has gone through major extensions of meaning, and through the definition of "cultural gene" and the "unit of cultural transmission, or a unit of imitation" it arrived at the general definition of practically any type of picture, text and/or audio based content which spreads in the internet (on memetics: Dawkins 1976, 1993; Dennett 1990; on cultural epidemiology: Sperber 1996; on the meaning change of the word *meme*: Veszelszki 2013).

Typical features of internet memes include the gossip effect (which is characteristic of orality), topicality and immediate reaction to events. An *internet meme* is a phenomenon, concept, text, image or the combination of text and image, which spreads in the internet similarly to fashion; its content can be joke, gossip, picture, web-site, web-link and news (or hoax). What is common in these things is their capability to spread rapidly through the internet (on folklore see Hoppál 1984; Balázs 2011).

The aim of producing internet memes can be entertainment, humour, the mocking of a socialpolitical phenomenon, caricature, and even advertising (when a content is shared on a social website, the author of the content is shown to the addressees who can therefore propagate the given page). In addition to this, memes also have community-building effect. These communities can place even multinational companies under pressure. Memes can mobilize computer users from societal, social or political aspects.

The most famous meme characters are various drawings depicting real faces (such as: Pokerface, Me gusta, Okay, Trollface, LOL, Challenge accepted, Forever alone, and Fuck yea; W11; *Image 24*), which can be placed practically into any context to make the content funny. One of the most famous faces is the "Y U No guy", a typical comics character spreading on the internet; *Image 25*), The rage face originates from a Japanese sci-fi manga or anime series. It first appeared to illustrate the text "I TXT U Y U NO TXT BAK?", that is "I text to you why don't you text back?". Subsequent mutations of this meme preserved the 'Y U NO" element, and they usually calls attention to certain phenomena with (grammatically) incorrect language or with abbreviations used in text messaging. Most of the 'Y U NO" pictures reflect on topical political, social or most commonly popular culture issues – so they can only be understood with sufficient background knowledge.

Image 24: Meme faces

Image 25: Y U NO guy

In virtual communication involving multiple participants (e. g. on internet forums or social networking websites), a *troll* is a person who intentionally causes disturbance and disagreement, a provocateur, "a CMC user who constructs the identity of sincerely wishing to be part of the group in question [...], but whose real intention(s) is/are to cause disruption and/or to trigger or exacerbate conflict for the purposes of their own amusement" (Hardaker 2010: 237). Trolls usually try to disturb communication with hateful or personal remarks, incomprehensible or irrelevant comments (Shin 2008; Veszelszki 2017). This destructive behaviour is called *trolling*. "Trollface" (also known as "Coolface") is a black-and-white face with a rascal smile used during trolling. Trollface usually comes with the popular catchphrase "Problem?", which is the punch line of many comics circulating on the internet. Trollface, the protagonist of such comics, sends a cheeky wink at the reader on these pictures; *Image 26*).

Image 26: Trollface

Internet memes in their original or more or less mutated form can survive on the internet for years, but most of them typically live for a few weeks or months (Veszelszki 2013; Horváth–Mitev–Veszelszki 2013).

4 Digilect (and its Effects) based on the Findings of Two Questionnaire Surveys

4.1 Questionnaire for primary school students (2008)

In September 2008 I surveyed the 5–8 grade students of a rural primary school and the 1–2 grade students of a grammar school which offers 6-year-long programmes (where grade 1–2 was equivalent with grade 7–8 in primary school). The survey resulted in 218 completed questionnaires. Most of the respondents were born in 1994 or 1995 (attending forms seven or eight at the time; see *Figure 10*). I selected this age group for the survey because previous surveys (among others: Kilian 2001; Storrer 2001; Thaler 2003; Schlobinski 2005; Schneider et al. 2005; Érsok 2007; Frehner 2008; Veszelszki 2005a, 2005c and 2007) focused on older secondary school students (aged 14–18) or college/university students (aged 18–25).

The aim of the survey was to discover whether the usage of mobile phones and computers influences the way primary students write, and if so, how.

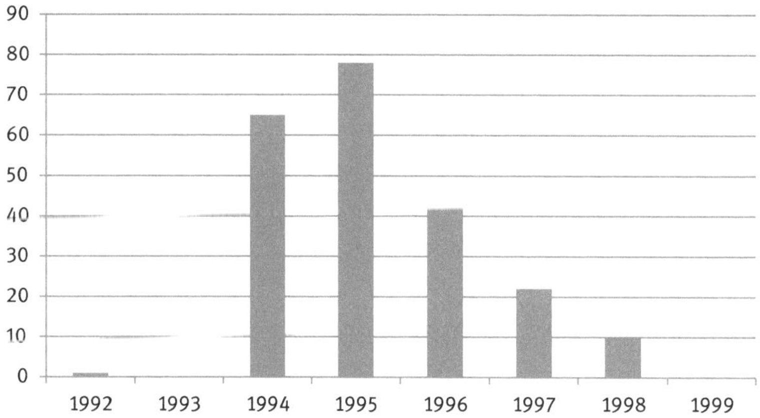

Figure 10: Respondents by year of birth

The respondents were prompted in the questionnaire to provide some personal data (grade, age, sex) and to answer altogether 12 questions (six closed, four semi-closed and two open-ended questions). The closed-ended questions were organized in tables so that similar questions on mobile phone and computer usage could be viewed parallel to each other (see *Table 15*).

DOI 10.1515/9783110499117-004

While closed- and semi-closed questions primarily related to the usage of communication technology, open-ended questions were mainly concerned with the attitude toward such technology. In the last part of the questionnaire the respondents were prompted to complete sentences such as: *If I had no mobile phone...*; *Since I have been using a mobile phone...*; *Computers are useful because...*; *Computers are criticized because...*; *E-mails differ from instant messaging in that...*; *Text messages differ from instant messaging in that...* At the end of the questionnaire respondents had a blank field to make any other remark.

Table 15: Excerpt from the questionnaire

1. Do you have a mobile phone?	2. Do you have a computer at home?
☐ no	☐ no
☐ < 1 year	☐ < 1 year
☐ 1–3 years	☐ 1–3 years
☐ 3–5 years	☐ 3–5 years
☐ 5< years	☐ 5< years

Only 6 % of the respondents stated that they had no mobile phone and only 4 % that they had no computer at home. Students studying in lower secondary education (forms five to eight) said that they had had a mobile phone for one to three years, and most of them had owned a computer for more than five years (*Figure 11*).

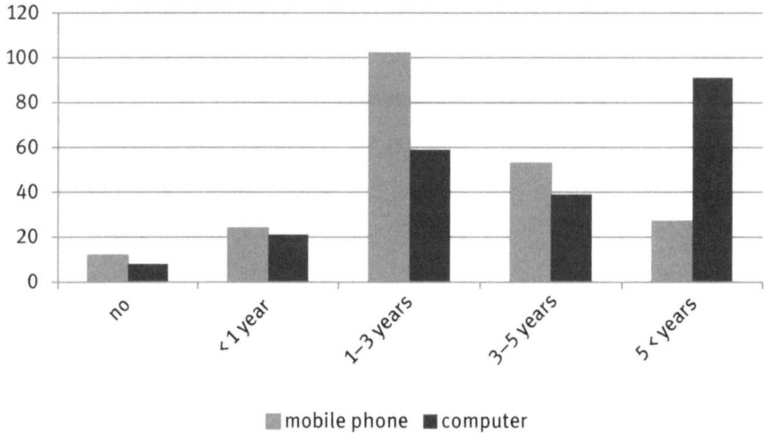

Figure 11: Length of mobile phone and computer ownership

4.1.1 Mobile phone and computer with the eyes of students aged 10–14

According to how the respondents completed the sentence "*Since I have been using a mobile phone...*", 94 % of mobile user students felt that mobile phone had had a positive impact on their life (ability to call friends and family, easier administration, sense of security, increased popularity among peers etc.). Only 3 % submitted a negative answer, which was "they have less money" in all of the cases. The remaining 3 % had a neutral attitude toward their device.

Interestingly, several students expressed the idea of being strongly attached to their phone: "*If I had no mobile phone... I would probably be dead/I would collapse/that would be an impossible situation.*" Attitudes reflected in the following responses also raise concern of the potential psychological and social consequences: "*...I would go to school alone*" or "*I should go everywhere in person.*" Responses like "*I would feel embarrassed*" and "*it would be bad because others have one*" indicate that mobile phones are prestigious accessories, and not having one is considered a shameful thing. Only 10 % of the students said that their life would remain the same without a phone, but the same percent said that they would buy or study better to earn a phone if they no longer had one. Responses like not having a mobile phone would be bad "*because I couldn't play with it*" or "*because I couldn't call my parents*" already relate to the topic of what mobile phones are used for.

Naturally, the most commonly used functions of mobile phones are making and receiving calls and texting. This fact is supported by responses given to open-ended questions: "*Since I have been using a mobile phone I can call my friends (even in the afternoon)* [37 %] or *my family* [23 %]." It is worth to note that mobile phones are widely used for gaming: more than half of the students aged 10–14 in the two schools regularly play games on their phones. It may somewhat distort the results that only making/receiving calls, texting, mobile gaming, WAP[54] and MMS were included as options in the questionnaire, while listening to music and using the camera were not (these functions were added by the students in the "other" field).

The students' attitude toward computers were also examined. The task to complete the sentence "*Computers are criticized because...*" approached the issue from an "adult point of view". 28 % of the students said that looking at computer screens damages (or are believed to damage) the eyes. Further negative attitudes toward computers: "*children sit in front of them a lot*"; "*play on them all the time*"; "*they distract students from learning*".

The students also mentioned lack of computer literacy as a cause for criticism: people criticise computers because they cannot use them (or in other

54 WAP (Wireless Application Protocol) is a quite outdated technical standard that allows users to access information via mobile phones.

words: "*because adults are evil xD*"). Maybe it is worth reiterating here why children learn how to use new technology easier and faster than adults. According to the research findings of Valéria Csépe (2002), procedural memory and implicit learning play a key role in learning how to use an instrument, that is children learn how to use devices through practice which requires many endeavours from them without any explicit instruction (2002: 171). Research showed that this form of learning is much more favourable for children than for adults.

Eleven students mentioned that people may become addicted to computers, and six students that "*there are websites on the internet which are inappropriate for young children*" (*Figure 12*).

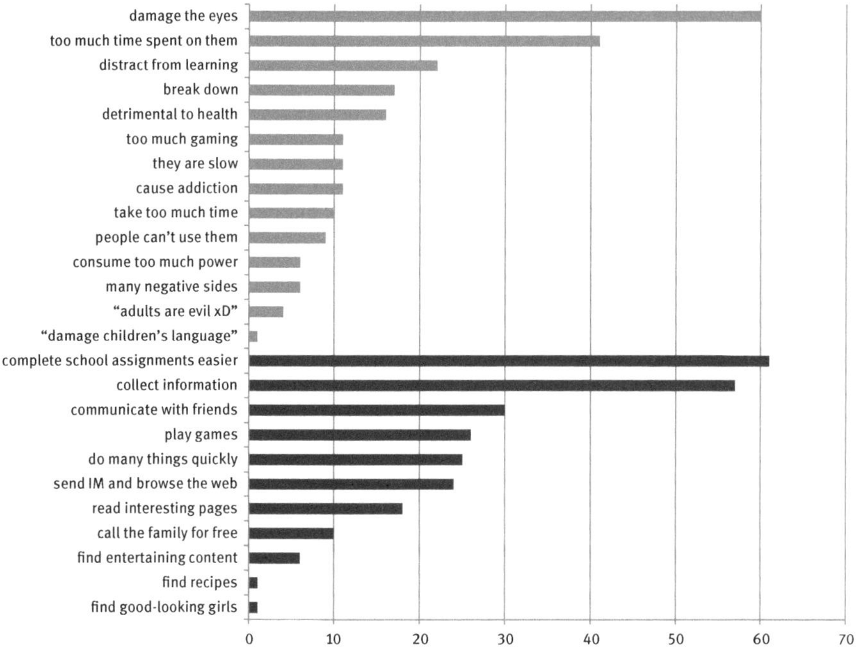

Figure 12: Advantages and disadvantages of computers as seen by students (Computers are criticized because... – marked with light gray; Computers are loved because... – marked with dark gray)

According to students aged 10–14, computers are useful primarily because they help complete school assignments and find information, secondarily they enable communication with peers, and thirdly they make a good platform for gaming. The informative function of the internet was mentioned by 54 %, while communication and gaming by 13–13 % of the respondents.

Helmut Brammerts (1996 and 1999) classifies the functions of the computer according to its role in education, therefore it can be a tutor, instrument, source of information and means of communication. My survey showed that students consider the computer's informative and communication function as the most important.

The questionnaire offered the following answers for possible computer functions: e-mail, instant messaging software (e. g. MSN), web browsing, social networking websites (e. g. iwiw, myVIP, Facebook) and online games. Two thirds of the respondents regularly used instant messaging software, around 60 % browsed the web or social networking websites, and the same percentage used computers for (online or offline) gaming (*Figure 13*).

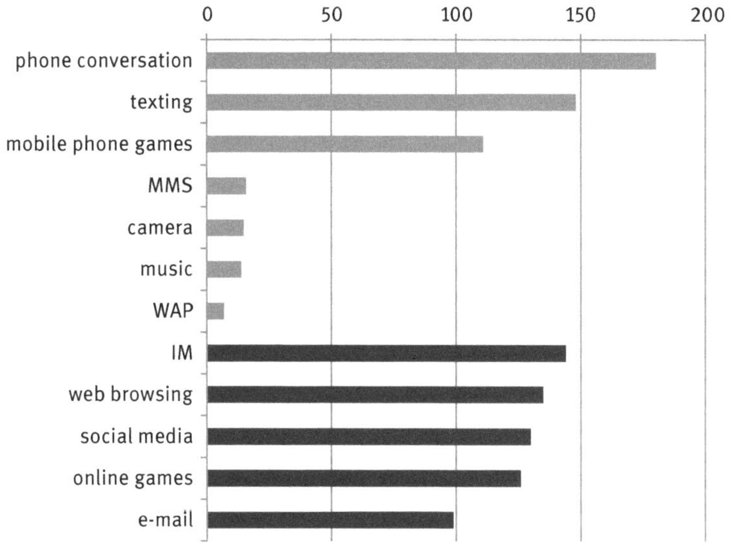

Figure 13: What do students use mobile phones and computers for? (Mobile phone – light gray; computer – dark gray)

Instant messaging (IM) dominates greatly over e-mails in this age group. Students see the main difference between e-mails and IM in that e-mails are not responded instantly and it takes a lot of time to answer them. In other words, e-mail communication "*is not real time*" interaction, that is people can take their time to compose their response as nobody is waiting at the other end of the line (this is how the students expressed the idea of the synchronous–asynchronous distinction between communications devices, which is considered as a basic principle in infocommunications research; cf. Storrer 2001; Thaler 2003; Schneider et al. 2005).

Approximately the same percentage of the respondents said that they used neither IM nor e-mails. 90 respondents said to write e-mails less frequently than on a weekly basis, and only five respondents stated to write e-mails for several hours a day. When comparing the two means of communication, the students found e-mails much more longer, tiring, boring, letter-like and "*generally more appropriate*" than IM. E-mails do not allow for viewing conversation history, but they can be read again anytime. While e-mails were considered as "*complex and coherent texts*", instant messaging were said to comprise "*highly casual sentences*". (Schneider's research group came to similar conclusions when they asked their respondents about the similarities and differences of face-to-face communication, traditional letters, public and private IM, texting and e-mails [Schneider et al. 2005: 53–65].)

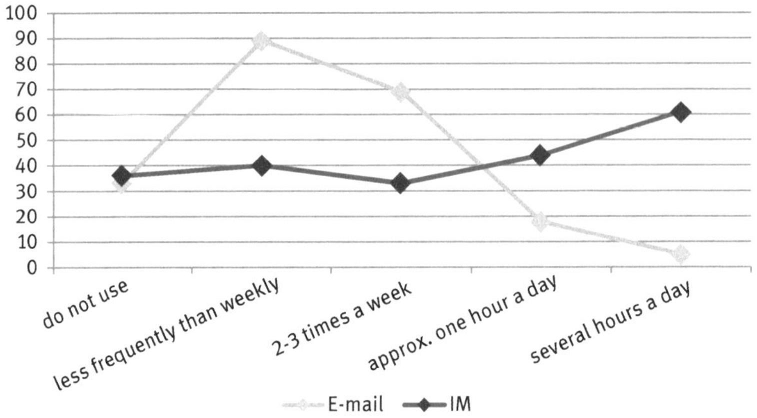

Figure 14: E-mail and IM usage frequency

As opposed to the relatively rarely used e-mails, IM is used several hours a day by nearly one third of the students, according to their responses (*Figure 14*). Therefore, it is reasonable to assume that the form of writing used by students during instant messaging strongly influenced them when writing (by hand) in more traditional settings. IM demands instant wording and enables sending messages by a keystroke, which makes it similar to spoken communication (cf. Főző 2006).

4.1.2 Differences in typing and "traditional" handwriting

The questionnaire showed that the students were aware of the fact that they wrote differently on the keyboard than by hand (in school). 80 % admitted this, and only 20 % said to write the same way on a computer than in class or when writing a traditional letter (*Figure 15*).

Those who write "differently", saw the main difference between "traditional" handwriting and typing in abbreviations. Nearly half of them made no distinction between upper and lowercase letters and 22 % never or only rarely used punctuation marks. Some also mentioned that they did not use diacritics and omit the space after punctuation marks. Further features of "computer writing", according to the age group of 10–14-year-olds, include the use of smileys (emoticons) and slang words. Some wrote on the computer "*the way it feels comfortable*", while others wrote in a more organized way then in class ("*my classroom handwriting is hardly legible*").

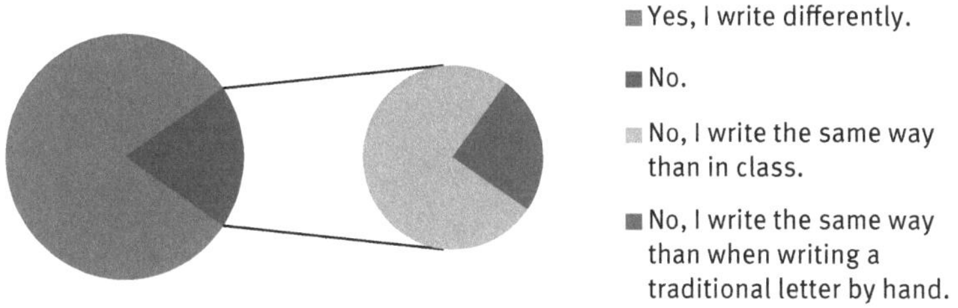

Figure 15: "Do you write differently on a computer than by hand?"

The main motives behind abbreviating words, writing with no capitals and omitting punctuation marks is the eagerness to be quick (82 % of those writing differently gave this reason). 55 % of those who communicate through typing do not pay attention to the typing process which sometimes results in typos. Peers also play an important role in this respect (for 35 %), which means that the way one writes has become a sign of group identity: students express that they belong to a group by using the abbreviations customary in that group. 20 % of the students consider correct spelling as unimportant so they pay no attention to it. Much less commonly mentioned causes included making intentional mistakes for the sake of their effect or the inability to write correctly due to technical issues. Some students also admitted that they "*like writing this way*" or that this form of writing is some kind of encrypted messaging for them (*Figure 16*).

4.1.2.1 Phonetic spelling

Nonstandard, phonetic spelling (~eye dialect, see: Bódi 1998; Balázs 1999: 141) and the usage of unique abbreviations or acronyms provide the opportunity for users to develop their own style. Phonetic spelling overrules word analysis, one of the principles of Hungarian orthography (e. g.: *mondjuk* 'say' > *mongyuk* or *monnyuk*; *játszani* 'to play' > *jáccani*; *Image 27*).

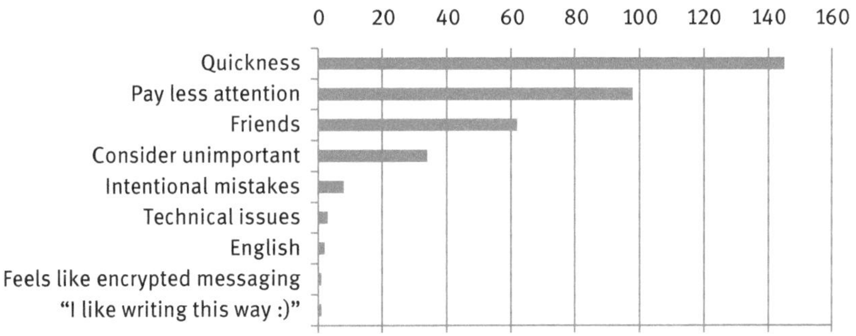

Figure 16: Students' reasons for writing "differently"

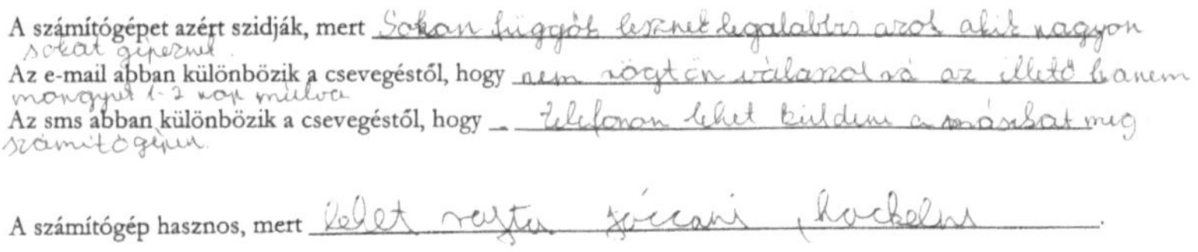

Image 27: Phonetic spelling in handwritten texts (*mongyuk* instead of "mondjuk", *jáccani* instead of "játszani") (Source: questionnaire for elementary school students; see *Annex 6*)

4.1.2.2. Abbreviations

In instant messaging abbreviations are used primarily to approximate typing speed to normal speech rate (cf. Storrer 2000; Fix 2001; see Chapter 3.1). It is also this pressure of time why participants in internet communication are more tolerant toward spelling mistakes and typos (for more details see: Storrer 2000).

The answers to the question "*Is there any abbreviation which you use only when writing on computers?*" showed that out of the 218 students who completed the questionnaire 60 (27.5 %) use only standard abbreviations, as opposed to the 154 students (70.6 %) who also use "substandard" abbreviations (2 % refused to answer). The respondents were prompted to provide examples as well: 89 students submitted altogether 126 distinct abbreviated forms, some respondents provided up to 6–8 examples. This group, which will be referred to here as "students with high preference for abbreviation", has the following social profile: 8th-grade female grammar school students who had owned a mobile phone for one to three years and a computer for more than five years; send one or two texts a day; use IM regularly for several hours a day; use social networking sites and browse the web frequently.

The most commonly mentioned abbreviations were *nm* (< *nincs mit* 'you are welcome', 29 records) and *h* (< *hogy* 'that', 6 records). The verb *vagyok* 'I am' was abbreviated by 15 students as *vok* (see also *Image 28*), and the expression *nem tudom* 'I don't know' was shortened by 11 respondents to *nemtom*.

Amióta mobilom van, boldogabb vok (vagyok xD)

Image 28: The word *vok* in handwriting

If we try to categorize all the "abbreviations" mentioned by the respondents (which are not always abbreviations but the students interpreted this term to include all word-forms deviating from the norm of conceptual literacy), we have to create many overlapping categories.

I use the term contraction (szóösszerántás) as a subtype of clipping where the middle of the word is omitted (*vagyok* 'I am' > *vok* and *wok*, *holnap* 'tomorrow' > *hnap* and *hn*, *szerintem* 'in my opinion' > *sztem*, *igen* 'yes' > *in*, *semmi* 'nothing' > *smi*, *volt* 'was/were' > *vt*, *nagyon* 'very' > *naon* and *non*). This category may also include word-forms which presumably derive from medial orality (less articulated pronunciation), therefore their spelling indicates a possible way of pronunciation: *nem tudom* 'I don't know' > *nemtom* 'dunno', *mit tudom én* 'I don't have the faintest' > *mittomén*, *azt hiszem* 'I believe' > *asszem*, *megyek* 'I'm going' > *mek* and *mék*, *miért* 'why?' > *mér*, *mert* 'because' > *mer*, *egészségedre* 'bless you!' > *egs*. The process from spoken to written language, however, is not necessarily unidirectional: it is also possible that linguistic forms used in conceptually spoken but medially written language (e. g. instant messaging) move to medial orality (spontaneous speech). Examples to this reverse process include *műxik* (< *működik* 'works'), *naon* (< *nagyon* 'very') and *szal* (< *szóval* 'so') which are increasingly common in oral conversations.

Though it shortens the word-form by one letter, the type of writing which is characterized by forms like *pux* (< *pusz* 'kisses') or *cya* (< *szia* 'hi') may not be considered real abbreviations but rather new elements specific to digilect. (The word-form *cya* allows for two possible interpretations: it is either taken as the non-standard variant of the Hungarian *szia* 'hi' or as the phonetic adaptation of the English *see you*). The Hungarian grapheme *sz*, which consists of two letters, is the most common one to be replaced by a one-letter grapheme: either with *x* (*lesz* 'will be' > *lex*; *pusza* 'kisses' > *puxa*, *szerintem* 'in my opinion' > *xt*) or with *c* (*szívesen* 'you are welcome' > *cív*, *köszi* 'thanks' > *köcce*). No shorter form is achieved by using *w* as a substitute for *v* (*szóval* 'so' > *szowal* and *xwal*) or *y* as a substitute for *i* (*szia* 'hi' > *cya* and *sya*) but this is also a characteristic feature of digital communication: it makes the style unique. The following letter replacements invoke baby talk (imitating lisp): *szeretlek* 'I love you' > *cer* and *cejizlek*; *hiányzol* 'I miss you' > *hiánzol*. The above mentioned replacement types can also be combined, such as in *szóval* 'so' > *xwa*. And it is worth mentioning that replacement may also result in longer forms: *szia* 'hi' > *xijjah*.

Only three instances were collected in the questionnaires for the very creative type of abbreviation where only those letters are capitalised whose names are pronounced (e. g. *emlékszel* 'you remember' > *Mléxel* [where the letter *M* is pronounced *em*], *jó estét* 'good evening' > *jóST*-t [where the letters *ST* are pronounced as *esté*], *ember* 'man' > *Mber* [where the letter *M* is again pronounced as *em*]). Perhaps the students did not consider these forms as abbreviations (or the pace of typing in IM renders the use of them impossible), but they are markedly present in my corpus compiled from the text messages of another age group (Veszelszki 2007).

Marking pronounced letters with capitals leads us to the next category of abbreviations. Shortening words without vowels resemble Arabic or Hebrew script where certain vowels are omitted. As this type of abbreviation generates ambiguity, it may only be used for frequently used words which are easy to recognize. Digital communication is (also) characterized by the frequent use of phatic expressions (*puszi* 'kiss' > *px*, *sorry* > *sry*, *ok* > *k*, *szóval* 'so' > *szvl*) and negative particles (*nem* 'no/not' > *nm*), but the word *szeretlek* 'I love you' > *szrtlk* also fits into this category due to its high frequency. Pronouns form a special subcategory here (*magam* 'myself' > *mgm*, *velem* 'with me' > *vlm*).

The special role of the English language played in communication technology is clearly shown by the fact that the respondents submitted 18 different abbreviations taken or derived from English words. The major part of internet-specific acronyms used in instant messaging (Fix 2001: 59; Teplan 2005: 51–53) have been borrowed from English. Haase et al. (1997) found that these IM-specific expressions can filter into oral conversations as well (to be discussed in detail in Chapter 5.1). Acronyms provided by the students are listed in *Table 16*.

Much less frequently but Hungarian expressions are also shortened to acronyms (*nem tudom* 'I don't know' > *nt*, *semmi baj* 'no problem' > *sb*, *nem baj* 'no problem' > *nb*, *nincs mit* 'you are welcome' > *nm*). According to the questionnaire, students seldom shorten words to initials (*hogy* 'that' > *h*, *vagy* 'or' > *v*, *csá* 'hi/bye' > *cs*), it is more common that they clip the end of the word (*pillanat* 'minute' > *pill*, *akkor* 'then' > *akk*, *szívesen* 'you are welcome' > *cív* and *sziv*, *szeretlek* 'I love you' > *cer*, *puszi* 'kiss' > *pu*).

Letters are often replaced with numbers ("*numbers in letters*"). Students aged 10–14 provided the following abbreviations which include numbers: The digit

1 (spelled in Hungarian as *egy*) can be used to create the following forms (and their variants): *egy* (indefinite article), *1szer* (< *egyszer* 'once'), *1szerű* (< *egyszerű* 'simple'), *1ütt* (< *együtt* 'together'), *m1ek* (< *megyek* 'I'm going'), *mind1* (< *mindegy* 'all the same'). The digit *2* invites an English abbreviation to Hungarian: *2* 'to' and 'two'. Five respondents mentioned the word *ötlet* shortened with the digit *5* (spelled as *öt*): *5let*. The digit *6* (spelled as *hat*) may replace both the stem *hat* 'influence' and the verb-forming suffix *-hat* 'can (modal auxiliary verb)'. *66* [< hathat 'can influence'], *6ás* [< hatás 'effect'], *+lát6od* [< megláthatod 'you can see it'], *+mond6od* [< megmondhatod 'you can tell it'], *felhív6sz, hív6sz* [< felhívhatsz, hívhatsz 'you can call me'], *i6* [< ihat 'can drink'], *minden6ó* [< mindenható 'almighty'], *taliz6* [< talizhat 'can meet'], *tud6om* [< tudhatom 'I should know']. *6+* is the abbreviation of an abbreviation (*hat. meg* < *határozd meg* 'define it'). The digit *7* (spelled as *hét*) mainly appears in words starting with *hét-* (*7en* [< héten 'this week'], *7fő* [< hétfő 'Monday'], *7v, 7vége, 7vg* [< hétvége 'weekend'], *a 7en* [< a héten 'this week']) but it can also replace the verb-forming suffix *-het* 'can (modal auxiliary verb)', as in *le7* [< lehet 'it may be']. The abbreviation *jó8* [< jó éjt 'good night'] and *w8* (w+eight = wait) plays on the similar pronunciation of the Hungarian *éjt* 'night' and the English *eight*, but the digit *8* may also be used to shorten the expressions *nekem nyolc* and *tök nyolc* both meaning 'all the same for me' (*nekem8, tök8*).

Table 16: English acronyms provided by the students

Acronym	Original full form
LOL	laugh(ing) out loud
omg	oh my god
afk	away from keyboard
nc	no comment
ty	thank you
bb	bye-bye
b	bye
re	return
np	no problem
thnx, thx	thanks
w8	wait
wtf	what the f*ck
vofl	Virtual Online Football League

Another space and time saving solution is to replace a series of characters with a single character based on homophony (replacement with symbol instead of letter) The prefix *meg-* and the conjunction *meg* may also be replaced by the + sign.

+besz [< *megbeszéljük* 'we'll talk it over'], *+cslom* [< *megcsinálom* 'I'll do it'], *+hívás* [< *meghívás* 'invitation'], *+kaptam* [< *megkaptam* 'I've got it'], *+lát6od* [< *megláthatod* 'you can see it'], *+mond6od* [< *megmondhatod* 'you can tell it'], *+mondom* [< *megmondom* 'I'll say it'], *+van* [< *megvan* 'got it'], *+vár6om* [< *megvárhatom* 'I can wait it']. Number and letter as replacement characters are combined in the word *1x* (< *egyszer* 'once') which was not only mentioned as an example but was also used in the open-ended questions by the respondents (*Image 29*).

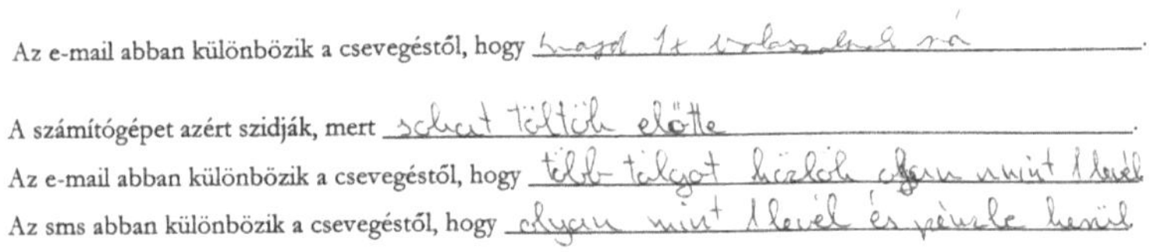

Image 29: Replacing series of letters or full words with numbers (*1x* < *egyszer* 'once', *1 levél* < *egy levél* 'one letter')

A possible explanation for the omission of the grapheme *ly* in words like *hüe* (< *hülye* 'stupid'), *mien* (< *milyen* 'what sort of') or *vmeik* (< *valamelyik* 'one of them') is that *ly* is the homophone of the intrusive consonant *j* which is not marked in writing (*Image 30*).

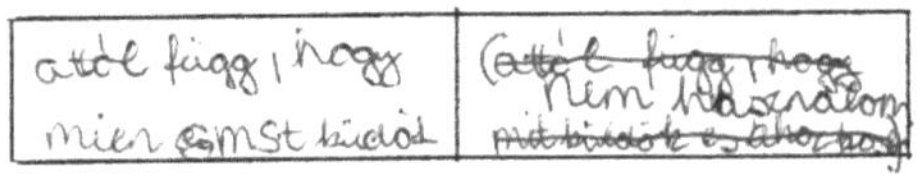

Image 30: Omission of *ly* (*mien*)

Though the following examples were provided by the students, they are not abbreviations but the unique elements of IM vocabulary, therefore they are not detailed here: *láwcsi* and *luwchy* 'I love you', *ja* 'yes', *okésh* and *oksih* 'okay', *ok* 'okay'. The description of diminutive forms mentioned by the respondents will also be omitted here: *lecó* < *lecke* 'homework', *megcsörizés* < *megcsörgetés* 'ringing sy's phone', *talcsizás* < *találkozás* 'meeting', *uncsitesó* < *unokatestvér* 'cousin', *üzi* < *üzenet* 'message'.

4.1.2.3 Emoticons

The students considered smileys as one of the most characteristic features of digital writing (*Image 31*).

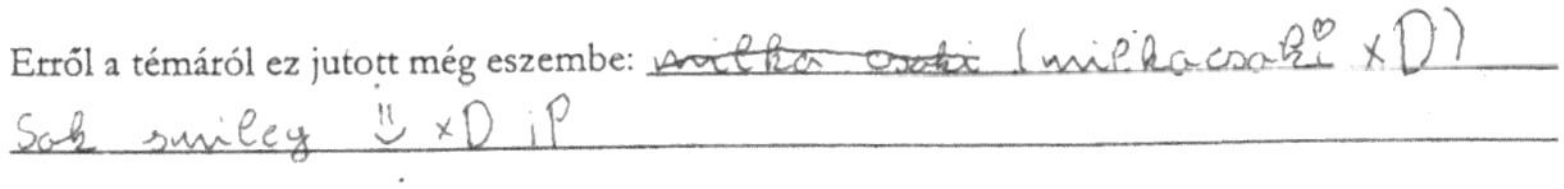
Erről a témáról ez jutott még eszembe: ~~mitka orbi~~ (mikacsaki xD)
Sok smiley :) xD :P

Image 31: Smiley as the most emblematic feature of digital technology

Despite the many completed fields, the answers given to the questions "*What smileys do you use in the following means of communication? Please give examples in the corresponding fields.*" could not be quantified but the scanned tables below clearly shows the number and diversity of the examples provided for each means of communication (texting, e-mail, IM [= MSN], other; *Image 32*).

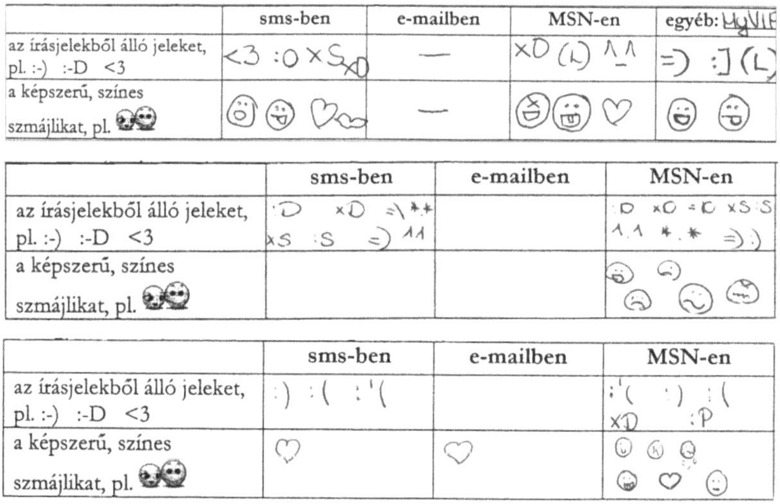

	sms-ben	e-mailben	MSN-en	egyéb: MyVIP
az írásjelekből álló jeleket, pl. :-) :-D <3	<3 :O xS xD	—	xD (L) ^^	=) :] (L)
a képszerű, színes szmájlikat, pl.		—		

	sms-ben	e-mailben	MSN-en
az írásjelekből álló jeleket, pl. :-) :-D <3	:D xD =\ ** xS :S =) ^^		:D xD =D xS :S ^.^ *.* =) :)
a képszerű, színes szmájlikat, pl.			

	sms-ben	e-mailben	MSN-en
az írásjelekből álló jeleket, pl. :-) :-D <3	:) :(:'(		:'(:) :(xD :P
a képszerű, színes szmájlikat, pl.			

Image 32: Smileys used in texting, e-mails and IM (according to the questionnaire)

The above tables also show that e-mail as a communication channel was not so important for students aged 10–14: many examples were given for texting and IM but none or just a few for e-mails. There may be several explanations for this: they did not use electronic mails; their e-mail client did not support inserted images; the typical recipients of their e-mails differed from the persons with whom they chat. To check the validity of the latter hypothesis, I have compared the age of e-mail and IM partners of students aged 10–14 but the comparison did not reveal any significant difference: it is only an indicative result that many of them choose IM when communicating with people aged 15–18 and e-mails when writing to people aged 26–40 (*Figure 17*).

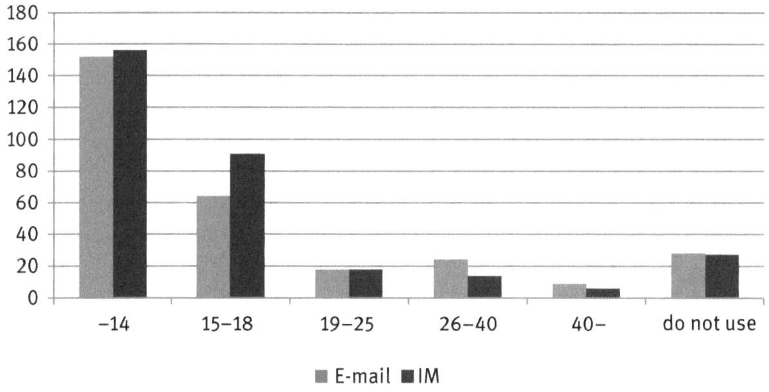

Figure 17: E-mail and IM partners of students aged 10–14 by age

More interestingly, the students automatically used emoticons when completing the open-ended questions by hand, as illustrated by *Image 33* below.

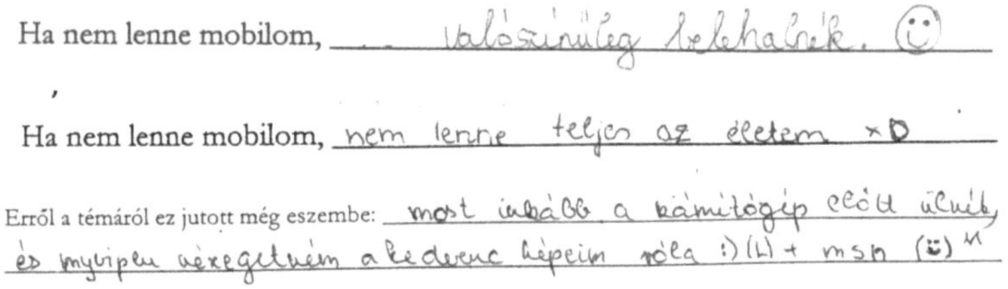

Image 33: Smileys in handwriting

Some respondents ended each of their answers with an emoticon either after the (often decorated) closing punctuation mark or instead of it (*Image 34*). As we can see, Borbála Keszler definitely had a point when she called emoticons “the new punctuation marks of internet communication” (Keszler 2004: 85).

Smileys cannot be considered as redundant items for supplementing or decorating the content, since they can modify or in fact negate the meaning of what has been written, such as in the following example. “*There are many violent computer games*”, says one of the respondents, but quickly adds a smiley to mitigate (or even negate) this statement which is more likely to reflect the attitude of parents (*Image 35*).

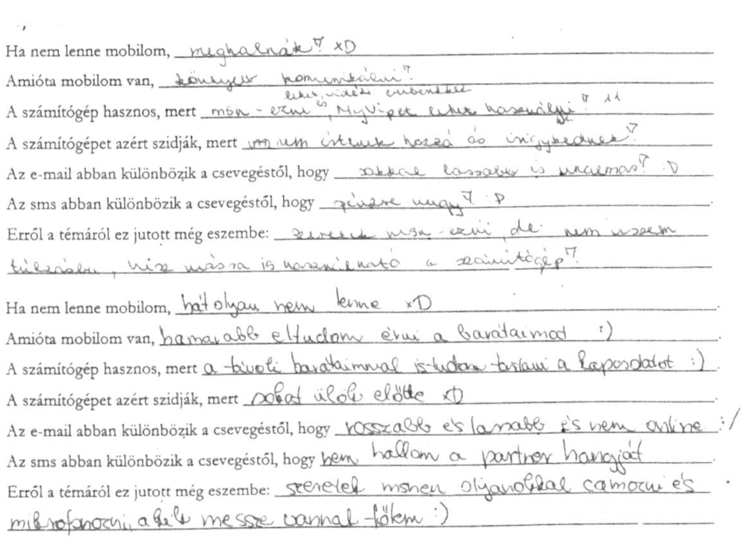
Ha nem lenne mobilom, meghalnák ∇ xD
Amióta mobilom van, könnyebb kommunikálni ∇
A számítógép hasznos, mert msn-ezni és Nyugipet (?) lehet használni ∇ !!
A számítógépet azért szidják, mert van aki sokat ül előtte
Az e-mail abban különbözik a csevegéstől, hogy sokkal lassabb és unalmas ∇ :D
Az sms abban különbözik a csevegéstől, hogy pénzbe van ∇ :P
Erről a témáról ez jutott még eszembe:
Ha nem lenne mobilom, hát olyan nem lenne xD
Amióta mobilom van, hamarabb eltudom érni a barátaimat :)
A számítógép hasznos, mert a távoli barátaimmal is tudom tartani a kapcsolatot :)
A számítógépet azért szidják, mert sokat ülök előtte xD
Az e-mail abban különbözik a csevegéstől, hogy rosszabb és lassabb és nem online :/
Az sms abban különbözik a csevegéstől, hogy nem hallom a partner hangját
Erről a témáról ez jutott még eszembe: szeretek msnen olyanokkal csatorni és mikrofonozni, akik messze vannak főleg :)

Image 34: Smileys as graphemes.

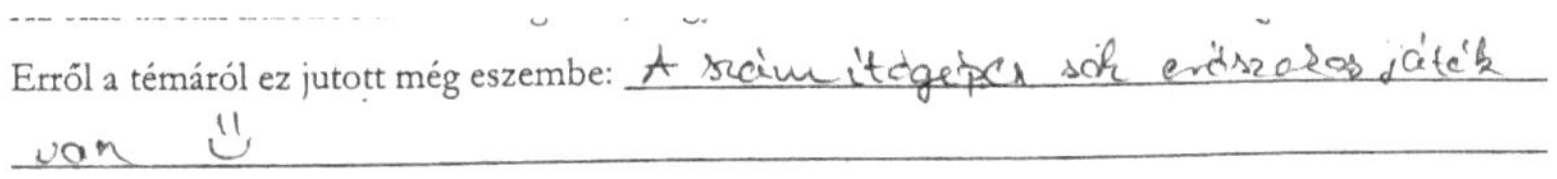
Erről a témáról ez jutott még eszembe: A számítógépen sok erőszakos játék van

Image 35: Emoticons as meaning-modifiers

4.1.3 Handwriting characteristics of primary school students according to the survey

The purpose of the questionnaire survey conducted in late 2008 was to assess the language use of primary school students in grades five to eight and grammar school students in grade one to two (equivalent to primary school grades seven to eight. Despite my anticipations, it was not the answers which delivered new information but the way the respondents completed the open-ended sentences. The characteristic features of digilect were mainly present in the handwriting of the 10–14 age group.

4.2 Questionnaire (2010)

I conducted an online questionnaire survey in 2010 summer (from 15 June to 15 August) using Google's survey programme (cf. Veszelszki 2013a). The online questionnaire was shared with potential respondents through e-mail and various websites. As the questionnaire was spread online, it could reach a very high number of respondents including complete strangers. As another advantage over paper-based surveys, the respondents of electronic forms are more honest and open, less influenced by social expectations and tend to submit longer answers to open-ended questions, according to research findings (cf. Köhler 1999: 188). Köhler (1999: 190), however, also warns that electronic forms are normally completed anonymously which potentially restricts the scope of respondents to those interested in the subject. This way active internet users are inevitably overrepresented in such surveys. Despite these considerations, though not at all disregarding them, my choice fell on online surveying.

The questionnaire was completed by 647 respondents until 15 August 2010. Only one of the completed forms had to be excluded from the evaluation process for containing consistently irrelevant answers. The questionnaire's register was intentionally formal, which was achieved by using the V form.

The first four questions prompted the respondents to enter socio-demographic details which were relevant for the survey (sex, age, highest level of education, occupation). Other sociological data, such as residence, were disregarded due to their irrelevance.

The second larger group of questions related to internet and mobile phone usage: 5. How many hours a day do you use the internet? (questions using Likert scale); 6. Do you have a mobile phone? (closed-ended, yes/no question). The respondents were also asked to reflect on how important they find sharing and receiving information (through landline calls, mobile phone calls, online calls, IM, Twitter, texting, e-mails, handwritten letters) and how often they use these functions when communicating with others. They were expressly prompted to specify usage frequency in the case of text messages and e-mails (0, 1–5, 6–10, 11–20, 21–30, 31– pcs/week). In the case of instant messaging, the respondents were asked to specify average weekly time (0, 1–5, 6–10, 11–20, 21–30, 31– hours/week) and the software they used (MSN – Windows Live Messenger, Google Talk, Skype, Yahoo, Facebook, IM interface of online games, online chat sites, such as gyaloglo.hu, chat.hu). The questions included two sociological ones with linguistic relevance: whether they engaged into instant messaging with strangers and in a foreign language.

The next group of questions dealt with social networking websites (e. g.: iwiw, Facebook, MySpace, MyVIP, Hi5), the frequency of their use (daily, two or three times a week, once a week, less frequently than on a weekly basis, never) and the functions used (searching the online profile of friends, contacting strangers, browsing new information shared by friends, sharing images and texts, commenting contents shared by others, using applications like games or quizzes, instant messaging, communication, learning).

Various studies have shown that the choice of communication channels depends on the communication partner and/or the purpose of communication. To verify this observation, the respondents were prompted to specify the most important communication partners (friends, partner, parents, children, siblings, relatives, colleagues, boss, employee, teacher) for IM, e-mails and texting. Subsequently, the respondents had to assign their preferred communication channel (phone call, postal mail, postcard, text message, e-mail, e-card, message wall on social networking website, IM) to the following four situation: 1. sending birthday/Christmas/New Year/etc. wishes/greetings; 2. arranging schedules; 3. doing official business; 4. talking about private or personal issues (if meeting in person is not an option).

The core of the questionnaire was comprised by the open-ended questions. These tried to find out whether the respondents know and use emoticons (*When do you use emoticons? What sort of emoticons do you use? In what situation would you NEVER use emoticons?*

The question *What abbreviations do you use in IM/texting? Please provide some examples.* aimed to elicit and collect the best known and most commonly used emoticons in instant messaging and texting. The respondents were asked to say expressly whether they have experienced interference between digilect and handwriting. *Have you ever used any internet or text message specific abbreviation or word form in handwriting? If so, please specify.*

The terminology of the questions *Do you think WRITING is being changed by internet and mobile communication?* and *Do you think ORAL CONVERSATIONS are being changed by internet and mobile communication?* needs some clarification. These questions were intentionally formulated to be comprehensible for lay, non-linguist respondents (even below the age of 14 years).

Hungarian terminology relating to written and spoken language does not distinguish between the same terms as for example German terminology. In the introduction of her work *Einführung in die Schriftlinguistik* Christa Dürscheid (2006: 19) gives a list of various expressions used to refer to the concept of *written language* and *writing*. The list is provided here without any detailed definition, just to illustrate this diversity: Schrift 'writing', Schriftzeichen 'grapheme', Schriftsystem 'writing system', Schrifttyp 'type of writing', schriftliche Sprache 'written language', geschriebene Sprache 'written language', Schriftsprache '(standard) written language', Schreiben 'writing', Schriftlichkeit 'literacy'. The questionnaire used the terms *handwriting/writing by hand* and *typing* which simply referred to the technique.

The following question was similar to a word association game: *What comes into your mind when you hear the following notions? Please, write your first thought!* (IM, chatting, iwiw, Facebook, emoticon, smiley, texting, handwritten letter).

The last question (*Any other comment on the topic:*) gave free space to the respondents to express their views (and indeed, many of them lived with this opportunity).

The answers were collected in the Google Documents system, raw data was processed by MS Excel, and cross-correlations were analysed by SPSS and/or manually.

The summary includes many quotations from completed forms. My aim was not only to document the current situation but also to provide the basis for comparative analyses in the future.

4.2.1 Demographics and internet/mobile phone usage statistics

Exactly 647 completed forms were submitted to the survey. More than two thirds of the respondents were female and only one third were male. This proportion is usually explained in sociology by women's greater readiness to help.[55]

The respondents' demographic profile was the following: the largest proportion (42 %) comprised university/college students (aged 19–25) who spent much of their time online. My own generation, that is young working adults aged 26–40, were represented by 32 %. Secondary school students (aged 15–18) accounted for 10 % (64 persons), adults aged 36–45 also for 10 % (65 persons), adults aged 46–55 for 3.9 % (25 persons), adults aged 56–65 for 2.3 % (15 persons) and adults aged 66 and older for 0.2 % (1 person) of the total.

More than half of the respondents (54 %) had university or college degree. As many of the respondents were university or college students, the percentage of respondents with completed secondary school education was also high (37 %). 6 % had completed primary education and 2 % (13 persons) had no completed education.

As regards occupation, 49 % of the respondents were white-collar workers (in the fields of economy, law, technology, humanities etc.), 36 % were university or college students, 11 % were secondary or primary school students, 4 % were blue-collar workers, and 0.6 % were pensioners.

55 Research has also shown that social networking websites have more registered female members than male ones. This may be explained by gender roles: traditionally women have been responsible for maintaining social networks (cf. Bonka–Kraut–Frohlich 2003).

The survey would probably have yielded different results if it had been conducted on a larger sample representative of age and occupation. Still, as it is apparent from the questions, the primary aim of the survey was not to collect quantifiable information but to obtain as many linguistic data and language user opinion as possible through open-ended questions, focused on qualitative results. Thus, a representative quantitative survey would not have provided any significant added value in this specific case.

The respondents with the above detailed demographic profile exhibited the following internet and mobile phone usage characteristics:

Nearly half of them used the internet for 1–3 hours, one third for 4–7 hours, 8 % for up to 8–12 hours and 9 % for less than one hour on a daily basis. 3 % of the respondents (22 persons) spent more than 12 hours a day online (the respondents were prompted to indicate actual internet usage time and disregard the time when their computer/device was online but unused).

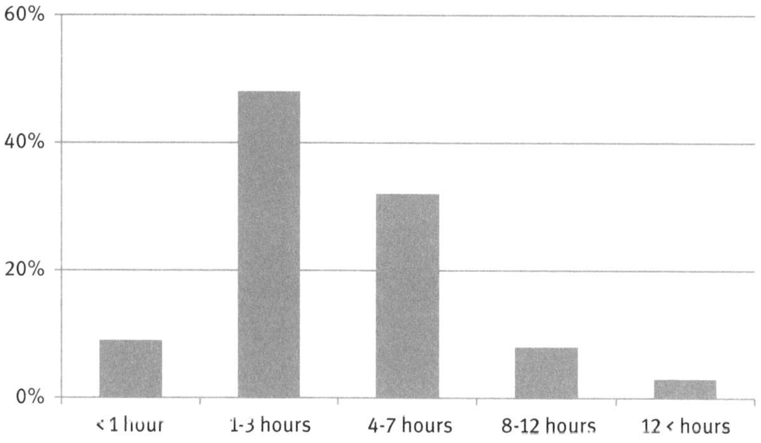

Figure 18: How many hours a day do you use the internet?
(Please indicate actual internet usage time and disregard the time when your computer/device is online but unused.)

The same question, that is the length of daily internet usage, was examined in relation to age. *Figure 18* clearly shows that the majority of the respondents spent one to three hours online a day and that the younger they were, the more time they spent online (the figure is, however, not representative as the different age groups comprise a different number of respondents, which is especially apparent for the 56+ generation which comprised relatively few respondents; *Figure 19*).

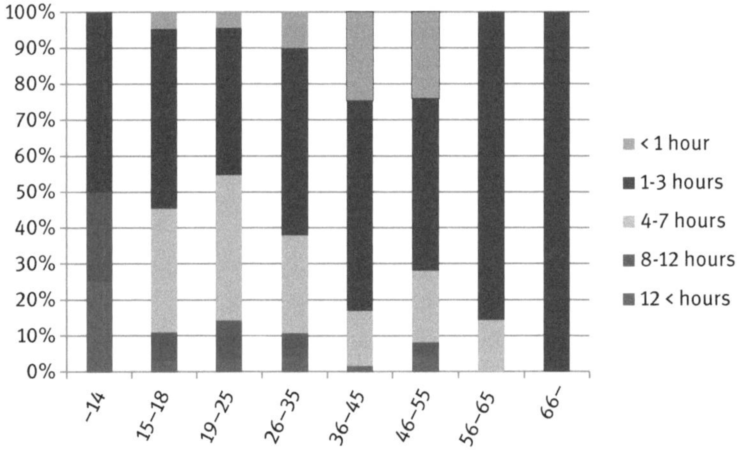

Figure 19: How many hours a day do you use the internet? Correlation between age and internet usage (contingency table)

The contingency table resulted in an unexpected result: men spent much more time online than women, that is the more time of internet usage we consider, the higher probability the respondent will be a man (*Figure 20*).

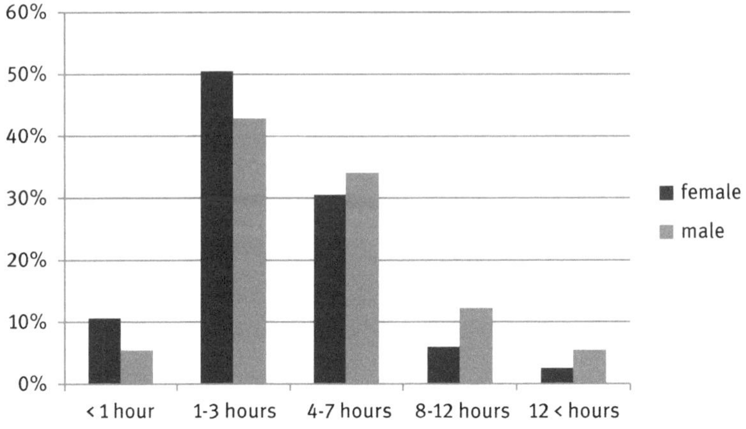

Figure 20: How many hours a day do you use the internet? Correlation between sex and internet usage (contingency table)

99.7 % of the respondents owned a mobile phone (only two out of the total 647 answered that they do not own one).

The respondents were also asked to specify how important they find different means of online and mobile phone communication; *Figure 21*). Apparently, the highest preference is for e-mails: half of the respondents consider this channel as very important and another one third as important, whereas only 1 % stated that e-mail is not important for them. Altogether two thirds of the respondents found mobile phone calls very important or important, as opposed to landline calls, which were considered as unimportant or rather unimportant by 45 % and 27 %, respectively, and very important only by 6 %. Text messaging is very important for 25 %, important for 33 %, neither important nor unimportant for 20 %, rather unimportant for another 25 %, and unimportant for 4 %. Instant messaging was considered unimportant by 30 % and very important by 16 %. The respondents were also asked about their preference for Twitter, which was relatively new in Hungary at the time. Not surprisingly, 87 % said that microblogging was unimportant for them. As opposed to this, the traditional genre of handwritten letters were regarded as unimportant by 45 %, rather unimportant by 30 %, neither important nor unimportant by 14 %, and important or very important by only 10 %.

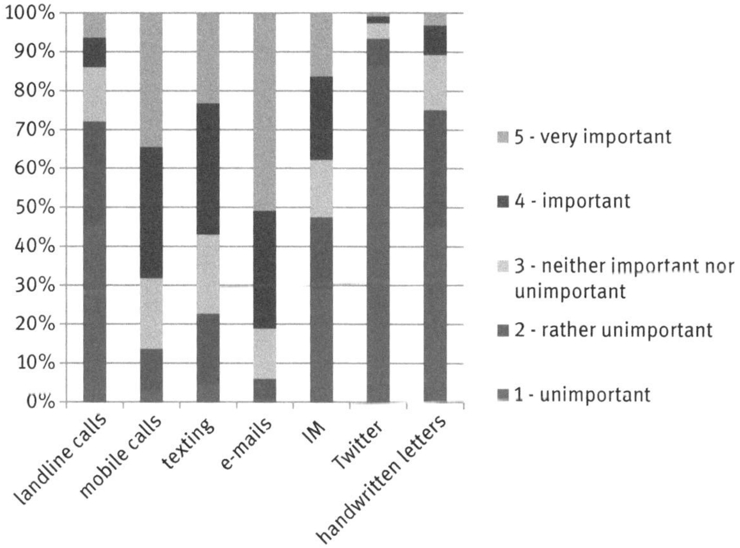

Figure 21: How important are the following means of communication for you?

The preference for different communication channels and the sex of respondents showed no significant correlation. There was, however, an anticipated correlation between preference for instant messaging and age: younger generations (aged 30 and under, especially below 18 years) consider IM more important than older ones (*Figure 22*).

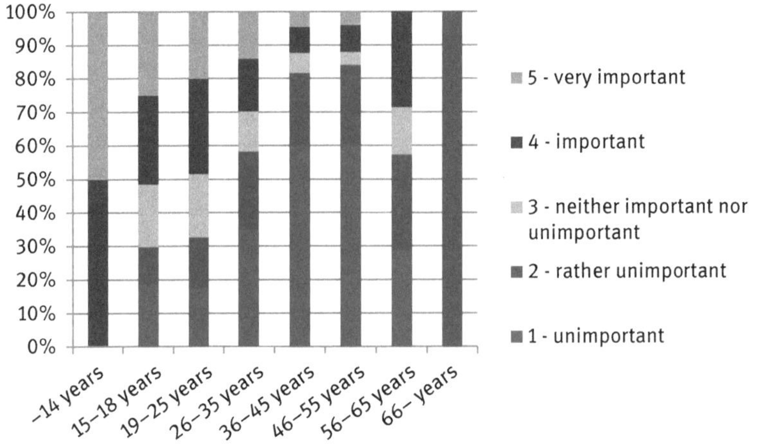

Figure 22: How important is IM for you when communicating with others? Correlation between age and IM preference (contingency table)

The same correlation characterizes the relationship between age and daily hours spent with IM. The younger the respondents were, the more hours they spent with instant messaging (even more than 30 hours a week; see *Figure 23*).

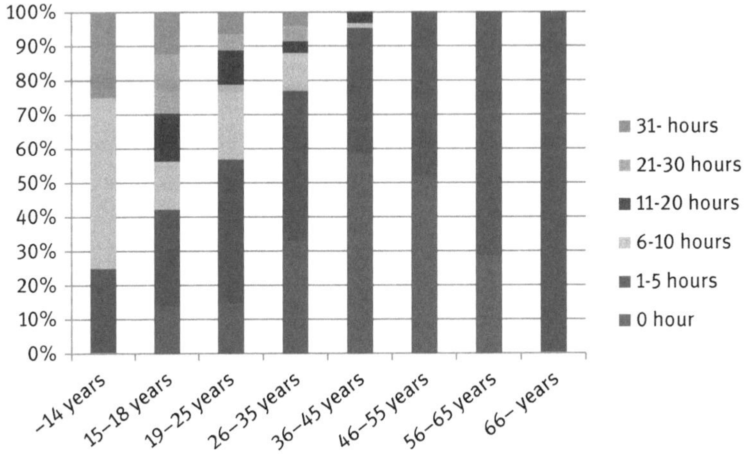

Figure 23: How many hours a week do you use IM? Correlation between age and IM preference (contingency table)

More than two thirds of the respondents write e-mails, and nearly half of them write texts and instant messages on a daily basis. The vast majority almost never write letters by hand, and only a few write occasionally, less frequently than on a weekly basis. More than 10 % said to write any type of text by hand less frequently than on a weekly basis, and a remarkable 5 % said that they never write anything by hand (*Figure 24*).

For handwritten letters, the world of multimedia creates extremely strong competition. The popularity of traditional letters is significantly less than that of texts or e-mails. Nevertheless, traditional letters are still used and have special impact specifically due to this "scarcity" (Höflich 2003a: 18). The fact that it was made by hand is what makes it stand out from the alternative means of communication, as an „object that is made by hand is not quite like any other object. It is unique, and carries the inescapable marks of the person who made it" (Cheal 1987: 158). In our very fast world, handwritten letters provide a remarkable counterpale: it has the power to slow us down (Höflich 2003a: 18).

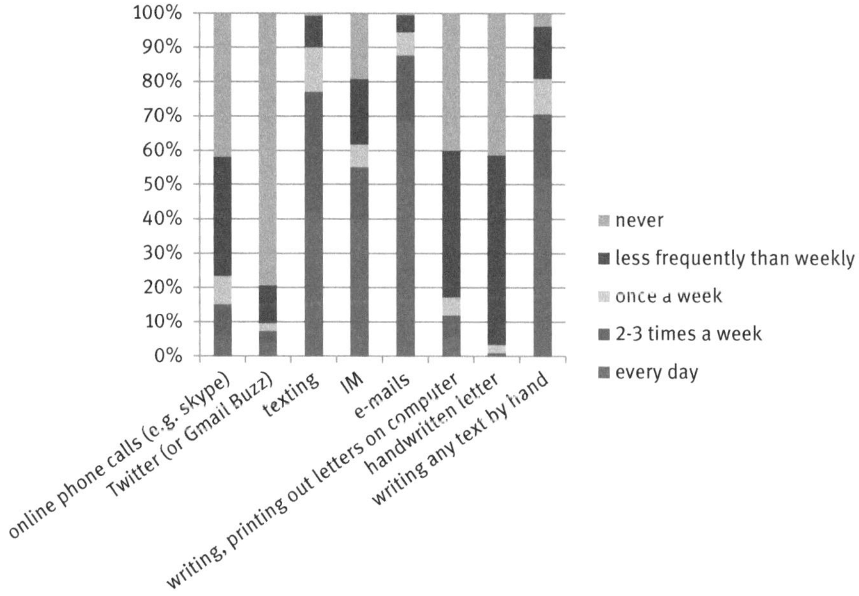

Figure 24: How often do you use these communication channels?

22 % of the respondents sent more than 31, 10 % sent 21–30, and 17–18 % sent 11–20 or 6–10 e-mails a week (*Figure 25*). Around half of the respondents sent one to five, one quarter sent six to ten, and one eight sent 11 to 20 text messages a week (*Figure 26*).

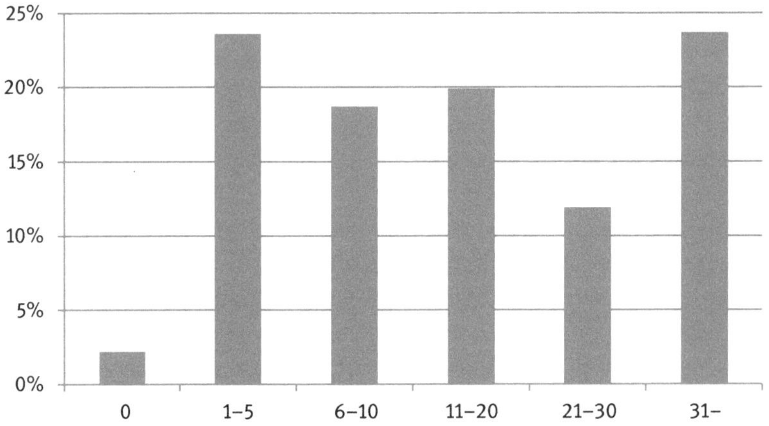

Figure 25: How many e-mails do you send a week?

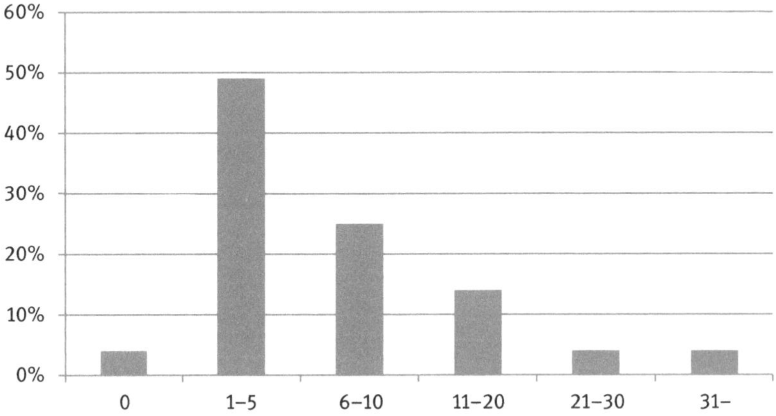

Figure 26: How many text messages do you send a week?

25 % of the survey respondents did not use IM, and 40 % used it relatively rarely (one to five hours a week). 13 % spent an average of one to one and a half hours a day (6–10 hours a week), 6 % spent two to three hours a day (11–20 hours a week) and nearly 5 % spent three to four hours a day (21–30 hours a week) with instant messaging. Only 4 % of the respondents said to spend more than four hours a day (31 hours a week) using IM (*Figure 27*). The most commonly used applications for instant messaging were MSN – Windows Live Messenger, Google Talk, the integrated messenger of Facebook, but many used special websites for this purpose

(e. g. chat.hu, gyaloglo.hu, omegle.com, chatroulette.com, dc++, IRC, mIRC, ICQ, iChat, Pidgin). Chatting through corporate applications (Microsoft Office Communicator) or the integrated interface of online games are also important channels of IM. Although relatively rarely, Skype, the messenger interface of Yahoo! and Lotus Notes Sametime were also used for this purpose.

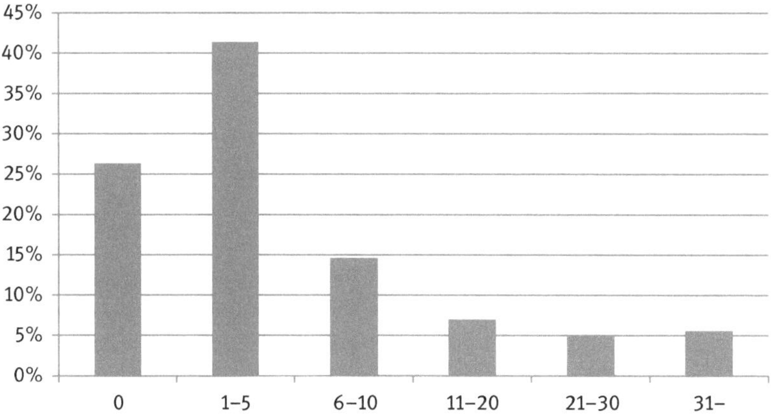

Figure 27: How many hours a week do you use IM?

According to the answers, 15 % of the respondents (93 persons) chatted online even with strangers, and 85 % only with friends, if at all. The respondents gave the following reasons for their preference to chat with people whom they knew from real life: "*for me IM is a way to maintain rather than make relationships*"; "*I don't feel like I want to chat with strangers*"; "*because I need to be related to the person somehow to have something to talk about* "; in other words, the majority of the respondents believe that: "*it's best to make friends face-to-face*". Many respondents explained why they refused to chat with strangers with the fact that they barely had time for their friends, let alone strangers: "*I've got enough friends and I'm happy if I have time for them*". The dangers of making acquaintances online and distrust for strangers were also mentioned: "*I think making friends online is unsafe*"; "*I used to chat with strangers but there are so many idiots on chat sites these days*"; "*because they pretend to be what they are not*"; "*I'm distrustful*"; "*I don't like being bothered*"; "*I like to know whom I'm chatting with, but it is so easy to conceal your true self on these sites*"; "*she may say she's only 12 but in reality he's a 40-year-old paedophile*". One of the respondents referred to a parental advice: "*my mother told me not to talk to strangers.*"

Negative experiences and disappointments may also discourage people from making friends online: "*yes I used to, but they were all complete idiots*"; "*because their second question was always whether I was a virgin or whether I used tampon*"; "*I used to chat with strangers, I used to make friends that way, but I no longer do so because many people wrote to me to have netsex*"; "*I have bad experiences with talking to strangers online (they were often not what they claimed to be)*". The following explanation has a linguistic character: "*it's too fast for me and I hate to type with mistakes*".

Those who were not deterred from talking to strangers online explained their attitude with the opportunity to make new friends, learn and exchange opinions. "*it's worth exchange opinions with strangers, you can learn interesting things*"; "*I chat with foreigners because I'm interested in other cultures*"; "*yes, with foreigners to practice the language*"; "*because of my work*"; "*to get to know others, make friends*"; "*it's easier to get to know others:)*" In some cases relationships established online may prove to be working offline too: "*This is how I got to know my classmates when I was a freshman. 10 % of the people in my year got to know each other through MSN.*" As opposed to those who find IM unsafe, several respondents found it much safer to talk to strangers online than to meet them in person: "*I'm curious about others, and this is one of the safest way to do this if you use it responsibly.*" In certain cases messaging with strangers is specific to a certain genre: "*it's a must when playing games online*"; "*only in online games, such as Honfoglaló*". Curiosity and entertainment are two of the main motives of instant messaging with strangers ("*curiosity*"; "*exciting*"; "*entertaining*"; "*relaxes me*"; "*good pastime*"; "*killing time*").

One half of the respondents (49.5 %, 320 persons) do, the other half (50.5 %, 327 persons) do not chat in foreign languages. Chatting in a foreign language is considered good to learn/practice the language, do work or communicate in online games (usually in English).

Currently, the most popular social networking website in Hungary is Facebook (in 2010 summer 54 % of the respondents used it on a daily basis, and the 2011–2016 Facebook statistics suggest that this percentage has only increased since the survey). Iwiw, which was a market leading social networking website in Hungary between 2004 and 2009, had lost much popularity by 2010–2011: only one third of the respondents visited it every day, and only one quarter two or three times a week. The once dominant role of iwiw is reflected in the fact that only 7 % of the respondents did not go to the website at all, while this proportion is 25 % for Facebook. The usage of other social networking websites (MySpace, Hi5, LinkedIn etc.) was insignificant compared to iwiw and Facebook (*Figure 28*).

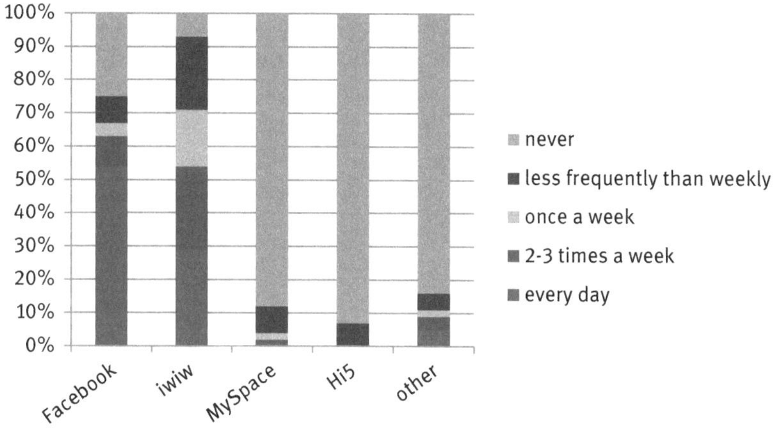

Figure 28: How often do you visit the following social networking websites?

4.2.2 Relationship between communication factors and preferred means communication

In 2002 Kovács–Krajcsi–Pléh examined people's preference for technical means for different social situations (cancelling a prearranged appointment, apologizing for failing to do something, arranging a meeting, dating). The survey, which polled 99 university students, showed that the more personal a situation is, the more important face-to-face communication becomes. The second place was taken by mobile phones (2002: 182). In an official/workplace setting, however, the most preferred form of communication is e-mail.

To verify this finding, my questionnaire prompted respondents to specify their most important communication partners (friends, partner, parents, children, siblings, relatives, colleagues, boss, employee, teacher) for IM, e-mails and texting. Responses were classified into three major categories: friends, family and formal relationships. The summary of the answers, however, did not bring any unexpected results. It showed that IM was used primarily with friends, secondarily with family members and least commonly with colleagues or in an official setting. Texting is an overlapping category as this form of communication is used both with friends and family members. E-mail is mainly used for official communication, less commonly for communication with friends and least commonly to contact the family (*Figure 29*).

The preferred communication channel not only depends on the partner but also on the communication purpose. I chose four situations: sending greetings, arranging schedule, doing official business and talking over personal/private issues. Unfortunately, several respondents misunderstood the question: despite being expressly prompted to imagine a situation with a person who lives at a distance from the respondents, they ranked meeting in person at the first place (these responses were disregarded in the figures below). Several respondents made a point by noting that *preference* means both *most willingly* and *most often*, but sometimes the two does not overlap.[56] That being said, here are the results:

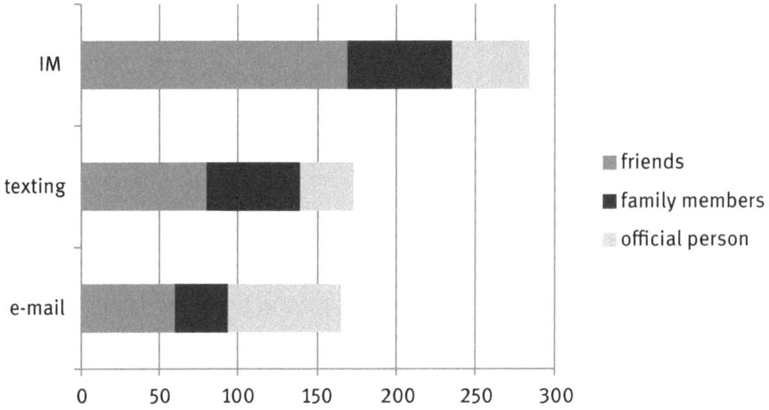

Figure 29: Most typical partners in IM, texting and e-mail communication

I surveyed approximately one hundred university students in 2006 asking them about birthday, Christmas, New Year etc. greetings (Veszelszki 2006), which showed that 84 % used text messages for arranging meetings and 74 % sends greetings in texts (in addition, 70 % of the respondents considered communication with friends as the most important function of texting). The 2010 survey supported that still many people sent text messages to their friends and relatives living far away from them (on the linguistic aspects of good wishes and greetings see Chapter 3.6; *Figure 30*). The second most common form of expressing good wishes was through phone call, followed by e-mail, e-card and social media. Several respondents noted that their choice of communication channel largely depends on the recipient: "*it depends on the person: I call or send text to close friends, write a message in a social networking website to other friends, and call less close relatives*"; "*this very much depends on the person to whom I want to*

56 As one respondent noted in relation to sending greetings: "*my preference, as far as my desires are concerned, is postcard, but my preference, as far as my habits are concerned, is text message (unfortunately, it depends on time and money).*"

express my good wishes. If it's a close friend or relative, I prefer phone calls, but if it's a more remote friend, I opt for a social networking website. I use the latter more because I have more remote friends."; "*if the person is important to me, I call/text/ visit him or her, if the person is not so important or these channels are unavailable, I write a message on Facebook.*" The following way to express good wishes is quite a unique one: "*I write a personal message with the Paint graphics programme on a digital photo taken by me and send it through e-mail.*" Interestingly, the questionnaire tickled one of the respondents' fancy to write postcards: "*I don't normally do such a thing, but maybe this year I'll surprise my friends with a postcard, as nobody expects that these days.*"

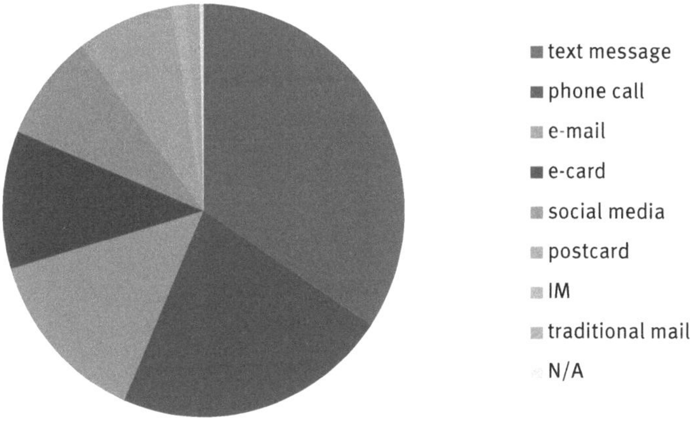

Figure 30: If you send birthday/Christmas/New Year etc. wishes to someone (who live far away from you), what is your preferred way of doing so?

Schedules are primarily arranged verbally (in the case of remote communication partners), more specifically through phone calls where immediate feedback can be given. As regards the digital genres, e-mails and text messages were considered as the most suitable channels for this purpose (*Figure 31*).

The preferred way of communication for doing official business is also phone calls which enables instant interaction, however, e-mail is also used for this purpose due to its formal character. This communication purpose is often fulfilled through traditional letters as well, which may be explained by the fact that authorities are often obliged to send notifications through postal mail. Although doing business online was not mentioned as an option in the questionnaire, several respondents mentioned that they met (or even used) this opportunity on the official website of government agencies (*Figure 32*).

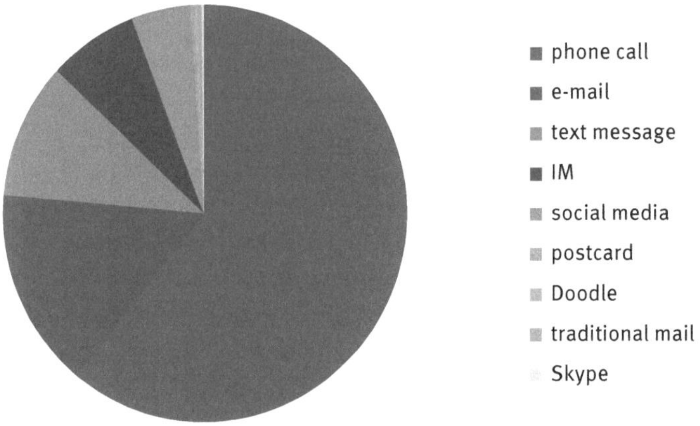

Figure 31: If you arrange an appointment with someone, what is your preferred way of doing so?

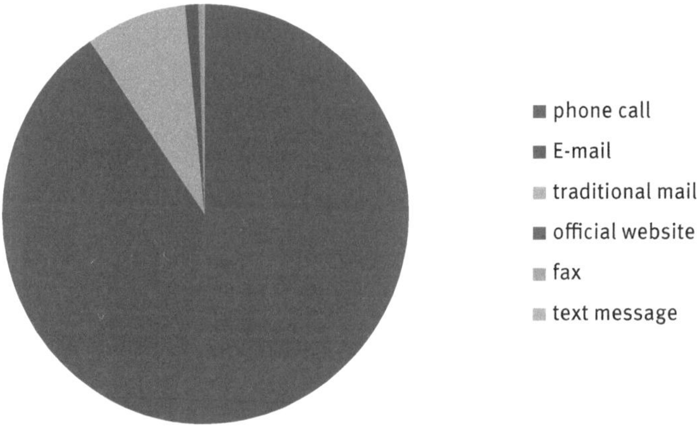

Figure 32: What is your preferred way of doing business?

When it comes to discussing private or personal matters, many respondents prefer personal meetings. But if they have to use technology for this purpose, the number one choice is phone or Skype. One respondent said to "*prefer the phone but typically use instant messaging:)*" for such conversations IM was found appropriate for discussing private issues by 17 % of the respondents, and e-mail by 13 % (*Figure 33*).

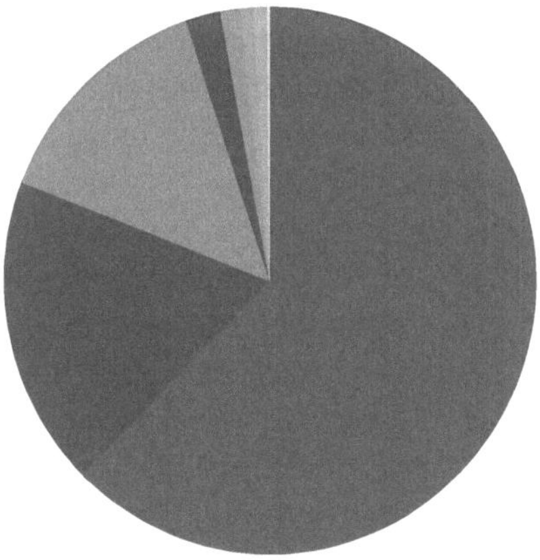

Figure 33: If you discuss private or personal matters with someone (who live far away from you), what is your preferred way of doing so?

4.2.3 Internet-specific expressions, abbreviations, word-forms

The question *What abbreviations do you use in IM/texting? Please give some examples.* received more than 620 different forms and expressions that were thought to be abbreviations. Chapter 3.3.1 includes a detailed collection and description of the most important character-saving and typing-speed enhancing techniques. These are the following: abbreviation with initials and related internet-specific acronyms; Arabic style writing by omitting vowels and its variant where the omitted vowel is marked with a capital letter; replacing letters with number, symbol or a different letter. Other types include contraction (e. g. *vagyok* 'I am' > *vok*, *szerintem* 'in my opinion' > *sztem*), clipping (*amúgy* 'by the way' > *am*), the euphemistic short forms of swear words, and unique techniques (see *Table 17*).

Table 17: Types of character-saving techniques

Character-saving technique	Possible subtypes
abbreviation with initials	internet-specific acronym
omission of vowels (Arab style writing)	marking the pronunciation of vowels with capital letters
substitution of a letter	substitution of a letter with number substitution of a letter with another letter substitution of a letter with symbol
other types	blending clipping (euphemistic abbreviation of swear words) other: other technique not falling within any of the above types

The full table of abbreviated forms categorized by abbreviation technique is included in *Annex 3*, so the table below features only those abbreviations used in digital communication which were mentioned by at least ten respondents (*Table 18*). The respondents may have been influenced by the acronyms listed in a previous question of the survey (which is not detailed here), as the forms *h* (*hogy* 'that'), *asszem* (*azt hiszem* 'I think') and *mind1* (*mindegy* 'all the same') were mentioned particularly commonly. The character-saving techniques listed above were among the most frequently mentioned ones. The so-called internet-specific acronyms, that is abbreviations typically used in internet communication and usually taken from English (Fix 2001: 59; Teplan 2005: 51–53; cf. Chapter 3.2.1) occurred in large numbers in the list, e. g.: *lol*, *omg*, *asap*, *wtf*, *brb*, *rofl*. Short forms containing only the initial letter are very often ambiguous solutions: for example *h* can be *hogy* 'that' or *ha* 'if' depending on the context; *v* can mean *vagy* 'or', *van* 'is/are', *vagy* 'you are' or *volt* 'was/were'; while *m* may stand for *mert* 'because' or *mint* 'as'. In general, it may be established that frequently used expressions have abbreviated forms (as they are easy to recognize and decode).

Table 18: Most typical abbreviations in texting and instant messaging given to open-ended questions

Type of abbreviation	Abbreviation	Meaning	Number of occurrence
abbreviation with initials	*h*	hogy 'that'; ha 'if'	326
phonetic spelling	*asszem*	azt hiszem 'I think'	124
substitution of a letter with number	*mind1*	mindegy 'all the same'	95
blending	*sztem (szerintem* 'in my opinion')	szerintem 'I think'	77
blending	*vok (vagyok* 'I am')	vagyok 'I am'	65
internet-specific acronym	*lol (laugh out loud)*	laughing out loud	62
blending	*nemtom*	nem tudom 'I don't know'	48
blending	*vmi*	valami 'something'	46
substitution of a letter with number	*mind2*	mindkét, mindkettő 'both'	39
substitution of a letter with number	*jó8*	jó éjt 'good night'	35
clipping	*pill*	pillanat 'moment please', azonnal jövök 'I'm back in a minute'	34
abbreviation with initials	*v*	vagy 'or'; van 'is/are'; volt 'was/were'	33
internet-specific acronym	*omg, OMG*	oh my god	30

Table 18: (Continued)

Type of abbreviation	Abbreviation	Meaning	Number of occurrence
abbreviation with initials	*nM, nm*	nem 'no'; nincs mit 'you are welcome'	29
blending	*valszeg (valószínűleg* 'probably')	valószínűleg 'probably'	29
blending	*hnap (holnap* 'tomorrow')	holnap 'tomorrow'	23
internet-specific acronym	*asap*	as soon as possible	22
substitution of a letter with symbol	+	meg (suffix, conjunction)	21
abbreviation with initials	*m*	mint 'as', mert 'because'; miért 'why'	20
abbreviation with initials	*p*	puszi 'kiss'	19
blending	*tom*	tudom 'I know'	19
blending	*szal (szóval* 'so')	szóval 'so'	18
blending	*thx*	thanks	17
phonetic spelling	*lécci*	légy szíves 'please'	16
internet-specific acronym	*wtf (what the fuck)*	what the fuck?	16
substitution of a letter with number	*7vége*	hétvége 'weekend'	15
clipping	*am (ami* 'which')	amúgy 'anyway'	15

Many respondents said that they avoid abbreviations because they do not like them. "*I usually use the full form of words and only use general abbreviations: de.* 'in the morning', *du.* 'in the afternoon', *kb.* 'approximately', *stb.* 'etc.'". The same idea with a pinch of humour: "*I don't use abbreviations. I support telco service providers.*". Many respondents said that abbreviations deteriorate the language and we should protect the language from them: "*I don't use abbreviations because in my opinion they disfigure our language which I find beautiful*"; "*The Hungarian language is very sophisticated so I use the full form of words.*" There are users who are even more redundant in texting than others: "*where the situation permits, I usually describe things in more detail than others as, according to my experiences, it is easy to misunderstand the message even without abbreviations, let alone using them.*" Typing with all ten fingers is a useful way to avoid abbreviations, according to several respondents: "*I don't use abbreviations.*" *Instead, I've learnt how to type fast. And I don't use diacritics either, as you can see in this sentence*"; "*as I can type with ten fingers, I find it easier to write everything down in its full form, I'm accustomed to it, I seldom abbreviate anything*"; "*I can type fast, I use the full forms, I don't like redundant abbreviations they are hard to read.*"

Abbreviations can also be avoided by opting for another word or expression, the respondents say: "*I consciously avoid abbreviations, it's rather that I formulate my message in a more compact and straightforward way*"; "*I rarely use abbreviations, I prefer finding shorter synonymous words or reformulate my message*"; "*I never use abbreviations in text messages, I would sooner reformulate my message if it's too long.*"

The usage of abbreviated forms can vary by partners as well: *"only in informal conversations with friends when we share experiences, because I know that they know what I mean by them. In any other case, I consider it irritating and offensive when somebody wants to tell me something important in a code language."*

Many respondents believe that the main reason behind abbreviations is the character limit of text messages. *"in text messages, when I'm running out of characters, I use them"; "It depends on whether I have enough characters left in the text message. If this is the only way to stay within the limit, I use for example* h (hogy 'that'*) or anything else that may come in handy."*; *"A* vmi 'something' *or* vki 'somebody' *here and there, sometimes maybe a* h 'that', *and more rarely a* sztem 'I think', *a* pill 'wait a minute' *or a* valszeg 'probably' *but nothing else, really. And I only use these in texts due to the character limit."*; "*if it is absolutely necessary, then, to save characters, I shorten* hogy 'that' *to* h, vagy 'or' *to* v *and* mert 'because' *to* m, *but I usually avoid them as I don't like them.*"

Some respondents provided fast type techniques rather than specific examples: "*I shorten word parts which are pronounced the same way as numbers:* jó 5let 'good idea', 6ásos 'effective'"; "*I sometimes omit vowels, like* mndg 'always'", "*we use unique abbreviations with my partner, typically by omitting the vowels (e.g. szrtlk (=szeretlek* 'I love you')"; "*I omit vowels (and maybe some consonants) but only if it remains comprehensible.*"

A unique technique draws on Orwell's newspeak as it uses + and ++ as intensifiers (taken from *plusgood* and *doubleplusgood*). Such abbreviations are, however, not very common, which was confirmed by the respondent as well: "*unfortunately, I've never met anyone who understood them, except for programmers as I symbolize what is good or bad according to the rules of value increase/reduction used in programming.*"

As a special way of abbreviation "*words can be written together, such as* nemár (ne már 'no way'), jólvan (jól van 'all right')". One of the respondents reflected on this the following way: "*writing together words falling under the same stress*", and another one: "*I sometimes write words together to be fast, and I rarely use capitals*".

People rarely admit that they intentionally misspell words to make them shorter, but there are examples for this as well: "*sometimes I write misspell words to make my message shorter*, e.g. télleg (tényleg 'really'), h (hogy 'that'), mind1 (mindegy ‘all the same’), hy (hogy ‘that’), vok (vagyok ‘I am’), re...)”; “*phonetic spelling (sometimes it is shorter to use J instead of LY* [these letter mark the same phoneme, and differ only in writing], *or BA instead of the suffix BAN*)”.

Too many abbreviations may definitely hamper communication, but not always by making the text more difficult (or impossible) to understand, but due to user attitudes: “*I don’t like abbreviations and I try to discourage my friends from using them too. The most common ones are* vok (vagyok ‘I am’), h (hogy ‘that’), sztem (szerintem ‘in my opinion’), teccik (tetszik ‘like’), *but in the long run I terminate communication with anyone who keeps flooding me with these words or look for an alternative way to communicate with them (e. g. phone),*” or: “*I won’t answer to anyone writing that way.*”

Example 66: Respondents’ examples for abbreviations

mind2 napon LmNtM bulizni, de közben vm @home, h ruhát váltsak:D kva jó nap v. örülök, h láttalak, puX. asszem mNM kLl. vigyimagi (mindkét napon lementem bulizni, de közben voltam otthon, hogy ruhát váltsak:) kurva jó nap volt, örülök, hogy láttalak, puszi. Azt hiszem mennem kell, vigyázz magadra! ‘I went to party on both days, but in the meanwhile I went home to change clothes:D It was a fucking good day, and I’m glad I met you. kisses. I think I have to go now, take care!’)

Am szeretem h lehetne ha akarnám.:) (Amúgy szeretem, hogy lehetne, ha akarnám. ‘By the way, I like them, so I could use them if I wanted to.’)

Nm hasznlk röv.t (Nem használok rövidítést ‘I use no abbreviations’)

The respondents were asked whether they used such digilect-specific abbreviations in handwriting as well. A total of 227 persons (35 %) replied “yes”, and 420 persons (65 %) said “no”. The difference is not significant between the age groups but younger generations tend to be more tolerant toward abbreviations and they use them—even in handwriting—more often than older generations.

The same topic was covered by the following question but from a different aspect: *Have you ever used any internet or text message specific abbreviation or word form in handwriting? If so, please specify (Figure 34).*

According to their own account, the respondents use abbreviations and unique word forms mainly in private texts created for their own purposes. “*When taking notes on lectures I often abbreviate words*”; “*When I put something down for myself, or when I make a draft before writing a formal letter, I often use abbreviations and even smileys*”; “*only in my own notes.*” Diaries were also mentioned in this context: “*yes, but only:) in my diary (which I write with my hand:)*”; “*yes but only in informal mails or diary entries*”; “*Yes, in my own diary I sometimes shorten* hogy (‘that’) *to* h *but nothing else, really.*” Abbreviations speed up note taking in class: “*during class, if I’m lagging behind, I use abbreviations in my notes*”; “*yes, when I'm taking notes, e.g.* +halt (meghalt 'died') *Though I think I had already used these when I was taking notes by hand as speed has always been a critical factor.*"; "*in my university notes I use virtually all the abbreviations I use in internet communication.*"

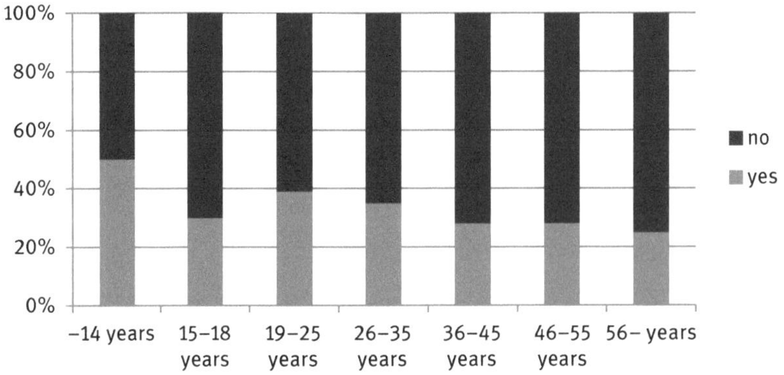

Figure 34: Do you use these abbreviation in handwriting? Use of digilect-specific abbreviations in handwriting by age

Many respondents admitted to use more abbreviations in texts written for themselves than in those which are or may be viewed by others: "*in class I use even more abbreviations than in instant messaging, because I may be understanding what I write in MSN but I cannot guarantee that the discussion partner shares my logic... in my notes, however, it is only me who will read my lines later on.*"; "*only in daily notes taken for myself*"; "*only in notes I made for myself, e.g.: the symbols shares my logic@ or &, or abbreviations like* htl *for* hotel"; "*only for own purposes (e.g. notes in class or at home) but never when writing to others (formal letter, test paper etc.)*"

The informal character is emphasized in the following comments as well: "*only in short letters or notes written to my friends or family members:* h (hogy 'that'), am (ami 'which'), tod (tudod 'you know')"; "*only when taking notes for myself or for a close friend:* v (vagy 'or'), h (hogy 'that'), asszem (azt hiszem 'I think')". The following respondent stresses the importance of common knowledge and background: "*longhand is perfectly fine when I'm communicating with the same people I'm texting or instant messaging with, as they'll understand the same vocabulary, only the communication medium is different. Naturally, when I assume the other party will not understand these expressions, or I'm writing in a formal matter, I try to approximate my style to the standard language. So, since it is only the medium of communication which varies, I use the same word forms as on digital devices.*"

As far as the scope of use is concerned: "*everything I use in texting or instant messaging, I also write in longhand letters and notes, if it fits the situation (obviously not in letters written to grandma)*"; "*Always in my own notes. Often in non-formal letters*"; "*in handwritten notes, memos, information sheets post-it notes and class notes*"; "*when I leave a message on a slip of paper.*"

Note how important it is in student-student communication to pass notes in class: "*When it's free hour in school but we're not allowed to be loud, we often exchange handwritten messages. We use loads of abbreviations in such messages to discuss everything silently within the 45-minute class*"; "*only when passing notes in class*"; "*when I'm exchanging handwritten messages with a friend during class, I like to abbreviate words.*" This form of communication finds its way into workplace meetings as well: "*yes, when I'm exchanging messages secretly with others during a boring meeting, usually in English:* U R (you are)".

The following comments were made specifically on internet specific acronyms: "*When the teacher says something funny in class, I write, for example a* lol *in my notes*"; "LOL *in my own notes, during in-class messaging*"; "*yes, when I exchanged messages secretly with a classmate:* lol, lmao,omg *and smileys.*" Emoticons are also commonly used: "*longhand messages (e. g. left on the kitchen table), greeting cards: smiley but drawn horizontally*"; "*Only in-class messaging comes into my mind; when taking notes in class I use smileys at most.*"

The use of abbreviations or emoticons in handwritten texts may even be sanctioned in some cases: "*under the effect of the texting fever of the time about five years ago, one of my former classmates replaced* hogy 'that' *with h and also spelled something phonetically in a literary (!) essay and got into trouble because of this*"; "*the teacher detracted some points because of this in high school (literary essay).*" "*Unfortunately, I'm so accustomed to instant messaging that I insert a few abbreviations or phonetically spelled words almost unintentionally, especially when I have to write fast. Test papers require particular attention from me in this regard, but in my own notes I use these abbreviations without problem*".

A certain degree of frivolousness is also attached to the use emoticons and abbreviations: "*During taking notes, when form is not so important, I sometimes abbreviate words. It's not by chance that I only wrote down such things when passing notes in class or writing for fun*"; "*It almost always happens with me when I'm writing just for the sake of it, for example:* nan (nagyon 'very'), asszem (azt hiszem 'I think'), h (hogy 'that'), sztem (szerintem 'in my opinion'), vok (vagyok 'I am')." Written text, according to some respondents, weigh more than spoken texts: "*Handwriting/letters always demand some kind of seriousness, in my opinion. Such word forms are fine in texting or instant messaging but a letter really deserves much more respect and I don't like the idea of handwriting losing seriousness.*" The following comments can be considered ironical or even self-contradictory: "*Not really, prbly:)*"; "*IMHO, no*" and "*prbly no*"

To sum up, slightly more than half of the respondents (331 persons) do not use digilect-specific word forms in handwriting, the rest of them do, particularly in informal texts (e. g. university notes taken for own purposes, memos, diaries, short messages or notes sent to a close friend, and passing notes in class or at the workplace). The word forms provided by the respondents were organized in the following table (*Table 19*).

Table 19: Character-saving techniques used in handwriting, probably influenced by digilect[57]

Usage	Type of abbreviation	Examples provided	Number of occurrence
no	–	–	331
yes	–	[no examples provided]	72
	–	[mentioned for previous questions]	27
	abbreviation with initials	***h*** (hogy 'that')	111
		m (mert 'because')	5
		n (nagyon 'very')	1
		úh., uh (úgyhogy 'so')	3
		v (vagy 'or')	10
	internet-specific acronym	*Asap*[58]	9
		lmao (laughing my ass off)	1
		lol (laugh out loud)	9
		omg (Oh, my God)	4
		***szvsz (szerény véleményem szerint* 'in my humble opinion')**	2
		wtf (what the fuck)	3
	marking the pronunciation of vowels with capital letters	***Lment***	1

57 The influence is not always obvious or justifiable. Highly probable cases are set in bold.
58 *As soon as possible* (borrowed from English).

Table 19: (Continued)

Usage	Type of abbreviation	Examples provided	Number of occurrence
	substitution of a letter with number	***1általán*** (egyáltalán 'anyway'), *4 szólamú* (négy szólamú 'four-part'), ***mind2*** (mindkettő 'both'), ***j7en*** (jövő héten 'next week'), + ***6ározó*** (meghatározó 'determining'), ***4zet*** (négyzet 'square'), ***mind1*** (mindegy 'all the same'), *1x* (egyszer 'once'), ***1ház*** (egyház 'Church'), *2esben* (kettesben 'just the two of them/us'), *4szemközt* (négyszemközt 'face-to-face'), ***7főn*** (hétfőn 'on Monday'), ***7vége*** (hétvége 'weekend')	30
	substitution of a letter with another letter	–	
	substitution of a letter with symbol	+ ***csináltam*** (megcsináltam 'I did it'), +***halt*** (meghalt 'died'), +***6ározó*** (meghatározó 'determining')	5
		ø (no)	1
		1x	2
	blending	***asszem, axem, aszem***[58] (azt hiszem 'I think')	21
		***hnap (holnap* 'tomorrow')**	3
		mek (megyek* 'I'm *going/I'm off')	12
		***mképp (másképp* 'in another way')**	1
		***nan (nagyon* 'very')**	1
		***naon (nagyon* 'very')**	1
		***szal (szóval* 'so')**	4
		***sztem (szerintem* 'in my opinion')**	14
		tkp (tulajdonképpen 'in fact')	1
		tod (tudod 'you know'), ***tom*** (tudom 'I know'), ***nemtom*** (nem tudom 'I don't know')	14
		tul.képp (tulajdonképpen 'in fact')	1
		uannyi (ugyanannyi 'the same amount')	1
		***valszeg (valószínűleg* 'probably')**	4
		***vok (vagyok* 'I am')**	6
	clipping	***akk (akkor* 'then')**	1
		***akko (akkor* 'then')**	1
		***am (ami* 'which')**	3
		***csina (csinálsz* 'you do')**	1
		lécci[59] *(légy szíves* 'please')	1

58 This form derives from orality but in writing it mainly appears in digital communication. Cf. *lécci* (*légyszi* 'please'), *szal* (*szóval* 'so'), *nemtom* (*nem tudom* 'I don't know'), *tom* (*tudom* 'I know'), *télleg* (*tényleg* 'really').

59 A form justifiably taken from orality.

Table 19: (Continued)

Usage	Type of abbreviation	Examples provided	Number of occurrence
		pill	2
		szomb (szombat 'Saturday')	1
		tört (történet 'story')	1
	other	***b/c*** (because)	1
		btw (between)	1
		cmq	1
		eg. (for example, taken from English)	1
		emoticons::) (Amúgy szeretem, hogy lehetne, ha akarnám. 'By the way, I like them, so I could use them if I wanted to.'):(:'(:O:D:-) < *3m*,[60] *XD*, *ikszdé*	42
		figyi (figyelj 'look!')	1
		kivi (kiváncsi 'curious')	1
		luv ya (taken from English)	1
		monnyuk (mondjuk 'let's *say')*	1
		lájk (English "like" adapted to Hungarian phonetics)	1
		ok (taken from English)	2
		tali (találkozó 'meeting')	1
		***télleg (tényleg* 'seriously')**	2
	+ conventional abbreviations	*ált*. (általában 'generally'), *de*. (délelőtt 'in the morning'), *du*. (délután 'in the afternoon'), *kb* (körülbelül 'approximately') [without fullstop!]; *pu* (pályaudvar 'railway station') [without fullstop!], *ún*. (úgy nevezett 'so-called')	6
		vki (valaki 'someone'), *vmi* (valami 'something'), *vhol* (valahol 'somewhere'), *vmilyen* (valamilyen 'somekind'), *bmi* (bármi 'anything ')	27
	unclassifiable	*kk*., *o*	2

To find out more about the generation gap in digilect, the respondents were prompted to write any word they think people older than them would not understand. The question could not be more specific as boundaries between generations are blurred, and it was also assumed that the questionnaire—being an online one—would be completed by digital natives, that is people aged 16–35. The question aimed to find out whether digitally inexperienced people had any difficulty understanding those who are adept in digital communication and technology.

60 Notes on the use of emoticons. "*I use smileys when grading test papers.*"; „*instead of saying thanks, I use:)*"

Based on the answers, the respondents were aware of generational differences and problems in understanding caused by the different vocabulary: "*I seldom communicate with older people on the internet, but when I write a letter to grandma I always try to be very clear.*"; "*I can say practically anything to grandma (aged 80) about the internet or computers, she won't understand it.*" "*It depends on what generation we are talking about: my parents (aged 50+) will have difficulties understanding them, pensioners will not understand a word*"; "*It's usually a problem for me that I can't use emoticons when I'm texting to my mom as I think she wouldn't understand them.*"; "*My mother could not decode, for example, these expressions:* jó8 (jó éjt 'good night'), omg, lol, tali (találkozó 'meeting'), vok (vagyok 'I am'), szmájli ('smiley'), LOL, sorry, lávollak ('I love you'), lájkolom ('I like it')."

The so-called generation gap, however, is opening in the other direction as well: a young respondent admitted that sometimes he had problems understanding even younger generations: "*I'm only 23 but sometimes I don't understand what my 13-year-old cousin is saying to me*"; "*I don't really know, maybe the words I use in connection with the university and the internet, but this applies more to my brother who is two years younger than me, sometimes even I don't understand what he's saying.*" "*There are people who are difficult to understand even for me. I feel quite old in such cases, though I'm only 26.*"

The parents of twenagers (young adults in their 20s), aged 40–50, is generally in a halfway house position: "*I'm 20 and I think my parents (aged 47) would not but people around 30 would surely understand the following:* pls, omg, tal (találkozó 'meeting')." According to the respondents, the 60+ age group generally does not understand the new vocabulary of digital communication: "*It depends on who much older people are concerned. Regarding people between me and my parents (aged around 28–50) it depends on their attitude and fields of interests. The age group of my parents (aged around 45–55) largely understands new words. But the age group of my grandparents have no chance to understand e.g. internet abbreviations, smileys, etc.*"; "*I think people aged 60–100 will not understand words like* instant messaging or skype."; "*Elderly people (grandparents) will probably have problems understanding words like IM, MSN, skype, the names of different apps and other words.*" Naturally, there are some examples in the 60+ age group as well: "*My 60-year-old mother is an internet addict.*"

Sometimes people in their twenties have to interpret between teenagers and their parents aged 40–50: "*usually I have to translate for my younger brother when he's talking to our parents.*" A parent sent the following response: "*Today's teenagers are hardly understood by their parents. I also have a teenager child who uses these abbreviations in texting. I don't really understand them, though I don't really want to learn them either. Instead of writing a long text message, I prefer calling the person on the phone.*"

Many respondents noted that being adept in digital technology—and thus using the special vocabulary—does not (necessarily) linked to a certain age group: "*I don't think this would depend on age… my father probably knows more of these than I do as he uses the internet more*"; "*in writing, during instant messaging, it took only ten minutes for my mother to get used to the above abbreviations, though she has experience in IM so she might be a wrong example*"; "*knowing these words primarily depends on the field of interest and the occupation, but e.g. my mother don't know even these words:* pendrive, skpye, e-mail." This observation, however, may not be generalized: "*my parents know what an e-mail is, but don't know the word* HD*:)*" Although the following answer lacks any substantial message, its form tells us many things: "*nem tom, mekkéne őket kérdezni* (nem tudom, meg kellene őket kérdezni 'I don't know, you should ask them')".

The words listed by the respondents may be classified into five larger groups: 1. English words (abbreviations, acronyms, "*simple English words*", and Hunglish[61] expressions); 2. Hungarian abbreviations; 3. (school) slang; 4. IT and related technical terms; 5. internet-related jargon.[62]

The following comments underline the importance of the English language: "*Many people do not understand the words, abbreviations taken from English which are sometimes used even in spoken conversations:* lol, omg, lájkolom ('I like it')"; "*the most frequent but not generally understood word is perhaps* lol *which is even used in speech.*"

Expressions from school slang include the following: *adja az ívet* ('is very successful'), *fasszentos* ('cool'), *faxa* ('cool'), *flashel* ('hallucinate'), *flamózik* ('eat'), *beolt* ('perform sexual action'), *kajakra* ('really'), *kira* ('cool'), *zsírsámán* ('cool'). Note that innovative diminutive forms were also considered to be incomprehensible for older people: *fullosan be vagyok baszcsizva* ('I'm totally drunk'), *ajcsi* ('gift'), *ari* ('cute'), *bari* (< barátnő, 'friend'), *cejetlek* ('I love you'), *harizik* ('is angry'), *harcsizik* ('is angry'), *ceriz* ('love').

The topic of the questionnaire might have directed the respondents' attention as they mentioned only a few school slang expressions, mainly IT terms and "internet slang based acronyms". This trend was expressly mentioned in a comment: "*I don't think this is true to today's slang in general, it's only the influx of IT expressions.*" According to the questionnaire, such expressions include: *divix* (DivX), *empeg* (MPEG), *fájl* (file), *iPod*, *mél* (mail), *peer*, *pendrive*, *skype*. Expressions

---

61 Or as the respondents put it: "the mixed English–Hungarian language used at multinational companies"; "IT/communication/business talk", such as: *callban vagyok* ('I'm engaged in call'); *ASAP írd meg a reportot* ('write the report ASAP'), *forecastol* ('send a forecast'), *meetingel* ('hold/attend a meeting').

62 Names given by respondents: "chat language", internet jargon, internet language, SMS language.

relating to the internet and digital communication: *hnap* (holnap 'tomorrow'), *ikszdé* ("XD" spelled phonetically), *irl* (in real life), *gógli* ('Google'), *megjelöl* ('mark as friend on a social networking website'), *laggol* ('lags'), *lájkol* ('likes'), *diszlájk* ("dislike" spelled phonetically), *láma* ('lame'), *legugliz* ('search on Google'), *lfmao, lol, low, meek, omfg, nézz meg a facebookomon* ('check my Facebook profile'), *pókol* ('to poke'), *rotfl, thx, szorri* ("sorry" spelled phonetically), *wow, wtf, xd.*

Video and online games form a special field: "*usually abbreviations, gamer slang, geek speak but this is not asked in the questionnaire, if I'm not mistaken. We can talk about WoW (World of Warcraft, an online game) or FPS games (first-person shooters) for minutes without anyone understanding us.:)))*," a respondent said. According to another respondent, outsiders do not understand the abbreviations used in connection with online games: *cs-zik* (play with "Counterstrike"), *codozik* (play with "Call of Duty", *travizik* (play with "Travian"), *gémel* (play a game), *wowozik* (play with *World of Warcraft*).

Example 67: Sentences which are considered undecipherable for "older" generations

Foursquare-en épp **becsekkolt** *a kedvenc kocsmájába ('He just checked in his favourite pub in Foursquare.')*
Facebook-on azt **posztolta,** *h nem* **lájkolja,** *amit a tesója* **share-elt...** *('He said in a Facebook post that he did not like what his brother had shared.''*

4.2.4 Use of emoticons

Of the 647 respondents, only 52 said that they never use emoticons in any form of text. "*I never use them, in fact, I hate it when somebody sends me such things, in particular if the person is male!!!!!!*" said one of the respondents in her reasons. Another opinion about emoticons: "*To be true, I almost never use them and it makes me angry when others do so... because they mock and stupefy the Hungarian language (in my opinion)*" Naturally, there are much more answers in favour of emoticons: "*Emoticons are integral part of our conversations.*"; "*almost always, I'm almost addicted to them:D < — you see?;)*".

Looking at emoticon usage by text type, we find the following (considering, however, that the open-ended question did not allow for obtaining precise statistics on emoticon usage): More than half of the respondents use emoticons in IM,[63] e-mails, text messages and on social networking websites. Online forums, blogs

63 Variations for the word *IM* in the responses: chat, chatting, electronic conversation, internet conversation, im, Facebook chat, Google Talk, torrentchat, msn, online game chat. The instant messaging feature of Skype may also be included in the list.

and Twitter were mentioned relatively less often. As far as non-digital text types are concerned, the most typical categories where the respondents used emoticons were handwritten (informal) letters and handwritten messages and memos. Many respondents mentioned notes taken in class, and a few mentioned notes passed in class and postcards. Interestingly, it is usually teachers who use emoticons in test papers (as suggested by the responses): "*when evaluating students' witty writings*", "*when evaluating test papers*", "*when compiling test papers*" (there were only three answers of this sort.) One respondent mentioned that he used emoticons in spoken communication (e. g.: *ikszdé* 'ex-dee; XD').

The questionnaire also confirmed that the primary function of emoticons is to express emotions ("*In general, when I want to add emotion or express mood, shock, agreement or sadness which would take long and complicated sentences to in words.*"), to add suprasegmental elements which are missing from written communication ("*when my emphasis is not clear in writing*"), and to compensate for missing body language ("*when I want to represent my facial expression in writing.*"). They can replace or complete punctuation marks: "*I almost always use them, mainly on MSN, as punctuation mark at the end of the sentence, indicating my attitude toward my statements.*" Emoticons are also used to make the message less severe or ambiguous or to cheer up the partner.[64] They are also useful to ease the impersonality of digital communication: "*I almost always use emoticons, because they strengthen this otherwise impersonal form of communication; it would be difficult to get through modality or irony without them.*" Emoticons allow for quicker reaction compared to verbal communication. "*to signify it to the other party that I'm actively taking part in the conversation. (To make up for facial expressions made in reaction.)*".

Naturally, different emoticons are used for different communicative intentions: "*When I want to emphasize that I mean no offense or, to avoid misunderstandings::) When I say something serious but want to make things vaguer:) When I meet something really funny::D When my message is weird, embarrassing, disturbing or maybe uncertain::S If I'm not amused by the message I received, or it's simply unexciting, or I want to express frustration or boredom: -.- Sometimes with a grin::]*"; and "*I use pretty much emoticons, they are almost integral part of my*

64 Explanation: "*I use them all the time in instant messaging and emails, what's more, when the guy comes to read the water meter I always leave a message to him with the reading and this message: Have a nice day:-) I don't know him personally, only by sight, we only meet when he comes to read the meter, but at such times he also writes a:-) in response.;-) This sign makes people happier and communication faster. It eases a tough situation. Long before I've met emoticons, when I wrote letters to my friends by hand, I made tiny joyful drawings on the envelop to brighten up the day of everybody laying their hands on it.:-), my favourite ones: XD,:@,:s:-(,:'(, -_-.*"

writings. Sometimes I even have to hold myself back. For example instead of writing "this is funny" or "that made me smile" I simply put a:) or:D. If we are joking, I use the emoticon with the tongue sticking out:P.[65] *These have variations as well, e.g. 8), =) or (=, (8 d: And another one which I use very often, the heart emoticon: < 3*".

The most frequently (339 times) mentioned emoticon was *:)* which has a complex but certainly positive meaning (including high spirits, happiness, joking, kindness and reaction to kindness, joy, indication that the message has been read, mitigation). The colon, which stands for the eyes, may be replaced by equals sign: *=)*. The second most frequently (256 times) mentioned emoticon was the character combination*:(*which signifies sadness, reaction to bad news and sympathy according to the respondents (its longer version, the *:-(*was mentioned 40 times). The laughing smiley*:D* was mentioned by 235 respondents (and*:-D* by another 40). With 113 records the fourth place was taken by *:P* (and its variant*:-P*) which refers to the gesture of sticking out the tongue, and means that the poster is joking or has nothing to add to what has been said. More recent emotions included *XD* and *xD* (100 records) which depict squinched eyes and mouth wide open laughing, and*:O* which expresses surprise, shock or, as one respondent put it, "big suck". The winking icon*;)* was mentioned by 97 respondents (and its variant, the*;-)* by another 20), while the emoticon *:S*, which means odd, bad thing, being sick or puzzled, was written down by 95 respondents. The original:-) smiley, which is the combination of colon, hyphen and closing bracket in this order, was mentioned only by 69 respondents. The *:'(* emoticon, where the apostrophe signifies a tear drop, expresses crying or even "unserious whining", the*:@* icon means anger and the*:$* blush or embarrassment. The *:** means kiss, while the*:/* and*:-/* character combinations mean being sceptical, undecided or uneasy. While these so-called European emoticons have to be rotated 90 degrees to be understood, the so-called Japanese or Asian emoticons depict human face without the need to rotate them: the ^^ represents joyful eyes, but may also mean sarcasm; the icon -.- means frustration, boredom. The heart as the conventional symbol of love is represented by the *< 3* character combination. Many respondents noted that they used animated emoticons in instant messaging which cannot be written down by ASCII characters. By typing the *(L)* character combination, for example, IM programmes automatically display the graphical image of a heart which is used for expressing love or saying goodbye. (The full list of emoticons mentioned by the respondents, as well as the frequency of their occurrence is provided in *Annex 5*.)

---

65 Impersonal structures, such as *poénkodás van* '~it's joking what we have here', *szmájli van* '~it's a smiley what we have here' (cf. Chapter 3.3.3).

The respondents associate emoticons with informal situations, therefore 90 % of them avoid using smileys in a formal setting (formal letter, test paper etc.). Half of the respondents use emoticons in handwritten letter and two-thirds when taking notes. Not surprisingly, in digital communication (e. g. message wall, e-mail, IM, text message) only 10 % avoid the use of smileys (*Figure 35*).

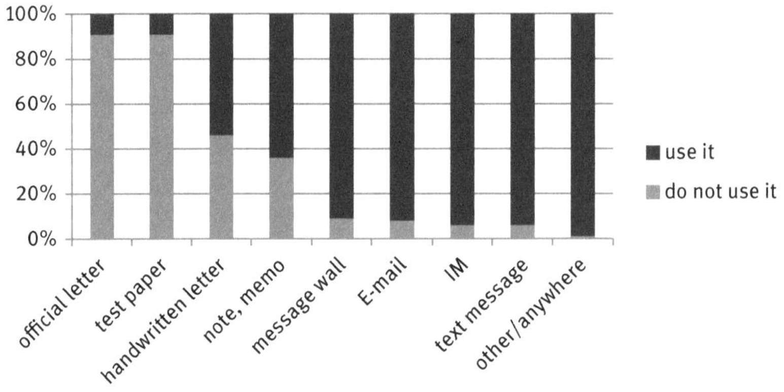

Figure 35: In what types of text would you NEVER use emoticons?

4.2.5 Digilect's impact on various written and oral linguistic forms, according to the questionnaire

4.2.5.1 What is "internet language" like?

Although some respondents said that "*there is no such thing as internet language*", as "*there are formal and informal ways to communicate on the internet, and there are languages relating to instant messaging, social networking sites and press. They all use different linguistic forms that are somewhere in-between oral and written language*" or "*The internet is wide enough to give space to both sophisticated and sluggish language. The pages I visit and the people I chat with lack any extreme linguistic forms*", the vast majority of them could identify characteristics which they attributed to internet communication.

The most commonly mentioned characteristics were simplification, (pursuit of) quickness and, in relation to this, the tendency to use abbreviations and more compact forms ("*keeping the number of characters at the minimum*"). Informal communication also affects the usage of T/V forms in Hungarian ("*many times even strangers write me letters using the T form*"). Informality and interactivity directly leads to, among others, the use of slang words in writing, the high number of diminutive forms and hypocorisms, as well as the complete or partial disregard for spelling rules ("*total ignorance of rules*"; "*freedom*"). Further characteristics, according to the respondents, include continuous and rapid change, variability.

The missing nonverbal features of oral communication are compensated for in writing by multiplied characters and punctuation marks (examples from the responses: "*ez naggggyon jóóóó* 'this is awesome'", "*tecccccik* 'I like it'", "*HAGYJ MÁR BÉKÉN!!!* 'Leave me alone!'"), as well as by emoticons and images.

Many respondents stressed the exclusivity of the internet-specific language variety which is incomprehensible to outsiders: "*outsiders won't understand it*"; "*It's a language that is not comprehensible to everyone, only to the members of the same group*"; "*internet users get along well with it, they understand each other. Outsiders are somewhat distrustful of the almost incomprehensible abbreviations and new words.*"

From a lexical perspective, internet texts are evidently dominated by the English language according to the respondents, namely through loanwords and calques. The following comment reflects well on this topic: "*English expressions are picked up without any change even by users who do not speak English.*" The respondents brought the following examples of macaronic language: *loadold be* 'load it', *downloadold* 'download it', *beloginolok* 'I'm logging in', *attendingol* 'attends'.

There were some evaluative comments as well. Creativity and word economy ("*linguistic economy*") were mentioned as positive attributes, and simplicity, superficiality and users' sluggishness as negative ones.

4.2.5.2 Written forms of text under change

The questions relating to written text types served a double purpose: they tried to elicit the respondents' observations on their own language use as well as on the language use of others.

The answers given to the question "*What kind of IM or SMS specific linguistic forms do you use in HANDWRITING?*" have been largely summarized in dőlt above (the respondents did not made a distinction between abbreviations and other linguistic forms and methods).

The yes-no question *Do you think WRITING is being changed by internet and mobile communication?* received only 30 negative, 200 unjustified and more than 400 justified positive answers, therefore the vast majority of the respondents believed that writing was being changed by digital communication. However, such statistics do not tell us much about the content of the answers given to open-ended questions. As the answers were often quite complex and lengthy, instead of a quantitative summary, I will provide a collection of responses showing possible trends and directions as seen by the mainly non-linguist respondents. The answers were quite diverse as the question was (intentionally) formulated to permit different interpretations.

Most of the respondents mentioned the change of spelling. Some viewed this process objectively but a large number of respondents considered it a negative trend. Many respondents are concerned about the form and spelling of written texts: "*the damaging effects are quite clear. Instant messaging and texting greatly contribute to the inability of younger generations to formulate thoughts and spell words properly.*"

Emoticons are filtering into traditional handwritten texts as well: "[Students] *often draw emoticons in handwritten messages; interestingly they rotate them by 90 degrees on paper as well as if they were still bound by the alphanumeric constraints of computers.*"

Many answers mentioned shortening and simplification as another tendency: "*my students are fond of abbreviations, such as: megmondtam/+mondtam*"; "*abbreviations, signs used* [on computers] *are appearing in writing*"; "*we are carrying short forms over to handwriting as well.*" One of the respondents even brought an example to this: "*abbreviations are gaining ground incredibly. To illustrate this, I copy here a comment from the website where I'm editor in chief. The poster is 16 years old, the topic is an account of concert experiences:* [Example 68]. *Punctuation marks are completely omitted, spelling rules are totally neglected, and abbreviations are so overused that they almost render the text illegible.*"

Example 68: Emoticons and abbreviations in an account of concert experiences (Source: questionnaire response)

*Ott voltam^^(a Spongyabobos táskámmal együtt:***Dszal** *észre lehetett venni) Nah mikor megtudtam hogy, a kisterembe lesz kicsit Sokkot kaptam* **o.o***,de aztán nem zavart annyira mert így közelebb lehettem az egész banzájhoz.De levegő az nem kell az hülyeség...-.- A SATEP az fasza volt jól nyomták a srácok:) csak a sok elmebeteg "táncikáló ember" vette el akedvem attól hogy, énis zúzzak na* **mind1.** *Azt hittem szarabb lesz aSuffokatemert nekem eddig nem igazán tetszett,sőt..De most megtetszett:Dlehet egy hölgyemény miattis:$nah de* **szal** *megtetszett a Suffokate! És amit nagyon vártam VEIL OF MAYAaz elején elvolt baszva a hangosítás kb. az első két számnál:/ de aztán megoldották:). Egy számot vártam nagyon a Namaste-tahw*-* s annyira jóvolt: O de az összes számuk az volt^^. S Végül a CarnifeXnemtudom mi a baj velük,sztem ultra faszás volt üvölteni a mikibe^^ s ultra közelről látni ahogy Scott "énekel":Dés áh az egész nagyon jóvolt^^ja és lett egy Carnifex-es dobverőm: D!*

Some respondents were convinced that formal and informal uses registers were becoming less distinct: "*I had a classmate who found it normal to write down textese even in test papers.*" However, several respondents had observed just the opposite of this, arguing that it would be increasingly important to distinguish between formal and informal styles: "*I don't think this change would affect official documents as they are based on standard forms*"; "*emoticons are quite common in non-formal documents*"; "*Abbreviations and smileys will certainly filter into non-formal literacy*"; "*Formal documents and papers are definitely not affected, but informal documents are, especially among the younger generations.*"

Similarly to the previous respondent, many others associated the change with certain generations: "*definitely, as far as younger people (aged 10–20) are concerned*"; "*this is mainly true to the younger generation whose members had their first mobile at the age of five.*"

Furthermore, the respondents also mentioned the loss of significance and changing function of handwriting compared to typing: "*The change is certainly apparent in that handwritten stuff are becoming increasingly rare. As it is no longer part of our everyday life (now that school years are gone), handwritten documents have gained a special value. There is something ceremonial in them. For instance, you'll still write a card by hand though you could send an online card with a few clicks. You'll do this precisely to express how much you value the other, that you invested paper, time, stamp and effort into the card.*" The following comment also reflects on this topic: "*My husband, for example, IS UNABLE TO WRITE BY HAND.*"

4.2.5.3 Change of orality-related text types (including, in particular, informal conversation)

The linguistic forms which were originally used in digilect but now also appear in oral conversations are summarized in *Table 20*. Expressions taken from the internet slang, that is the vocabulary of digital communication, include in particular the verbs *lájkol* 'like' and *meglájkol* 'like', as well as the noun lájk 'like' (or as a respondent put it: "Lájkolni – *this expression comes from Facebook. Since people have been texting and instant messaging, this is the first expression I use even in oral conversations.*" The expressions relate to Facebook which gained immense popularity in Hungary in 2010. The verb *műxik* 'works', which is a slang variant of the verb *működik* 'operates', probably originates from the SMS communication of the early 2000s. Interestingly, the word was coined to save characters in text messages, but is now used in orality as well. The phonetically spelled English words *dunno* and *luv ya* were possibly taken over from spoken English rather than the English digilect, and are widely used expressions in computer-mediated communication.

The interference between digilect and oral conversations is, however, quite clear when internet-specific acronyms are pronounced letter by letter in orality. These "*expressions which have gained independent meaning*", as one respondent put it, such as *lol* or *omg* are often used in oral conversations: "*sometimes saying* LOL *is OK when I'm with the right people*"; "*though I've heard many people saying* lol *I prefer not to use it*"; "*Yes, it's happened to me that I said* vétéef (wtf), lol *or* oh-my-god." **blush*:)*". Yet, there are people who feel offended by hearing these forms in oral conversations: "*I shiver with cold when somebody says OMG*", though such comments strengthen our assumption that this phenomenon is not at all rare. The acronym *OMG* 'oh my God' can be heard both with Hungarian ("*hát ez óemgé*") and English spelling ("*óemdzsí*").

Pronounceability and comprehensibility are important requirements for acronyms: "*Only what you can pronounce. But definitely not the*:) *type ones (it would be weird to say colon–parenthesis), or the types* exerint (ezek szerint 'consequently') *or* ë8 (jó éjt 'good night'). *But words like* SZVSZ (szerény véleményem szerint 'in my humble opinion'), sztem (szerintem 'in my opinion'), IRL (in real life), IMHO (in my humble opinion) *are quite useful ones.*"

According to the respondents, there is also a direct relationship between emoticons and oral conversations: "*smileys names such as* xD (ikszdé 'ex-dee')"; "*My 14-year-old nephew often uses the expression* XC"; "*Acronyms are becoming increasingly common, and sometimes I can also hear an* ikszdé (ex-dee, XD) *or a* kettőspontdé (colon-dee,:D)". The usage of *XD* in oral communication was mentioned by 19 respondents.

The last question prompted respondents to reflect on how digital communication makes oral conversations change *(Do you think ORAL CONVERSATIONS are being*

Table 20: Digilectic forms used in oral conversations according to the respondents' own account (see full table in *Annex 3*)

Type	Example	Number of occurrence	Note
internet slang, digilect form	*bug*	1	< English
	dunno	1	< English
	lájk	3	<English *like* (adapted to Hungarian phonetics)
	lájkol	8	'likes'
	láv van	1	'I love you'
	lávol (lávollak)	1	'I love you'
	luv ya	1	< English
	meglájkol	1	'mark with the like button'
	műxik	1	'works'
	ráírni valakire	1	'send a message to someone'
	rákeresni valamire	1	'use a search engine to find something'
internet-specific acronym	*asap*	13	< English
	brb	1	< English
	btw	1	< English
	lol, LoooL	89	< English
	ROFLMAO	1	< English
	rtfm	1	< English

Table 20: (Continued)

Type	Example	Number of occurrence	Note
	omfg	2	< English
	omg	29	< English
	szvsz	1	< *szerény véleményem szerint* 'in my humble opinion'
	wtf, vétéef	15	< English (the latter is a variant of the former spelled with Hungarian phonetics)
abbreviation, blending or clipping (presumably) created in digilect	[no examples provided]	16	
	am	1	< *ami* 'which'
	vok	1	< *vagyok* 'I am'
	köcc, köcce, köce	3	< *köszi* 'thanks'
	majd besz	1	< *majd beszélünk* 'talk to you later'
	mek	1	< *megyek* 'I'm going/I'm off'
	micsi	1	< *mit csinálsz?* 'what are you doing?'
	naon	1	< *nagyon* 'very'
	nemtu	1	< *nem tudom* 'I don't know'
	nömö	1	< *nm* < *nincs mit* 'you are welcome'
	pill	4	< *pillanat* '(wait a) minute'
	szal	4	< *szóval* 'so'
	sztem	3	< *szerintem* 'in my opinion'
	thx	1	< English
	tod, tom	2	< *tudod* 'you know', *tudom* 'I know' (maybe from spoken language)[66]
	tulképp	1	< *tulajdonképpen* 'In fact')
	valszeg	10	< *valószínűleg* 'probably')
	vok	5	< *vagyok* 'I am')
emoticons pronounced	[no examples provided]	1	
	ikszdé	19	< *ex–dee* 'XD'
	kettőspontdé	1	< *colon–dee* ':D'

66 Several respondents suggested that certain reduced word forms were created in digilect, though they originate from spoken language. See for example the following teacher's note: "*I try to use the correct linguistic and grammatical forms and I also encourage my students to do so. Therefore, I always tell them to use "utolsó" (*'the last one'*) instead of "uccsó".*"

changed by internet and mobile communication?). Nearly half of the respondents said "yes"—though without giving reasons—and only slightly more than fifty persons stated that internet communication did not affect oral conversations at all. Many noted that internet-based linguistic forms do appear in oral communication but this trend was not so substantial or striking as in written communication.

They thought that the change of oral conversations was specific to certain age groups, similarly to written communication, but many of them noted that the change was by no means a matter of the future, but was already in progress, if not completed.

It may be regarded as a sociological change with linguistic aspect that personal conversations are often replaced by online chatting: "*They prefer making friends online much more than personally. This gives confidence and a sense of security for many people.*"; or supplemented by it: "*Online social life does not replace but supplement offline activities, but people at a Friday evening party will touch on topics which they would know nothing about without the internet. On the other hand, Facebook timeline and Twitter enables us to share uncensored thoughts with our friends so we are constantly aware of others, so a meeting is not about telling news and exchanging information any more. This is good because you don't have to listen to monologues about the past four months, but also bad because those who are not involved in the Facebook buzz will miss personal topics*". The following comment also compares instant messaging and orality: "*many people can chat a lot online but are unable to speak a word in person.*"

The vocabulary of conversations has undergone the most spectacular change according to many respondents. "*popular abbreviations and expressions have oozed into spoken language*" The respondents illustrated this with English-based neologisms as well, such as *lájkol* 'like', *attendingel* 'attend', *dzsoinol egy klubba* 'joins a club', *képet addol* 'add a picture', *kommentel* 'comment'.

The use of emoticons and acronyms was a recurrent topic here as well: "*I also use expressions taken from the internet language (e. g.* ikszdé 'XD', óemgé 'OMG')"; "*I've already heard* lol *and* XD, *but I don't think they would significantly and permanently affect oral conversations*", "*We use* LOL *and similar words in oral conversations as well, instead of laughing*"; "*we utter words (*lol*) we used to use only in writing*". The emoticon *ikszdé* 'XD' is also a popular item in conversations: "*I've heard many people saying XD in an oral conversation (pronounced as "ex-dee") This seems very odd to me, though I'm young*".

I found the following comment particularly interesting: "*LOL and other abbreviations are linked to the internet communication because in live face-to-face communication you don't really need them. (Why would you say I'm laughing my head off when you laugh?:)).*" As opposed to this comment, saying the internet-specific acronym *LOL* during an oral conversation indeed replaces laughing, i. e. replaces nonverbal communication.[67]

All in all, the respondents considered the changing sociological circumstances of communication and the occurrence of digilectic elements (e. g. *lol*, *ikszdé*, *óemgé*), as the most important trends affecting oral (spontaneous) conversation.

67 See headword “LOL” in *Annex 10*.

5 The Impact of Digilect on non-digital Media According to Corpus Analysis Findings

Written orality (or with other words: spoken language in writing), as its name suggests, is primarily based on spoken language. For example, the formal reduction of *azt hiszem* 'I think' to *asszem* has taken place in orality but the reduced form often appears in written digital texts (though its function is not always the same, as shown in Chapter 3.3.6).

The process from spoken to written language, however, is not necessarily unidirectional: linguistic forms used in conceptually spoken but medially written language (e. g. instant messaging) tend to filter into medial orality as well (e. g. into spontaneous speech).

"*A Fácsén meglájkoltam a kvízedet, mert megalol volt.*"[68] Sentences like this flourish in students' everyday conversations but their meaning is not evident for everyone. The difference between internet citizens (netizens) and those who do not use the internet is not only in the possession of electronic literacy but also in the knowledge of the digital language variety (digilect). This chapter aims to present how the digilect influences other language varieties, including in particular spoken and written texts.

Several authors (e. g. Uwe Wirth 2005; Christa Dürscheid 2005: 93; Martin Haase et al. 1997; Gyula Laczházi 2004; Kelle 2000: 111) have forecast that internet communication will have an impact on other forms of communication, such as face-to-face (F2F) conversations and non-digital literacy. These suggestions, however, have not been supported by concrete descriptions or empirical studies. For this reason, the present chapter endeavours to document and record such interactions in motion through selected text types. The data of the text analyses detailed below are also supported by survey findings described in Chapter 4.

5.1 The impact of digilect on spontaneous speech and conversation

Many authors have dealt with spoken language from a grammatical perspective, including some empirical studies as well (for example, in English: Tannen 1982;

---

68 *A* ***Fácsén meglájkoltam*** *a kvízedet, mert* ***megalol*** *volt.* In English: 'Your quiz on Facebook was megaLOL, I gave it a Like.' *Fácse*: the playful name variant of the social networking website Facebook; *meglájkol*: recommend a content in social media by clicking on the "Like" button; *megalol*: absolutely funny (mega+LOL 'laugh out loud').

DOI 10.1515/9783110499117-005

McCarthy 1998; Krauthamer 1999; in German: Behaghel 1927; Hoffmann 1998; Fiehler 2000; Schwitalla 2003; Imo 2007a; Ágel–Hennig 2007; in Hungarian: Szalamin 1978; Keszler 1983 and 1992; Kuna 2007; Veszelszki 2008; detailed bibliography: Lanstyák 2009), but we are still in want for a thorough, comprehensive grammar of spoken language (Deppermann 2006: 44; Imo 2007a: 3 and 2007b: 4; cf. Chapter 3.3).

This subchapter focuses on observations made in respect of and examples taken from dialogues in spontaneous spoken language. As the examination could not rely on any systematically built database (i. e. an extensive spoken language corpus containing recorded instances of spoken language from 2009–2011), it was restricted to my own observations, as well as to those supplied to me by my colleagues, friends and students. This sporadic form of data collection, however, precluded the possibility of performing an extensive analysis. The examination of the interference between digilect and spoken language also reflects on the findings of the 2010 questionnaire survey discussed in Chapter 4. For the full collection of examples broken down to linguistic log entries see *Annex 10*.

Haase et al. (1997) also found that IM-specific expressions can filter into oral conversations as well. Examples to this reverse process include *műxik* (< *működik* 'works'), *naon* (< *nagyon* 'very') and *szal* (< *szóval* 'so') which are increasingly common in spoken language.

5.1.1 Internet-specific acronyms in spontaneous spoken language

A situation encountered in Budapest in 2010 is an illustrative example to this linguistic trend: Two teenagers were travelling on the bus. One of them told a joke to the other but the response was merely a laconic "LOL" without any facial expression, let alone "laughing out loudly" which the abbreviation stands for.

The acronym *AFK* means *away from keyboard* in computer-mediated communication. When mentioned in an oral conversation as *áefká* (pronounced with Hungarian phonetics), it means that the person concerned is no longer at the given place, he/she is not present.

According to students, the acronym *ASD* is a form of pause or expression of boredom in oral communication. In gaming, where the acronym comes from, *ASD* is used when one cannot make any meaningful comment so he/she just hits these three neighbouring letters on the keyboard. Thus, in writing the acronym has a phatic function: it means that the chatter is present but cannot contribute to the conversation. In spoken language the same three letters (pronounced in Hungarian as *áesdé*) are used when hesitating, approximately with the meaning: "well, whatever". Another interpretation says that the three letters have the combined meaning of the emoticons :A, :S and :D. A third interpretation holds the acronym as an expression indicating boredom. (The usage of *ASD* was found in three different situations involving different age groups.)

When chatting with strangers online, usually for the purpose of making friends/dating, the conversation often starts with typing the acronym *ASL*, meaning 'age, sex, location'. A high school student has used this expression even in face-to-face communication when talking to girl new to him (his remark: "*of course, S was quite obvious, I just wanted to be funny*").

The expression *BB* (bye-bye), as another English acronym used in IM, can also be heard from high school students when saying goodbye after school. It is pronounced as a Hungarian acronym: *bébé*. A variant of this expression is *BB the Q* (pronounced as *bébé tö kú*) where *Q* stands for *kockák* (literally: 'cubes', figuratively: 'computer nerds'), playing with the [kju:] sound which is the initial phoneme of the English *cube*. Interestingly, *BB* and *Q* is pronounced with Hungarian phonetics (*bébé* and *kú*), but the pronunciation of the definite article sounds more like English (*tö ~ ðə*). Thus for high school students the expression "bébé-tö-kú" means: 'goodbye, nerds'.

A high school student borrowed a pen from his classmate and when he received it he said thank you with the following expression: *téháiksz*. The internet-specific acronym *THX* (pronounced in Hungarian as *té-há-iksz*) is the short form of the English *thanks*. The English acronym made its way to the Hungarian digilect, and from there it filtered into Hungarian spoken language. (This acronym is used by university students as well in the same "thank you" meaning.)

The usage of the acronym *vétéef* (*WTF* with Hungarian spelling) was recorded in three different contexts. The following dialogue took place in a university setting: Teacher: "*What is your opinion about the novel you had to read for today?*". Male student: "*This is a big WTF*" *WTF* is also an internet-specific English acronym with an obscene character, meaning "What the fuck". According to my data collected from high schools students, it is also used to denote impatience.

Further (chiefly English) acronyms of digital communication which now appear in the oral conversations of high school students include: *gégé* (*GG* 'congratulations'), *géel* (*GL* 'good luck'), *idékâ* (*IDK* 'I don't know'), *enem* (*NM* 'you are welcome'), *enpé* (*NP* 'no problem'), *estéefú* (*STFU* 'shut the fuck up'), *óemgé* or *óemdzsí* (OMG 'oh my God').

5.1.2 Emoticons in spontaneous spoken language

As a similar technique, a funny story is sometimes ended with the pronounced form of a smiley: *And then ex-dee!* The smiley *XD* (pronounced as *ex-dee* with

English and *ikszdé* with Hungarian phonetics) depicts a face with squinched eyes and mouth wide open laughing.

In Hungarian it also exists as a verb: *ikszdézik* (literally: 'does the ex-dee') meaning 'rolls with laughter' which is equivalent with the meaning of the smiley *XD*. Similarly to *XD*, the emoticon*:D* also motivated the creation of a verb: *kettőspontdézik* (literally: 'does the colon-dee') was formed with the *-z* verb-forming suffix and means the same as the smiley*:D*: 'rolls with laughter'.

5.1.3 Internet-specific vocabulary

The homophony of internet-specific and every-day words may also serve as source of humour. The following joke, for example, was taken from the real life dialogue between two teenagers: "*Tudod, mi a közös a Twitterben és a Facebookban? –??? –Hát a **fészek**.*" The joke is untranslatable as it is based on a pun. It plays with the homophony of *face-ek* (pronounced as *fészek*), the Hungarian plural of the English word *face* (meaning 'faces' and also referring to the first part of the compound *Facebook*) and *fészek* 'nest' referring to Twitter's logo which is a blue bird. The pun can only be understood with both Hungarian and English language knowledge, and with at least a basic level knowledge of the online world.

The word *lájkol* 'like' comes from the social networking website Facebook (for the detailed description and etymology of the word see: Chapter 3.3.2). Its primary meaning is 'recommends a content in social media by clicking on the Like button' (e. g.: "*Lájkoltad már a képemet a Facebookon?*" 'Have you given a Like to my photo on Facebook?'), but it has also developed a secondary meaning in orality: 'likes someone' (e. g.: "*Lájkolom Gabit nagyon.*" 'I like Gabi very much.').

People playing video and online games have developed a special vocabulary (Balogh 2012, 2014). These expressions are often used in spontaneous oral conversations both in their original and altered meaning. The verb *lenerfel* 'weaken' is an example for the latter, as used in the following account of a survey respondent: "I know a boy who was obsessed with online role-playing games. He took to the slang of these games so much that he would use them even in the offline world. So when his girlfriend told him off in my presence, he whispered to me: 'She really needs some *nerfing*'" (in the original: *Le kellene már nerfelni az asszonyt.*). The verb *lenerfel* was derived from the video game slang *nerf* 'weaken'.

The original meaning of the expression *offon van* 'to be offline/unavailable', while figuratively it refers to an attitude of deliberately avoiding contact. *XY is észrevehetné már, hogy **offon vagyok**, egyszerűen hagyjon békén.* ('Why can'*t XY notice that I'm off(line) and leave me alone?*') Its alternative meaning: 'ignore someone', for instance: *Peti **offon van**, nem is szólok hozzá.* (*'Pet is off(line), so I won't talk to him.'*)

The verb *kontrolcékontrolvézik* ‘to copy-paste’ comes from the character combination Ctrl+C (pronounced as *kontrolcé*) and Ctrl+V (*kontrolvé*) which serve as shortcuts to the copy and paste functions in many computer programmes. This expression has not been recorded in writing, only in spoken language.

The following example is not really a lexical item, but rather the verbalized form of a technique used in instant messaging for making (self-)corrections: in an informal conversation a young woman said that “*Nem találkoztunk össze*” (‘We did not meet together’). Then she realized that the sentence was not grammatically correct (*találkozik, *összetalálkozik*), so she added with a smile: *Mínusz össze.* (*‘Minus “together”.’*) The woman corrected herself in speech just as she would in instant messaging, as it is customary in IM to correct something with the minus or plus sign respectively for deleting something from or adding something to the previous line.

To sum up the usage of acronyms and emoticons in spoken language, let me insert here a student’s opinion mentioned in the (unscientific) article *A tinik titkos nyelve* (*The secret language of teens*).

Example 69: A teenager’s views on using acronyms in spoken language

Tünci (17): *If instead of smiling somebody says LOL or EX-DEE without a trace of emotion on his face, that’s too much for me. And if someone is too lazy to write “I love you” to his girlfriend and finds the two-character emoticon < 3 perfect for the purpose, that’s utterly disgusting. My mom, for example, interprets the heart icons in a text as two balls of ice-cream, and she has no idea what I mean when I send a blushing smiley to her. Half of our messages are already in English, but even the Hungarian part includes more and more abbreviations, but this is the way we can save characters: “De most* ***sry****, rohanok, mert* ***tal*** *a* ***bnőmmel f9kor 1etemtér****;):* < 3”* [*where sry means sorry, tal means találkozom* ‘I meet’, *bnőmmel = barátnőmmel* ‘my (girl) friend’, *f9kor = fél kilenckor* ‘at half past eight’, *and 1 etemtér =az egyetem* (*előtti*) *téren* ‘at the square in front of the university’]

5.2 The impact of digilect on handwritten texts

The digilect is primarily linked to typewriting and digital media but some of its characteristics have already appeared in orality and, according to my observations, even in other types of written text (students’ notes, test papers, formal letters).

According to Dürscheid (2006: 55), the two basic types of writing (“manuscript”, that is the result of handwriting, and “typoscript”, that is the result of printing) was completed by a third one in the middle of the 20th century: “compuscript”, that is the result of writing on a computer. The latter category includes all texts written or edited on computers. These are typographically edited, editable and re-editable texts ("everyone writing on a computer has become a typographer"), and the product displayed on the screen can be modified, overwritten, rewritten, rearranged and reformatted. Electronic writing influences the process of text production (Dürscheid 2006: 55) but even our way of thinking about language.

5.2.1 The impact of digilect on handwritten notes

Peter Sieber (1998) analyzed the mid-term and final examination test papers of Swiss secondary school students. After comparing this corpus of texts from the 1990s and another corpus of final examination test papers from the period between 1881 and 1991, he assumed that young people were developing a new writing style which is closer to conceptual orality than to conceptual literacy in the system of Koch and Oesterreicher. He called this phenomenon *parlando*, a term borrowed from music theory. While recording the various manifestations of parlando, Sieber came to the conclusion that written communication patterns were changing, or it would be more appropriate to talk about a socio-communicative linguistic change. The reasons behind the change are related to changing communication needs and conditions.

School and university notes, which record spontaneous speech, typically have share some characteristics with spoken language (Krauthamer 1999: 4; Vallent 2008).

I conducted an empirical research to prove that certain characteristics of the digilect can be found in school notes as well. The analysis primarily focused on the exercise book notes prepared by the grade 7–12 students of Bibó István Grammar School in Kiskunhalas, Hungary between 2007 and 2009. The corpus consisted of altogether 2409 pages (115 exercise books and 45 final examination notes) and all students consented to the research. This corpus was completed with the lecture and seminar notes of university students studying at Eötvös Loránd University, Budapest, which also accounted for hundreds of pages. The 115 exercise books were provided by 99 female and 16 male students, most of them attending grade 9 and 10 (64 and 40 persons, respectively), and a few students attending grade 11 or 12 (8 and 3 persons, respectively). Looking at the exercise books by subject, the majority of them was dedicated to Hungarian Language and Literature (41), Geography (16), History (14) or Physics (14), and some to Biology, Chemistry, English, Music or Italian.

When examining the exercise books, main texts were discriminated from texts noted on the margin (as well as from the teacher's written comments).

The special character of the notes (incompleteness, individual note taking techniques) mostly restricted the collection to formal and lexical characteristics. A syntactical examination of this text type would have produced only very general results. In addition, the lexical aspect is difficult to distinguish from the grammatical one. This is partly because lexis and grammar is increasingly seen as a continuum rather than separate categories, and party because grammatical features relating to words could be classified both as grammatical and lexical feature. For this reason, the analysis of the notes did not include a detailed examination broken down by various aspects, only global conclusions were drawn.

The following characteristics may be attributed to notes in general (and may not be considered as an influence of digital communication): incomplete sentences, key words instead of sentences, logical connections are indicated with symbols rather than words, dominance of coordinate clauses.

Common background knowledge is essential to understand the notes. For example, Math students combine the symbol ∀ 'for all' or in Hungarian *mind* 'all' to generate ∀1 (*mind* + *egy* < *mindegy* 'all the same') or ∀2 (*mind* + *kettő* < *mindkettő* 'both') in the notes, while the symbol ℑ or the letter J means 'exists'. These short forms are only understood by a restricted number of persons.

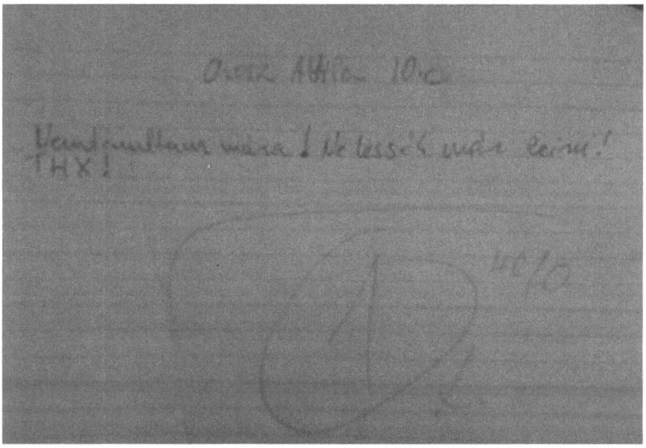

Image 36: *THX* in a high school test paper (Source: W9)

The examination of *notes* from a formal-technical aspect revealed the following main characteristics: replacement of letters with numbers (251 examples), initialisms (154 examples), replacement of letters with symbols (133 example), contraction (41 examples), emoticons (35 examples), phonetic spelling (10 examples), replacement of letters with other letters (6 examples), using capitals to indicate omitted vowels (4 examples), blending (4 examples), internet-specific acronyms (2 examples; cf. *Image 36*), other, unique technique not listed above (14 examples). The different types are described below from the point of view of how probable it is that they were affected by the digilect (from lower to higher level of probability; see *Annex 7–8*).

Replacement of letters with numbers was the most dominant formal characteristic in the notes. This is no surprise as the technique had been a long-established one in note-taking, far preceding the emergence of digilect, so it was not created but (probably) merely encouraged by the language of digital communication. The most typical examples included: *3szög* (< *háromszög* 'triangle'), *6vány* (*hatvány* 'power [of a number]'), *mind2* (< *mindkettő* 'both'), *1ik* (*egyik* 'one [of]'), *1enlő* (< *egyenlő* 'equal'), *1enes* (< *egyenes* 'straight [line]'), *6almi ágak* (< *hatalmi ágak* 'branches of power'), *7végére* (*hétvégére* 'by the end of the week'), *6ó* (*ható* 'having an effect'), *szent3ság* (< *Szentháromság* 'Holy Trinity'), *9ed* (< *kilenced* 'one ninth'), *4zetre emeljük* (< *négyzetre emeljük* 'raise it to the second power'), *1.sorban* (< *első sorban* 'in the first place'), *meg6ározott* (< *meghatározott* 'determined'*)*. Most commonly, the indefinite article (*egy* 'a/an') is replaced by the homonymous number *1* (pronounced as *egy* 'one').

The most typical example for replacing letters with symbols was the + sign standing for *meg-* (verbal prefix) or *meg* (conjunction meaning 'and' or 'plus'), such as *+határozott* (< *meghatározott* 'determined'), *+jelenés* (< *megjelenés* 'appearance'), *kapta +* (*kapta meg* 'received it'), *+hat* (*meghat* 'move [sy]'), *azé a föld, aki +műveli* (< *megműveli* 'cultivates it'). Another common symbol in this function is *x* standing for the multiplicative suffix *-szor/-szer/-ször*, for example: *3x* (< *háromszor* 'thrice'), *még1x* (< *még egyszer* 'once more'). Some students used the conventional heart symbol to replace the word *szív* 'heart' in their notes: *♥izom* (< *szívizom* 'myocardium') and *♥belhártya* (< *szívbelhártya* 'endocardium'). The following form is redundant: *háromΔ* (< *háromszög* 'triangle' where the symbol stands for *háromszög* 'triangle' which already contains the word *három* 'three').

Although many students used this technique, usually the same words were abbreviated with their initial letters. The abbreviation of the conjunction *hogy* 'that' with *h* or *h.* appeared in almost all of the texts examined, even in dictation exercises (an example for both short forms: *szerencséd* ***h*** *öreganyádnak szólítottál; szerencséd,* ***h.*** *öreganyádnak szólítottál*). Further initialisms: *cs* (*csak* 'only'), *m* and *m.* (*mert* 'because', *mint* 'as') and *v* (*vagy* 'or'). All three of the above techniques appear in the following example: "*modernsége abból fakad,* ***h. +jeleníti 1nagy*** *lélek sebzettségét*".

While the above mentioned formal characteristics were the result of traditional note-taking techniques, contraction and phonetic spelling are definitely linked to digital communication: Most probably, the form *uccsó* (< *utolsó* 'the last [one]') with nine occurrence is the result of less articulated pronunciation in spoken language but now is highly characteristic to digilect and note-taking as well. The letter *ly* is often omitted as if it was an intrusive consonant (probably as it is homophonous with *j* which can function as an intrusive consonant in speech), as in *mien* (< *milyen* 'what sort of') *ien* (< *ilyen* 'this sort of'), *oan* (< *olyan* 'that sort of'), *vmien* (< *vmilyen* < *valamilyen* 'some sort of'), *bármien* (<

bármilyen ‘any sort of’), *ameik* (< *amelyik* ‘the one which’). What is more, even the letter *j* can be omitted for the same reason though it is not acting as an intrusive consonant: *mesél a barátnőének* (< *barátnőjének* ‘to his girlfriend’), however, as there is only one example to this, it can be regarded as a spelling mistake as well). Contracted forms include: a *hnap* (< *holnap* ‘tomorrow’), *naobb* (< *nagyobb* ‘bigger’), *naon, non* (< *nagyon* ‘very much’; the latter being mentioned in the following side note: *még mindig* ***non*** *szejesselek ;)*), *sztem* (< szerintem ‘in my opinion’), *valszín, vszínűleg* (valószínűleg ‘probably’), *vt* (< valamit ‘something’). This technique was also used for shortening the title of a poem: *Petőfi: Hun wok* (Sándor Petőfi: Magyar vagyok ‘I am a Hungarian’ where *Hun* stands for *Hungarian* and *wok* for *vagyok* ‘I am’). Examples for pseudophonetic spelling: *ahhó* (< ahhoz ‘for/to it’), *1 ucca* (< utca ‘street’), *má* (< már ‘already/yet’), *hangulati eccség* (< egység ‘unit’), *hagy* (< hadd ‘let sy’), *mingyárt* (< mindjárt ‘almost’), *mibő* (< miből ‘from what’), *meccik* (< metszik ‘[lines] intersect’) and *láccottak* (< látszottak ‘[they] could be seen’). Clipping as a character saving technique could be found only in the word form *akk* (< akkor ‘then’), though this form occurred in several notes.

Replacement of a letter with another letter is a very important characteristic of digilect, despite the fact that it saves only one character: For example: *egycerű* (< egyszerű ‘simple’), *vixxahúzódik* (< visszahúzodik ‘draws back’), *öxe* (< össze ‘together’), *öxeesküvés* (< összeesküvés ‘conspiracy’), *öxejövetelek* (< összejövetelek ‘meetings’).

Before the emergence of digilect (and in particular text messages), the replacement of omitted vowels with capital letters could be found only in word puzzles: Examples include: *Lipszis* (< ellipszis ‘ellipse’), *számelmélet alapTtele* (alaptétele ‘principle [of]’), *prímTnyezős felbontás* (prímtényezős felbontás ‘integer factorization’), *maraDkot* (maradékot ‘remainder’).

Only two examples were found in the notes for internet-specific acronyms: *a Föld korong alakú* (***LOL***), *szabálytalan alakúak* (*krumpli* ***LOL***).

The occurrence of emoticons in students’ notes clearly shows the influence of digilect, in particular if we consider that the smileys in the notes were rotated by 90 degrees, the way they are typed on a keyboard. The most typical emoticon is the smiling face, probably as a comment on the teacher’s remark, e. g.: *keménységük: puha* ***:)***; *a hamuból humusz lesz* ***:)***; *Épp mikor mentem át az úton majd nem elütöt egy karácsonyfát szállító autó vagy legalábbis fenyő félék voltak rajta.****:)***; *a reneszánszba ez a menő mert az ókort utánozzák és ott is ez volt a cool* ***:)***; *régen vót minden házná koca* ***:)***. It may also reflect on self-correction: *imágó* [*crossed out*] *mégis jó* ***;)***. Examples for the variant of the emoticon containing equals sign: *A gázcsere színhelye a levél.* ***=)***, *Vanília a kedvencem!* ***=)***; *ódai* ***=]***. The smiley XD was also very common: *Kereszteshadjáratok lásd később* ***XD***; *csapat csata* ***XD***;

elgazosodik **xD**; *robbanás* **xD**; *jönnek az UFÓk* **xD**; *találnak egy kis kavicsot, gazt azt múzeum lesz belűle* **xD**; *pálesz* **xD**. Japanese smiling emoticons: *6–10 dia – 9 ^^; Délhús, Pick, Debrecen ^^; Rózsa Szeged ^^.*

Due to their informal (and sometimes dialogic) character, marginalia, that is comments written in the margins of the exercise books, share many characteristics with dialogue letters which are discussed in Chapter 5.2.2. The texts were collected from the already mentioned exercise books. Marginalia showed an uneven distribution: some exercise books (especially those that had to be submitted to the teacher for assessment) had none, others had drawings, short writings or decoration (made perhaps out of boredom) almost on every page. Of the categories listed above emoticons were the most common elements in the side-notes (280 occurrences), followed by replacement of letters with numbers (12), other letters (10) or symbols (4), phonetic spelling (5) and initialisms (5).

Initialisms only occurred in the form of *h* (< hogy 'that') or *m* (< mert 'because'). For example: *hülye* [name] *stíröli a 11-eseket* **m** *nyitva van nekik is az ajtó és tök felűnően lesi szegények*; [name *meg asszem* **h** *én*; *lehet* (**h** *ha mész*) **h** *ma elmegyek veled.*

Examples for replacement of letters with numbers: *5ös* (< ötös 'five [best grade in school]'), *1es* (< egyes 'one [worst grade in school]'], *1000jó* (< ezerjó 'thousand good'), *Zombikként felfaljuk 1mast* (< egymást 'each other'). The following side-notes contained the English abbreviation *4ever* (forever): *Kimi Ferrari* ***4ever****!!!, Ati* ***4ever***; even with Hungarian phonetics: *Szimcsi* ***4 evör*** *:P*. The two-letter grapheme (or digraph) *sz* is replaced with the one-letter grapheme *x* in *nem* ***hixem*** *L...*; ***perx-e-perxe-perxe****... =)* and ***Xar!***. Further examples for letter replacement: *cejetlekkk* (< szeretlek 'I love you'), *pijulok* (< pirulok 'I'm blushing'); *punxerelem* (< punk szerelem 'punk love'); *Tomee* (< Tomi 'Tommie'). Symbols are used for abbreviation purposes in the expressions *még 1x* (< még egyszer 'once more') and *+jegy* (< megjegy < megjegyzendő 'nota bene'), and for decorative purposes in *Điđu$* (< Didus '[name]'). Examples for contraction: *uccsó*, [name] *meg asszem h én* (uccsó < utolsó 'the last one', asszem < azt hiszem 'I think'); *Fúúú az hüe név!* (hüe < hülye 'stupid'); *múlt órán azt mta, hogy már csak 10 perc van ebből* (mta < mondta 'said').

LOL is the only internet-specific acronym which appeared in the marginalia: *„elcigányosodik"* **:D** *lol*; *go kill lol.* English abbreviations appear in the following (fully English) examples: *R U SURE?????* (*Are you sure?* and *I 1 2 C U night B4 9 o'clock;)* (*I want to see you night before nine o'clock*).

The side-notes also provide examples for euphemism: *Performáció: Vazze!*; *fogd be hülye k...v@.*(k...v@ < kurva 'bitch').

As an interesting spelling issue, students wrote separate words together even in handwriting if they considered them as a single intonation unit: *Persze csak szeretnédte azt!!* (*szeretnédte* < *szeretnéd te* 'you want').

Here are a few examples of the altogether 280 emoticon occurrences found in the marginalia: emoticons rotated with 90 degrees included :(,:), =D, ^^, ;), =), =/, **), :S, < 3, xD, :-), :D, while standing emoticons included: -.-; ^.^; *.*. Several humorous notes, comments, witticism were recorded in the marginalia together with emoticons: *„Egy őrült tudós fejéből pattantam ki...” by Juhi* ***xD***; *Heavy Medál* ***-)***; *T.ú.: Az utca másik része mintha kivilágosodott volna. Oli: De miért dudált?* ***xD***. Commentary on classroom events: *nem éri meg kitermelni* ***:S***; *Összefoglalás* ***:'(***; *vándorlás, mint motívum* ***:S***; *alacsony hőmérséklet* ***:S***; *Qrva érdekes volt* ***XD***; *Mainzi Hordó?* ***XD***; [*name*] *tanár úr órája* ***^^***; *jó kis óra ez!!!* ***XD :P***; *3 perc az hosszú idő* ***:P***, *neki biztos* ***XD***; *tud ám ** órát tartani* ***XXP***. Students used emoticons to mark their attitude toward school subjects as well: *Biology* ***:P***; *History****:)***; *Maths!* ***=D***; *Phsyics****< 3***. Examples for Hungarian and English macaronic language_ *First hercegnő =); Second hercegnő ^^; I'm the second hercegnő ^^; Ancsa királykisasszony The Only ^^; Liló hercegnő The First ^^*. Many instances were found where emoticons are used for the expression of love or affection: *Hercegem :); H:) E:) R:) C:) E:) G:) E:) M:); Máté és Judák :); Patrik :); Iza ♥ árpi :); Dominyc! < 3 :):$; Dani < 3; Gábor < 3*. Examples for similar function: *Szeretlek Cica X); láb jú xD; I love U:$* (*L*); *I love U nagyon-nagyon =); Zita = husombogyi I love you :P; Я люблю! :$* (*L*); and an impersonal structure: *Szeri van Me too:)*. MSN emoticon codes are often inserted in messages written l'art pour l'art without any real addressee: *X)* (*L*) *:D Pimpó* (*L*) *=D :) =) XD*. Note how the following word is written *Picyciwem = Jocó Jocó = Looser :D :D :D* (Picyciwem < pici szívem 'my sweetheart').

The following letter-like note on the margin of an exercise book, which was left without any answer, leads us to the next corpus containing notes passed by students in class.

Example 70: Notes on the margin of an exercise book

You will hate me for this. But hope you'll not XD Guess what, I've bought a hairgrip xD Cool isn't it? By the way I haven't, I only wanted to write something stupid. I'm curious when you'll notice this! xD 25 February 2009

5.2.2 The impact of digilect on handwritten letters (dialogue letters)

5.2.2.1 Instant messaging vs. communication in letters

There has been a general consensus in the literature of digital communication that the language of instant messaging cannot be placed in the orality–literacy dichotomy (Benczik 2001; Érsok 2003, 2006; Thaler 2003), it oscillates between the two categories (Wirth 2005). It is worth reiterating the theory set up by

Koch and Oesterreicher (1994) about conceptual and medial literacy/orality: in this theoretical framework instant messaging is medially linked to literacy but conceptually to orality, on the basis of the categories of temporal proximity and spatial remoteness. This novel form has many names, including among others: written quasi-orality (Wirth 2005: 71), written-spoken language (Bódi 2004), written chatting (schriftliches Plaudern, Wirth 2005: 72), virtual literacy (Érsok 2003) and secondary literacy (relying on the system of Walter J. Ong 1982).

Instant messaging is usually defined as the communication form of "remote presence" (Wirth 2005: 67). From a linguistic point of view the novelty of this lies in the fact that instant messaging is the first in the history of communication where writing is used for simultaneous, direct and situation-bound verbal communication without being preceded or followed in time by spoken language communication (Storrer 2001: 462).

Reinhard Nickisch (2003) argues that text messages and e-mails are simply the continuation of traditional letter communication. This striking statement is challenged by the following arguments: e-mails differ from traditional letters in the medium (electronic data vs. tangible paper), in appearance (screen vs. envelope and letter paper), in delivery time (e-mails are delivered substantially faster), permanence, relationship between the communication partners, and conventional formal characteristics.

According to Dürscheid, while electronic mails share certain characteristics with traditional letter communication or "snail mail" ("Schneckenpost", Dürscheid 2005: 93), instant messaging has nothing in common with it (Dürscheid 2005: 87). In reality, however, instant messaging can not only be related to face-to-face and phone conversations but also to traditional letter communication in terms of mediality and communicative dynamics. Although traditional correspondence is an asynchronous form of communication, a private letter is conceptually more closely related to orality which makes it similar to instant messaging. For Gottsched (1973: 145) private letters are the alphabetization (Verschriftlichung) of oral conversations, while Gellert argues that private letters do not equal to but replace real conversations (Gellert 1989: 111; Höflich 2003b: 39). Thus, traditional letters imitate spoken language conversations (Wirth 2005: 72) and this feature creates a kind of imaginary presence of the communication partners (Wirth 2005: 73).

The differences of letters (more specifically epistolary novels) and instant messaging were collected by Wirth (2005: 81): From a semiotic perspective, letters mark the speaker's state of mind implicitly through symptoms, while instant messages mark it explicitly through conventional signals (Wirth's examples – *ach!* vs. **empfindsamsei** – in English: *Ahhh...* vs. **deeply affected**). Self-descriptions and replacement signs do not only supplement for missing nonverbal information but also function as comments (e. g.: >*X is happy*< explicit description of emotion, *laugh* inflective, *LOL* acronym, *:-)* emoticon).

5.2.2.2 The corpus

The exercise books handed over by the high school students for analysis contained 29 pages of dialogue letters, typically notes passed secretly between students in class. (As these dialogue letters were found in the exercise books handed over voluntarily by the students and identification of the authors was impossible, the analysis raised no ethical concerns.) This corpus was later completed with 18 other dialogue letters (*Annex 9*).

An analysis of the texts revealed that they share more characteristics with instant messaging than with traditional letters. One of these similarities was quasi-synchronicity: while in oral conversations one party may interrupt the other while speaking, in instant messaging and dialogue letters any reaction is only possible after taking turns.

However, as a major difference from IM and traditional letters, dialogue letters assume the physical presence of both communication partners (so both temporal and spatial proximity are fulfilled according to the categories of Koch and Oesterreicher), but oral communication is impossible (as it would disturb class work). This characteristic feature was the main reason why I named these notes as *dialogue letters*. Dialogue letters therefore do not match the prototypical characteristics of written communication as, according to Günther (1983: 33), it would assume that "the reader is not physically present while the text is being written, and the writer is not physically present while the text is being read."

Dialogue letters further differ from instant messaging but resemble traditional letters in their permanence: while instant messages are written to the moment and are lost after the end of the communication (if they are not saved), both traditional and dialogue letters are written by hand and are therefore more permanent.

Christa Dürscheid observed that "certain forms of expression specific to the internet and text messages, including emoticons, inflectives and abbreviations" are increasingly common in non-electronic text types as well, such as in notes passed by students in class (Dürscheid 2005: 94). This phenomenon, however, was not described in detail, only illustrated with some examples by Dürscheid. In fact, she mentioned that no empirical research had been conducted to explore the interference (Dürscheid 2005: 95). I tried to remedy this deficiency by surveying elementary school students for their attitude in terms of language sociology and language use, the results of which was presented in Chapter 4.1. The direction of this survey is continued in this subchapter but this time the examination mainly focused on the writings of secondary school students.

In the following, I will examine how digilect has influenced handwritten private letters from four main perspectives: formal-technical, lexical, grammatical and textological-pragmatic.

5.2.2.3 From a formal-technical perspective

Different speakers in the texts are marked with different colours, as in most dialogues on MSN (Windows Live Messenger; *Image 37–38*).

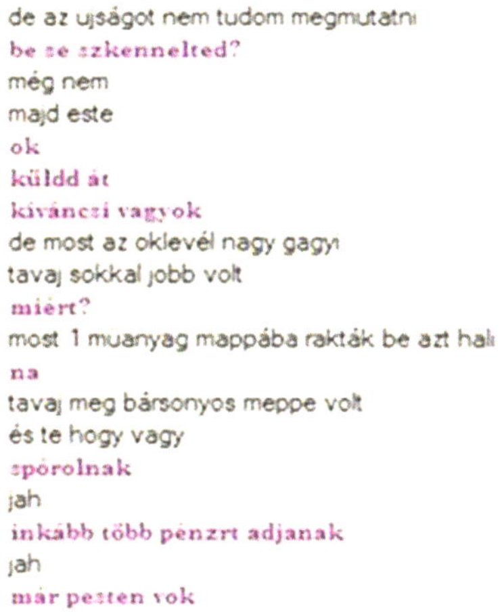
de az ujságot nem tudom megmutatni
be se szkennelted?
még nem
majd este
ok
küldd át
kíváncsi vagyok
de most az oklevél nagy gagyi
tavaj sokkal jobb volt
miért?
most 1 muanyag mappába rakták be azt hali
na
tavaj meg bársonyos meppe volt
és te hogy vagy
spórolnak
jah
inkább több pénzrt adjanak
jah
már pesten vok

Image 37: Example for instant messaging on MSN (speakers marked with different colours: black and purple; different fonts; no capitals)

Image 38: Two dialogue excerpts

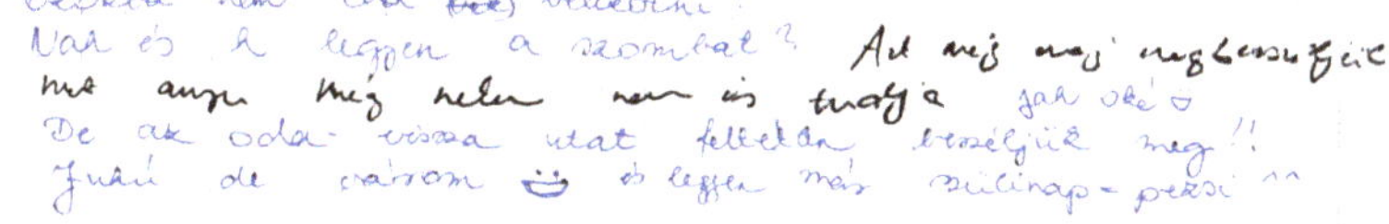
Nah szted az kihető ha mondom h családi okok
miatt nem...? Inkább verseny miatt... de a családi
okokra nem tud (tok) beleszólni!
Nah és h leggen a szombat? Azt még [illegible] megbeszéljük
[illegible] anyu még neki nem is mondta a jah oké ♡
De az oda-vissza utat feltétlen beszéljük meg!!
Jukú de várom 😊 és leggen már szülinap-pezsi ^^

Image 38.1: Letter excerpt No. 1 (each speaker marked with different colours: blue and black)

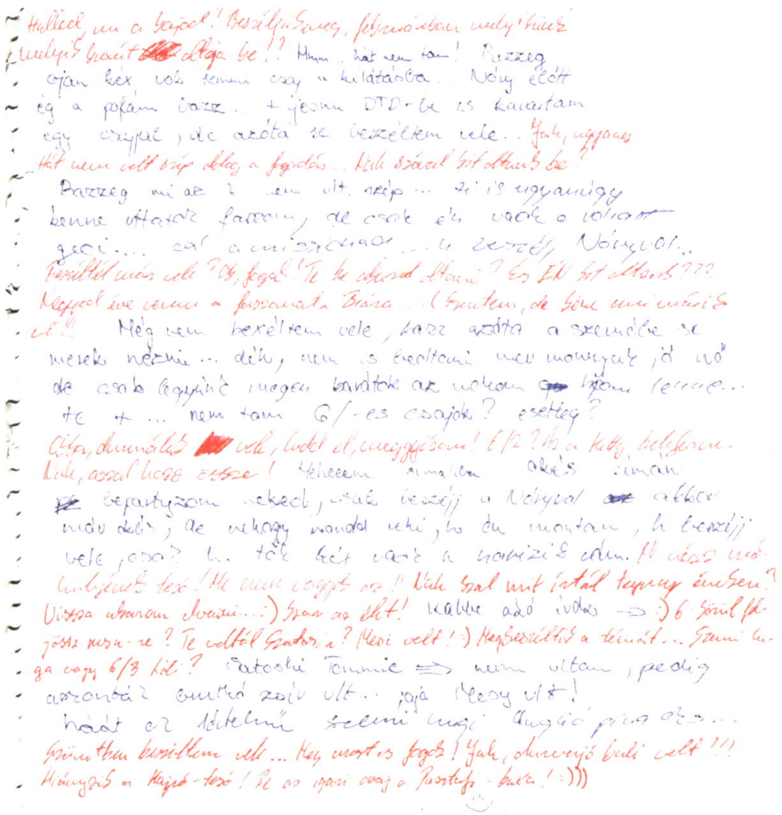

Image 38.2: Letter excerpt No. 2 (each speaker marked with different colours: blue and red)

Gyula Laczházi wrote as early as in 2004 that "considering how quickly and powerfully emoticons gain ground everywhere we may come to the conclusion that a new cult of emotions is on the rise. The importance of emotions and the increased attention directed at them are noticeable in many areas of life, including various scientific areas (philosophy, neurobiology, history, literary criticism) [...]. This avid interest in the postmodern era is primarily connected to the disappointment in instrumental intelligence and the realization of the limits of rationalism" (Laczházi 2004; similarly: Meier-Seethaler 1997). This subject philosophic approach is only strengthened by my examples. The following three phenomena clearly indicate how important it is for the writers of dialogue letters to express their feelings:

1. overuse of letters and characters to express emotions and intensity;
2. punctuation marks expressing emotions (*Ancsáék megint balhéznak... :/*; *Az agyam eldobom!!*; *Mit????????*; *Szeretlek szó benne volt!??*), and
3. use of emoticons instead of/in addition to punctuation marks.

The intensity of feelings can be expressed by the overuse of punctuation marks and iteration (Keszler 2006: 66). This can be achieved easily on a keyboard (by keeping a key pressed down) but it requires much more effort and time in hand-written dialogue letters.

Example 71: Overuse of characters in dialogue letters

ÉN megmondtam ***igeeen...***
Neeeeeem *:O*

Linguists were divided on the issue of the necessary number of punctuation marks in the 19–20th century (Keszler 2006: 103): one group argued for fewer, the other for more punctuation marks (see for example: István Gáti [19th century]; Károly Tamkó Sirató: Új írásjeleket [We need new punctuation marks]). "Virtual literacy [however] produced some new punctuation marks" (Keszler 2006: 66): emoticons are graphical signs which are suitable to express emotions and are primarily connected to new media and communication devices (more on emoticons: Veszelszki 2005b; Bódi–Veszelszki 2006). Gyula Laczházi (in 2004!) did not exclude the possibility that these signs would soon appear in conventional forms of communication, more specifically in paper-based notes.

According to my observations, mainly those punctuation marks are omitted in instant messaging and similar dialogue letters which have only segmentation function but no emotional intonation: commas, for instance, delimit units of thoughts rather than clauses in the conventional grammatical sense. Full stop as a sentence-ending punctuation mark are omitted in most of the cases. Exclamation marks, which mark emotional intonation in writing, and question marks, which mark modality, are both used in dialogue letters. Sometimes all sentences (or rather sentence-like units) are ended with an emoticon (*Example 72; Image 39*).

Example 72: Emoticons in dialogue letters

– Ohh értem ***:D*** *de nekem nehogy beteg legyél! Am fogadok, ha talizol Vele akk jobban leszel* ***;)***
– Max vasárnap taliznánk de nem akarok... ***:S*** *Am buliba biztos megyek* ***=)*** *Még 40°-os lázzal is* ***XD***
Tudod van az a Gere Scence ***XD*** *vagy ki És a blogján minden j helyett y írt.* ***XD***
– Szerintem se csalás ***XD*** *Ő is kirakhatná* ***XD***
– Ja úgy van h ahányan megnézik.
Ez nem cs filmekben van ***XD***
De leszarom, pasizni fogok, hogy legyen tiszta ideg ***:@***

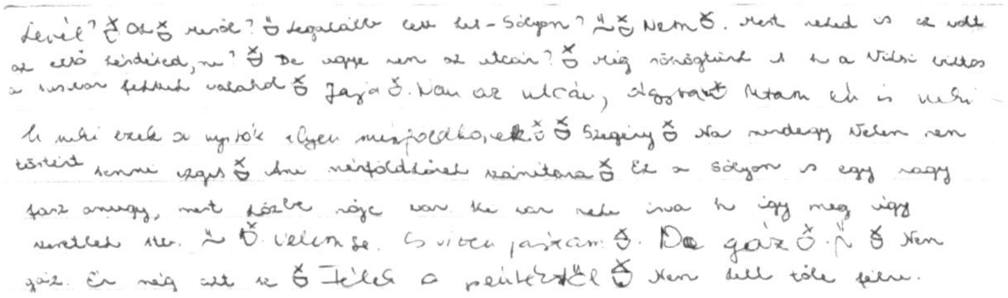

Image 39: Letter excerpt No. 3 (each sentence ends in an emoticon)

5.2.2.4 From a lexical perspective

The lexical characteristics of digilect appear in handwritten letters as well. The most commonly used ones were the following: foreign language influence, slang, verbal aggression, pro-sentences with phatic function.

Foreign (especially English) language influence is particularly strong in Hungarian internet communication. English words crop up in dialogue letters as well, sometimes spelled with Hungarian phonetics.

Example 73: English influence in dialogue letters

Hát, vagy megvárom h szakítsanak és akk ***heppyend****, vagy kidobom poszeidont mielőtt én esek pofára :S* (*heppyend* 'happy end')
cal a ***missionod****...* (missionod 'your mission')
Letter*?* ***Ya.*** (*Ya* 'yes') *Válaszolj rendesen XD*

The language of the letters is so close to spoken language that even *slang* elements occur in them (probably according to their own set of rules of communication).

Example 74: Slang in dialogue letters

És igazából azért ***vágtam pofákat****, mert közbe apu* ***osztott****, h késő van és h menjek aludni...* (*vágtam pofákat* 'made faces', *osztott* 'told me off')
Full nemvágok semmit *:S* (*full* 'totally', *nem vágok semmit* 'I don't understand a thing')
Beszéljük meg, februárban melyikünk melyik ***buxát oltja be****!?* (*buxát oltja be* 'lay a girl')
okés ***simán bepartyzom*** *neked* (*simán* 'easily/surely', *bepartyzom* 'arrange it for you')
aszonták ***brutkó zsír*** *vlt...* (*brutkó zsír* 'absolutely cool')

We can also find *swearwords* in dialogue letters, but as opposed to instant messages the rules of chat related netiquette ("chatiquette") are completely ignored: there are no euphemistic forms (*Image 40*).

Example 75: Swearwords in dialogue letters

az a ***kretén*** *meg otthagyott msn-en* (*kretén* 'idiot')
de hát nem ***basztathatom*** *ezzel minden egyes alkalommal...* (*basztathatom* 'keep messing with')

csak h ***kapja be*** *h otthagyott és h hiányzik :(:(:(* (*kapja be* 'fuck off')
ilyen ***faszságai*** *vannak.* (*faszságai* 'annoying habits')
Kabbe *akó írok* (*kabbe* 'fuck off')
Bazzeg *mi az h. nem vlt szép... ti is ugyanúgy benne vltatok* ***faszom***, *de csak én vaok a* ***rohatt geci...*** (*bazzeg* 'fuck', *faszom* 'shit', *rohatt geci* 'wretched jerk')

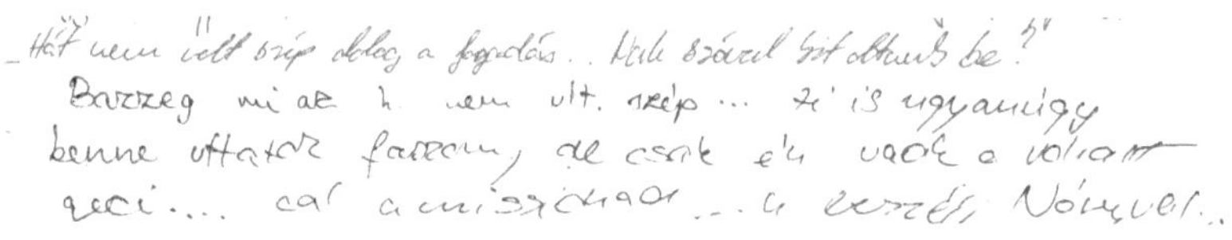
Bazzeg mi az h. nem vlt. szép... ti is ugyanúgy
benne vltatok faszom, de csak én vaok a rohatt
geci...

Image 40: Letter excerpt No. 4 (manifestation of slang and verbal aggression)

As one of the main reasons for passing notes in class is to maintain relationship, dialogue letters abound in interjections (indulatszók), more specifically *pro-sentences with phatic function* (interakciós mondatszók). In the last example the expression *A-A* is a metalinguistic one: it refers to the sound produced when shaking the head. (The same expression appeared in a Facebook comment, see Chapter 3.5.1.3.) Writing influences spoken language due to its optical-visual character. Dürscheid (2006: 40) illustrates this by compounds which refer to letters as icons. *V-kivágás* 'V-neck (sweater)', *S-kanyar* 'S-bend (racing track)', *X-láb* 'X-legs', *O-láb* 'O-legs' and the English *T-shirt*, but this interaction is reflected in the expression *I say it quotation marks* which is accompanied by a characteristic gesture. Certain paraverbal elements—e. g. *hihi* (laugh), *haha* (laugh), *öhöm* (agree), *aha* (agree), *a-a* (disagree)—can also be independent words if written down. These words then reappear in spoken language: for example in response to a statement a speaker may ironically say: *hihi* (Dürscheid 2006: 40–41).

Example 76: Pro-sentences with phatic function in dialogue letters

Nah *szal ügyi voltam ^^ //* ***Nah jah...*** (*nah* 'well', *jah* 'yes')
Jah, ugyanez. (*jah* 'yes', *ugyanez* 'just like me')
Heheeem... ('hehe')
A-A. ('no-no' [the words said when shaking the head])

5.2.2.5 From a grammatical perspective

Reduced grammatical structures move dialogue letters closer to spoken language conversations: context and common knowledge is required for understanding the speech products.

Example 77: Ellipsis in dialogue letters

I'm arranging my things and learning a bit and then I'm off to have a shower :D
I'll washing my hair and all that :P

just talk to Nóry and it will be fine
– And are you sure you don't want to?
– Now I am, yes... Yesterday's conversation...

The following forms of internet literacy/digiliect probably originate from medial orality and came to existence through looser pronunciation: *nem tudom* 'I don't know' > *nemtom* 'dunno', *mit tudom én* 'I don't have the faintest' > *mittomén, azt hiszem* 'I believe' > *asszem, megyek* 'I'm going' > *mek* and *mék, miért* 'why?' > *mér, mert* 'because' > *mer, egészségedre* 'bless you!' > *egs, valószínűleg* 'probably' > *valszeg*. The primary motivation behind these reduced forms include simplification, faster typing and humour/invention.

In Chapter 3.3 I compared neologisms used in text messages. While the words *müxik* (< *működik* 'it works'), *aszonta* (< *azt mondta* '(s)he said') and *nemtom* (< *nem tudom* 'I don't know') are only the formal variants of the full expressions, the word *asszem* has gone through some pragmatic change as well. *Nemtom* 'I don't know' is the merger of *nem* 'not' and *tudom* 'I know', imitating the pronunciation of the expression. No other (epistemic or pragmatic) meaning is attributed to *nemtom*; it is the abbreviated and merged form of *nem tudom*, a logical-semantic relationship between a negative particle and a verb (Kugler 2000b: 289). The change of *asszem*, however, is not restricted to mere formal modification but it extends to function as well. Now it seems that word is shifting from the main clause of a subordinate structure toward becoming a modifier which indicates epistemic modality.

Example 78: The occurrence of *asszem* and *nemtom* in dialogue letters

Nah ***asszem*** *du lesz min gondolkoznom...*
Garbais... de ***nem tom*** *milyen szakra XD*

Authors studying the language of the internet, such as Androutsopoulos and Ziegler (2003: 253), mention abbreviations and acronyms among the constructive characteristics of instant messaging. The dialogue letters examined contained a remarkably high number of *abbreviations*. Nearly all the abbreviation types and character-saving techniques used in instant messaging could be found in the dialogue letters (*Image 41*).

Example 79: Character-saving techniques in dialogue letters

nekem anya ***nm Ngedi h*** *használjam* (*nm Ngedi h* < *nem engedi hogy* 'does not allow me to')
Szal *tegnap beszéltünk msn-en... és* ***akk*** *beszéltük* ***h*** *talizunk hétvégén.* (*szal* < *szóval* 'so', *akk* < *akkor* 'then', *h* < *hogy* 'that')
ezért mondom, ***h mind1, h*** *mit veszel fel :D* (*h mind1, h* < *hogy mindegy hogy* 'that it's not important what')
Szóval tudja, hogy ki vagyok ***sztem...*** (*sztem* < *szerintem* 'in my opinion')

> *Fogadok* ***h 1000x*** *szívesebben találkozna veled!* (*h 1000x* < *hogy ezerszer* 'that a thousand times')
> ***am*** *komolyan bejött volna* ***vld****?* (*am* < *amúgy* 'by the way', *vld* < *veled* 'with you')

The abbreviations used in instant messaging and dialogue letters are compared in *Table 21*. The examples of digilect were taken from my earlier studies.

Table 21: Types of abbreviations in digilect and dialogue letters

Type of abbreviation	Example from the digilect	Example from the dialogue letters
initials only	*hogy* 'that' > *h. and h; vagy* 'or' > *v; csá* 'hi/bye' > *cs*	*hogy* 'that' > *h and h.; csak* 'only' > *cs*
acronym-like Hungarian abbreviations	*nem tudom* 'I don'*t know'* > *nt; semmi baj* 'no problem' > *sb; nem baj* 'no problem' > *nb; nincs mit* 'you are welcome' > *nm*	[no occurrence in the examined texts]
internet-specific English acronyms	*LOL; ASD; WTF*	*LOLi; WTF*
omission of vowels	*puszi* 'kiss' > *px; bocsi* 'sorry' > *sry; ok* > *k; szóval* 'so' > *szvl; hogy* 'that' > *hgy; szeretlek* 'I love you' > *szrtlk; magam* 'myself' > *mgm; velem* 'with me' > *vlm*	*veled* 'with you' > *vld; volt* 'was/were' > *vlt*
clipping	pillanat 'minute' > pill; akkor 'then' > akk; szívesen 'you are welcome' > cív and sziv; szeretlek 'I love you' > cer; puszi 'kiss' > pu	*akkor* 'then' > *akk; amúgy* 'by the way' > *am; mert* 'because' > *me*
omission of *j/ly* felt to bc intrusive consonants	*hülye* 'stupid' > *hüe; milyen* 'what sort of' > *mien, valamelyik* 'one of them' > *vmeik*	*milyen* 'what sort of' > *mien*
letters in mixed case the uppercase letters are pronounced as independent phonemes full (e. g. *M* as *em*)	*emlékszel* 'you remember' > *Mléxel; jó estét* 'good evening' > *jóST-t; ember* 'man' > *Mber*	*engedi* 'lets' > *Ngedi*
Blending	vagyok 'I am' > vok and wok; holnap 'tomorrow' > hnap and hn; szerintem 'in my opinion' > sztem; igen 'yes' > in; semmi 'nothing' > smi; volt 'was/were' > vt; nagyon 'very' > naon and non; szóval 'so' > szal	*szóval* 'so' > *szal; szerintem* 'in my opinion' > *sztem; vagyok* 'I am' > *vok* and *vaok; voltatok* 'you were' > *vtatok; tudom* 'I know' > *tom; tegnap* 'yesterday' > *tnap*

Table 21: (Continued)

Type of abbreviation	**Example from the digilect**	**Example from the dialogue letters**
substitution of a letter – with a different letter	*pusz* 'kisses' > *pux; szia* 'hi' > *cya; lesz* 'will be' > *lex; pusza* 'kisses' > *puxa; szóval* 'so' > *szowal* and *xwal; szerintem* 'in my opinion' > *xt*	*olyan* 'such' > *ojan; ilyesmi* 'something like this' > *ijesmi; kész* 'ready' > *kéx; beszéltem* 'I've talked to' > *bexéltem; szóval* 'so' > *cal; szemét* 'rubbish' > *xemét*
– with numbers	*egyszer* 'once' > *1szer; egyszerű* 'simple' > *1szerű; együtt* 'together' > *1ütt; megyek* 'I am going/leaving' > *m1ek; mindegy* 'all the same' > *mind1; 5let* 'idea' > *ötlet; hatás* 'effect' > *6ás; felhívhatsz* 'you can call me' > *felhív6sz; tudhatom* 'I should know' > *tud6om; héten* 'this week' > *7en;*	*egyébként* 'by the way' > *1ébként; mindketten* 'both of them' > *mind2-en* and *mind2en; egyértelmű* 'clear' > *1értelmű; dugható* 'pluggable' > *dug6ó; mindegy* 'all the same' > *mind1*
– with characters standing for a series of letters	*hétfő* 'Monday' > *7fő; hétvége* 'weekend' > *7v* and *7vége* and *7vg; nekem nyolc* 'all the same for me' > *nekem8; tök nyolc* 'all the same for me' > *tök8*	*meg* [*perfective prefix*] > +
– with number and character	*megbeszéljük* 'we'*ll talk it over'* > *+beszél-jük; +besz; megcsinálom* 'I'*ll do it'* > *+cslom; megvan* 'got it' > *+van; holnap* 'tomorrow' > *hol¤; Emese* [*female name*] > *M§E; pihen* 'rest' > *ΠhN egyszer* 'once' > *1x*	*ezerszer* 'a thousand times' > *1000x*
abbreviation/ blending indicating looser pronunciation	*nem tudom* 'I don't know' > *nemtom* 'dunno'; *mit tudom én* 'I don't have the faintest' > *mit-tomén*; *azt hiszem* 'I believe' > *asszem*; *megyek* 'I'm going' > *mek* and *mék*; *miért* 'why?' > *mér*; *mert* 'because' > *mer*; *egészségedre* 'bless you!' > *egs.*	*majd* 'later' > *maj; nem tudom* 'I don'*t know'* > *nem tom; nem tudok* 'I can'*t'* > *nem tok; mert* 'because' > *mer; mondjuk* 'say' > *monnyuk; megint* 'again' > *megen; mondtam* 'I said' > *montam; szerinted* 'in your opinion' > *szented; gondol-tam* 'I thought' > *gondótam*
baby talk[69]	*szeretlek* 'I love you' > *cer and cejizlek; hiányzol* 'I miss you' > *hiánzol*	*hiányzik* 'I miss him/her/it' > *hiánzik*

69 Not necessarily abbreviations but the word-forms typically become shorter.

Table 21: (Continued)

Type of abbreviation	Example from the digilect	Example from the dialogue letters
word formation with diminutive suffixes	*megcsörgetés* 'ringing sy's *phone'* > *megcsörizés; találkozás* 'meeting' > *talcsizás and talizás; unokatestvér* 'cousin' > *uncsitesó; üzenet* 'message' > *üzi*	*pezsgő* 'champagne' > *pezsi; haragszik* 'is angry' > *harizik*
other	*működik* 'it works' > *műxik*	[no occurrence in the examined texts]
techniques going against the trend of abbreviating words (purpose: to be unique)	*szia* 'hi' > *xijjah*	*szia* 'hi' > *cijja; föci* 'Geography [class]' > *föcee; Ancsa* [female nickname] > *Ancsaaa*

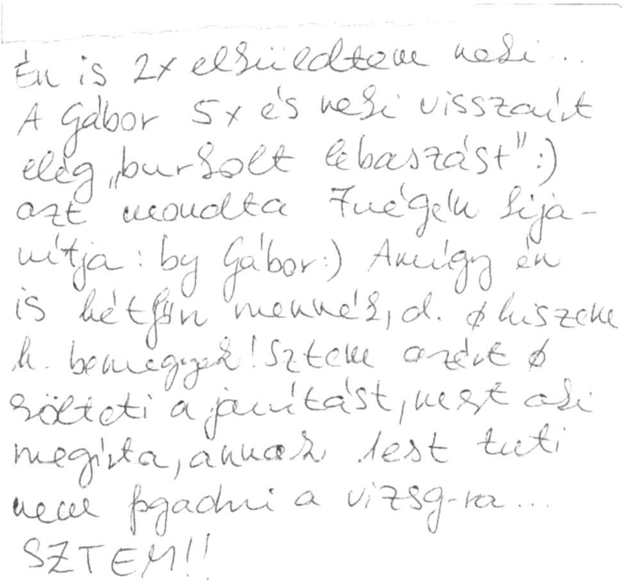
Én is 2x elküldtem neki..
A Gábor 5x és neki visszaírt
elég „burkolt lebasztást" :)
azt mondta Fuégeln kija-
vítja: by Gábor :) Amúgy én
is kétfjn mennék, d. ø kiszem
h. bemegyek! Sztem azért ø
kötötti a javítást, mert aki
megírta, annak lest tuti
nem fgadni a vizsg-ra...
SZTEM!!

Image 41: Letter excerpt No. 5 (abbreviations)

It is also a type of abbreviation when the spelled form of a word reflects the pronounced (and typically less articulated) form rather than the standard rules of spelling. *mondjuk* 'say' > *mongyuk* or *monnyuk*; *játszani* 'to play' > *jáccani*). Adapting the phonetic characteristics of oral conversations to writing results in *stylized orality*.

Example 80: Phonetic spelling in dialogue letters

csak ***beszéjj*** *a Nóryval, de nehogy mondd neki, h én* ***montam****, h* ***beszéjj*** *vele* (*beszéjj* < *beszélj* 'talk to', *montam* < *mondtam* 'I said')
aszonták [< *azt mondták* 'they said']
mer monnyuk *jó nő de csak legyünk* ***megen*** *barátok* (*mer monnyuk* < *mert mondjuk* 'because say', *megen* < *megint* 'again')

This writing technique, which strongly resembles orality, can go as far as *spontaneous writing*.

Example 81: Spontaneous spelling in dialogue letters

Bazzeg ojan kéx vok semmi csaj a kilátásba... (with standard spelling: *Baszd meg, olyan kész vagyok, semmi csaj a kilátásban...* 'Damn, I'm so frustrated, there's no girl on the horizon')
Nóry előtt ég a pofám bazz... + ijesmi DTD-be is kavartam egy csajjal, de azóta se beszéltem vele... (with standard spelling: *Nóri előtt ég a pofám basszus, meg ilyesmi.* 'I'm so ashamed to meet Nóri, and things like that')
csak h. tök kéx vaok h harizik rám (with standard spelling: *csak hogy tök kész vagyok, hogy haragszik rám* 'the thing is that I'm so frustrated that she's angry with me')

5.2.2.6 From a pragmatic-textological perspective

I agree with Wirth (2005) in that the main motivating factor in internet communication—besides collecting information—is raising attention, or with harsher words: narcissistic self-representation. This explains why phatic elements and linguistic rituals (greetings, starting and ending communication, saying thanks) play a distinctive role in the dialogue letters examined.

Example 82: The distinctive role of the phatic function in dialogue letters

Nah szóval akkor mizujs? ('Well, then what's up?')
– *Sziia Ancsaaa!!!:) Mizujs?* ('Hi Ancsa! What's up?')
– *Megvagyok, jólvagyok, ölelést szeretnék:((*('I'm doing fine, I'm OK, I need a hug')

The written dialogues mainly center around relationships, meeting, friendships, weekend programmes (parties) and school assignments. Even the IM software MSN (which was merged in 2011 with Skype) appears explicitly as a metatopic in the letters, such as in the following example (turn-takings are marked with en dash; *Image 42*).

Example 83: MSN as metatopic in dialogue letters

– *Figy... Elmondtam Neki a Cicás ügyet, és szente nem gáz, ha úgy hívunk egy pasit! Szented köxönjek Neki MSNen úgy hogy: Cijjja CiCa!? :D* ('Look, I explained this Kitty thing to him, and he don't find it odd if we call a guy like that! You think I should greet him on MSN like: Hi there, Kitty?")

– *hát hjam, végülis ja :D simán :D* ('Well, how should I say it, well, yes, surely!')
– *Kíváncsi vok, h mit xólna rá! Attom lenne!* ('I'm curious what he would say. That would be awesome') :D

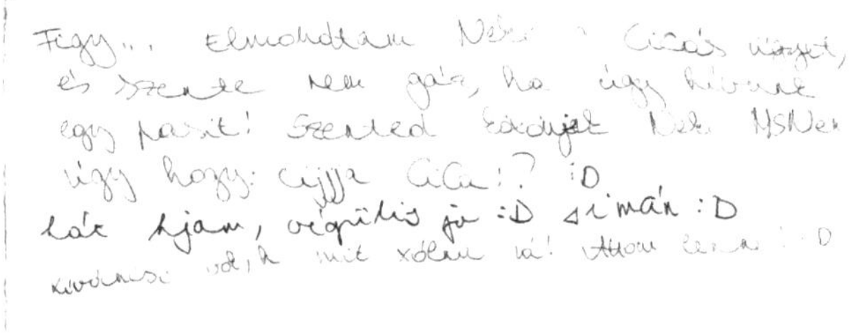

Image 42: Letter excerpt No. 6 (the speakers used different colours: pencil and black pen; MSN as metatopic)

The dialogue letters are characterized by leaping turn-takings and interweaving thematic plots. As opposed to spoken language communication, there are no interactive linguistic signs of turn-taking in dialogue letters (e. g. different intonation, pause, word order, addressing, etc. For more details see: Hoffmann 1998: 5); the parties take turns simply by passing the note. However, the parties use a distinct sign (//) if they comment on two separate topics in the same turn. This is not an idiolectic feature of one or two students but a common sign which appears in several texts (*Example 84*, *Image 43*).

Example 84: Marking thematic leaps in dialogue letters

De szeretem! Megvárom, h mi lesz. // Az nem kompromisszum, h csak te hozol áldozatot.
('But I love him. I'll wait and see. // That's not a compromise that only you make sacrifice.')
Najó ,majd cs lesz vmi XD // Én meg félek h mi lesz velünk. // Mibe jössz szombaton? :D
('Well, it will be fine anyway XD // As for me, I'm afraid of our future. // What will you put on on Saturday? :D')

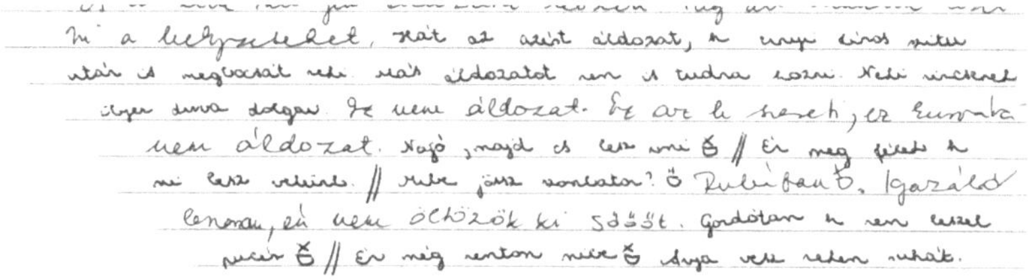

Image 43: Letter excerpt No. 7 (marking thematic leaps with the // sign)

5.2.2.7 From digilect—toward digilect

Traffic on the way from and toward digital communication is going in both directions: not only digilect is affected by other forms of communication but digilect also influences spoken language and (spoken language like) literacy. The influence of digilect on non-electronic literacy is evident in the notes passed by students in class (dialogue letters). To sum up, I copied an excerpt from a dialogue letter below which flourishes with the linguistic and non-linguistic characteristics of digilect (*Image 44*).

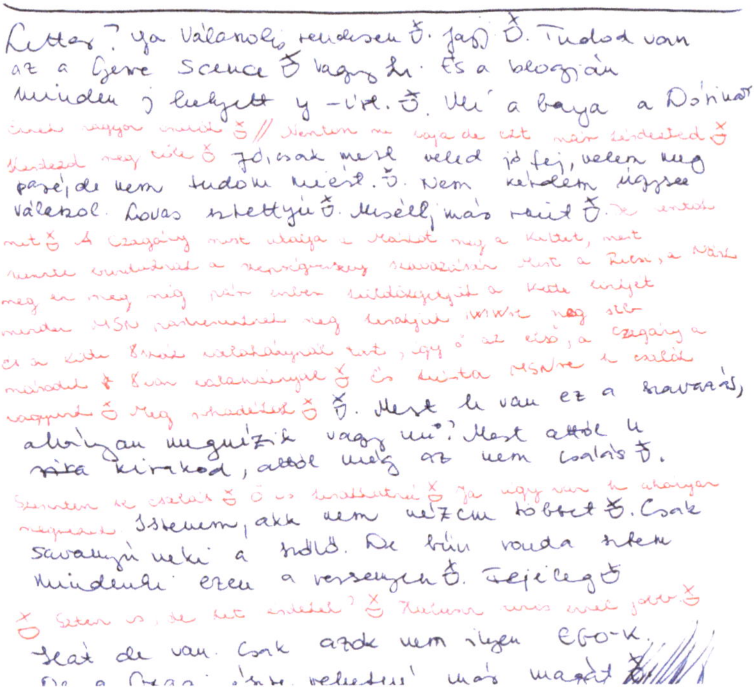

Image 44: Letter excerpt No. 8

5.3 The influence of digilect on printed texts, in particular advertisements

The way a text is produced affects the language of that text, this is why have to distinguish between handwritten and printed texts (Ludwig 1983). The present Chapter 5.3 examines how the language of digital communication influences digitally created but not digitally perceived, that is printed texts, with special regard to advertisements.

5.3.1 The characteristics and language of advertisements

The definition set forth in the Hungarian Code of Ethics for Advertising (W10), advertisements are commercial information which, on the one hand, calls and encourages consumers to buy products or use services, and, on the other hand, informs customers about the properties and function of the advertised goods.[70] Thus, advertisements both help and influence, entertain and inform consumers.

The Hungarian scholar in communication theory, Jolán Róka (1998: 352; also: 2003), defines advertisement as paid non-personal communication of persuasive information on products or ideas whose sponsors can be identified. The first sentence of David Ogilvy's book on advertising is also a definition of the term: "I do not regard advertising as entertainment or an art form but as a medium of information" (Ogilvy 1992: 7).

Advertising is not a new genre: László Faragó (1931/4) quotes Herbert N. Casson's infamous statement that the important thing is not "what" but "how" we advertise. This more than 80-year-old observation seems to be valid even today. The elements of the "how" are as follows: wording (which should be brief, easy-to-understand and easy-to-remember), consistent usage of brand name and slogan (which creates stable brand image), and visual design (which helps interpretation and ensures quick impact) (Jakusné Harmos 2005: 419–436).

As I have observed, the characteristic features of digilect are gaining ground in advertisements. What could be the reason for this? "The perception and acceptance of the message of an advertisement largely depend on how much it is adapted to the language of the target group" (Móricz–Téglássy 1999: 126). According to Eszter Bártházi, advertisements use foreign and novel expressions because they have the same psychological impact as rare things in general. This observation can be extended to the linguistic elements of digital communication as well: "If an advertisement's style is formal, sophisticated, elegant and, as a consequence, it gives the impression of exclusivity, it automatically makes the consumer want to be part of the exclusive audience—primarily by possessing the advertised product. Normally, advertisements targeting the general audience tend to have a plain, everyday language. By contrast, the language of advertisements popularizing prestige products contain more unusual, rarer or even novel expres-

70 ...any information which promotes the sale or other utilisation of any product, service, real property, right or obligation (hereinafter: goods), popularizes any company's name, brand or activity, and/or raises awareness of goods or trademarks (hereinafter: advertisement). Section 2 of Act LVIII of 1997 on Commercial Advertising Activity.

sions or foreign words which aim to add to the extraordinariness and uniqueness of the product" (Bárthází 2008: 460).

The three basic ways to influence (persuade) people (*ethos*: making the audience believe that the speaker has a good moral character; *pathos*: influencing emotions; *logos*: providing logical arguments) have been consciously used in rhetoric since Aristotle (Adamik et al. 2004). Advertisements typically adopt the latter two approaches (Kupper 2006: 207): affective and rational persuasion.

Rational advertisements are built on facts and give the consumer an objective, reasonable explanation why the product should be used. Thus type of argumentation generally focus on the advantages and positive characteristics of the product, so it tries to be objective. Emotional advertisements, on the other hand, are not related to the objective characteristics of the product, they influence consumers' emotions and mood, they make consumers want to identify themselves with the brand (Kiss M. 2005: 196; Deli 2005: 121; Bauer–Berács 1998).

If a product has different characteristics than its competitors, consumers tend to be affected more by cognitive arguments, so verbal persuasion is the best choice. But if a product is very similar to its competitors, affective persuasion (through images, slogan and references) works better (see also: Síklaki 1994; Szikszainé Nagy 1998). "The vast majority of people buy with their heart, or—if you like—with their emotions. Of course, they are also interested in logical reasons: why is the product good, what would be even better. And then they make an emotional decision: I like that one. I like this one better, I think it will do" (Maurice Levy cited by: Papp-Váry 2007: 97).

It is precisely the power of affective persuasion, the group identity effect of language why advertisements tend to contain digilectic elements today.

So what should the language of advertising be like? Kupper (2006: 244) summarized the main criteria as follows: the language of advertising should attract attention and interest, should be understandable and appealing; should influence the consumers' behaviour and possibly encourage them to buy. The list can be completed with the criterion that an advertisement should adapt to the style of the medium.

Popcorn and Marigold found that the two most important persuasive elements in advertising are visibility and recency (Popcorn–Marigold 2001: 132). Advertisements drawing on the characteristics of digilect apply exactly these two strategies of persuasion. Digilect is quite a recent language variety: strictly speaking it is little more than twenty-five years old (not so strictly speaking it is forty-five years old at the most: the first e-mail was sent at the end of the 1960s, the first e-mail with the @ sign in 1972, the first smiley in 1982 and the first text message in 1992; W2). Digilectic texts are normally very visible as they use many visual signs, such as emoticons, text colour or text format.

5.3.2 The influence of digilect on advertising

It has been more than twelve years (since 2004) that I have been collecting advertisements (on billboards, packaging, printed medium and websites) which feature one or more characteristics of digilect.

The corpus, which currently comprises more than forty different adverts, is continuously and rapidly growing. To illustrate this I provide below the excerpt of an admail received while I was writing the present subchapter. It contains two smileys (*Image 45*).

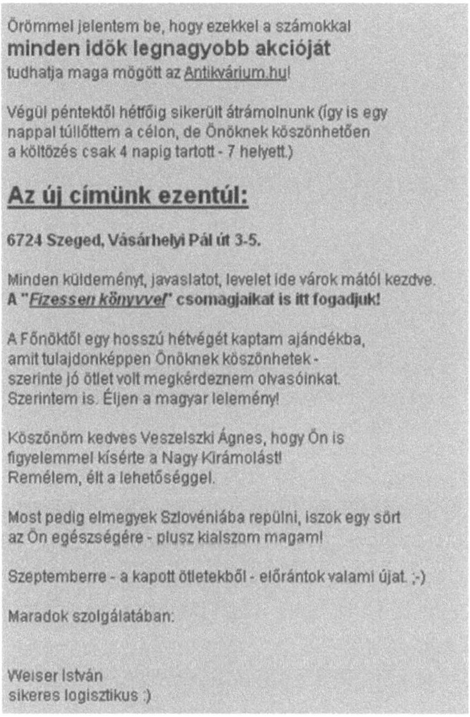

Image 45: Admail with two smileys (2010)

The texts presented in this book were all examined from four perspectives (formal, lexical, grammatical and textological-pragmatic). However, the advertisements in my corpus exhibited only the formal-technical characteristics of digilect.

The only exception was the advertisement of a sanitary napkin (*Image 46*). In addition to the yellow smileys used in IM, it also uses the abbreviations (e. g. *h* < hogy 'that', *vmi* < valami 'something') and characteristic vocabulary (e. g. *LOL, hi*) of digilect, especially in the first and the last sentences which carry the phatic function of the message: *„**Hi, h** vagy? Minden **OK**? Van itt neked **vmi**: Az Always Ultra-t most*

Dri Plus védőzóna teszi még biztonságosabbá, így segít hogy a szivárgás helyett végre vidámabb dolgokkal foglalkozhass. ***LOL, mi?*** *Hogy jobban teljenek azok a napok".* ('Hi! Howdy? All right? We've got sth for you: Always Ultra is now made even safer with Dri Plus protective zone. It helps U forget about your period and do things that are more fun. LOL, init? Making those days more fun '.) This advertisement page was published in a teen magazine and it imitates digilectic texts both visually and linguistically.

Image 46: Advertisement with the formal and lexical characteristics of digilect (Bravo 2009/11: 2)

The largest part of the corpus is made of ads containing smileys. The electronic (colon-hyphen-bracket) smiley is thirty-five years old in 2017. The image of the yellow smiley was originally not related to electronic communication but has carved a highly successful career since its birth in 1963: it has become a symbol of digital communication (cf. Chapter 3.5.1).

The smiley on the billboard advertisement in a rural city seems to be used in its original function of cheering up (*Image 47*). In another example, in the newsletter of a Hungarian electric utility the sockets have turned to yellow, orange and blue smileys (*Image 48*).

Image 47: Billboard with yellow smiley (Dunaújváros, Hungary 2010)

Image 48: Newsletter with smileys (Budapest, Hungary 2010)

Both the yellow smiley and the *Plug&Lay* on the convert poster below are elements of digilect (*Image 49*). The IT term *plug&play* refers to a computer function which enables immediate use of an external device when connected without installation or calibration. The advertised event used a playful distortion of this expression as a slogan. *Image 50* shows a street poster in Leipzig which advertises flats with smileys looking out of apartment block windows.

Image 49: Leaflet with yellow smiley (Budapest, Hungary 2010)

Image 50: Street poster advertising flats with smileys (Leipzig, Germany 2009)

"Baba", the long-established Hungarian soap brand, was advertised with animated smileys on a social networking website (iwiw). The banner, however, could only be inserted into the text as a static screenshot (*Image 51*).

Image 51: Internet banners (Screenshots, 2010)

The cable television advertisement in *Image 52* is a simple collage of images, but the whole picture gives shape to smiling face.

Image 52: Newspaper advertisement with smiley (2006)

The colon-bracket type emoticon may form part of the brand (and logo), such as in the case of the English language school "beHappy". The school's slogan is also closed with a smiley, as we can see it on their website and on subway posters (*Image 53*).

Image 53: Smiley as part of a logo and closing element of a slogan (2010)

The colon-bracket type smiley is incorporated into the brand name of the Ukrainian mobile service provider “life:)” as well, and the company’s SMS and MMS advertisement explicitly uses several types of emoticons (*Image 54*).

Image 54: Smiley as part of a logo; emoticons in mobile advertisements (Munkachevo, Ukraine 2010)

Image 55: Emoticons written on post-it notes in newspaper advertisements (2010)

The newspaper advertisement of a TV show (*Image 55*) depicts a notice board with pictures and post-it notes with two types of emoticons: the *:)* and the *:P* version.

Smileys also appear in the advertisements of telecom companies (*Image 56*). As a special service, the Swiss mobile phone service provider “Swisscom” offers to teach older generations how to use new technology, especially mobile phones: „Internet, E-Mail und SMS – die Möglichkeiten der neuen Kommunikationsmittel sind riesig. Wir zeigen Ihnen, wie Sie sie am besten nutzen. Swisscom ist es ein Anliegen, dass alle kompetent und sicher mit den neuen Kommunikationsinstrumenten umgehen können. [...]" (Swisscom advertisement). The text of the advertisement ("Wer bin ich?" 'Who am I?') is completed by a large question mark eyed smiley (or vice versa, the emoticon is accompanied by the text).

Image 56: Question mark eyed smiley on a flyer (Bern, Switzerland 2004)

It is easy and fun to find accommodation on the advertised website—at least this is suggested by the smiling face appearing on stickers, car labels and street posters (*Image 57*). The pseudo-phonetically spelled *Figyejjé má!* 'Hey, listen!' contains a tongue-sticking smiley on a street poster in *Image 58* (the purpose of this advertisement is unclear).

Image 57: Emoticons after a website URL on a car sticker (Balatonfüred, Hungary 2010)

Image 58: Emoticon on a billboard (Eger, Hungary 2010)

The following banner of conference on direct marketing shows that advertising professionals have also realized the importance of emoticons in advertisements (*Image 59*).

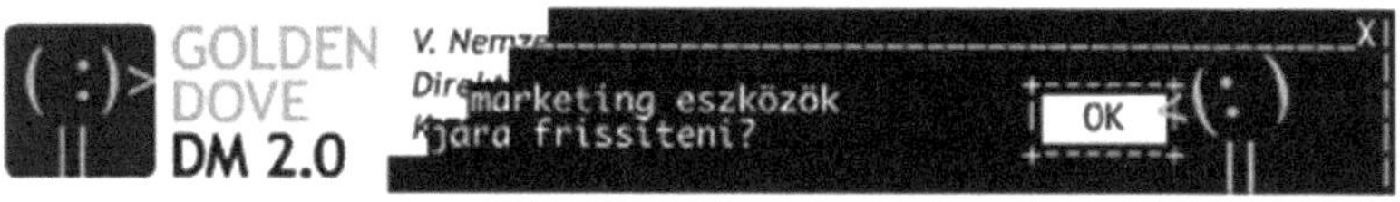

Image 59: Emoticons on a banner (2010)

The poster in *Image 60* tries to address and draw the attention of the appropriate audience by a smiley.

Image 60: Emoticons on a billboard (Kiskunhalas, Hungary 2009)

Image 61: Emoticons on a poster in a shop (Budapest, Hungary 2010)

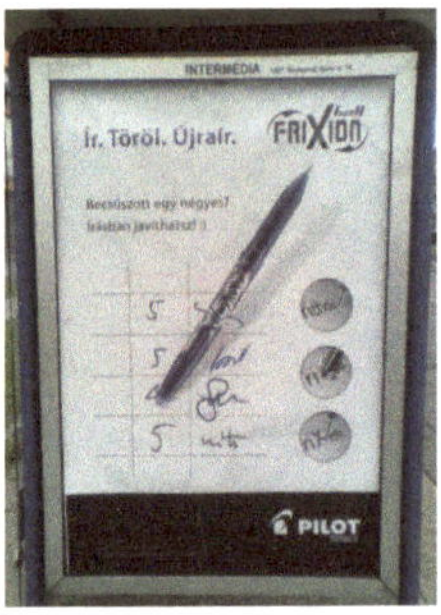

Image 62: Emoticon as sentence closing element on a street poster (Budapest, Hungary 2010)

The poster in *Image 61* targets school children (and indirectly their parents) by putting a smiley after the "Iskolakezdés 2010" ('Back to School 2010') label. The advertisement of an erasable pen in *Image 62* also adapted its language to the target audience by completing the "Becsúszott egy négyes? Írásban javíthatsz!" ('~Make up for your bad grades in writing.') slogan with an emoticon.

Image 63: Emoticon as slogan closing element on billboard and website (2010)

The slogan of szivesseg.net website also contains a smiley: *adni jó, és kapni sem rossz :)* ('it's good to give, and it's not bad to receive either :)'). They use this form both on their website and a billboard ad. In addition, the menu item "*tetszik? – segíthetsz*" ('Like it? You can help!') is also closed with a smiley (*Image 63*).

Image 64: Emoticon as part of a logo (1: Flyer of a political party, 2010) and as slogan closing element (2: Pop-up advert of a political party in a Facebook app, 2010)

Political parties also took advantage of the popularity of emoticons in their campaign: LMP used yellow smileys, Fidesz used ASCII character smileys in their advertisements (*Image 64*).

For a few days in May 2010 the newly formed government welcomed visitors on all government websites with the following message: *OMG* (*Image 65*).

Image 65: OMG (government websites, 2010)

Image 66: Emoticon as slogan closing element on billboard and website (2010)

Intersport displayed the message *Nyár :-)* ('Summer :-)') on both its website and billboards in the summer of 2010 (*Image 66*). Interestingly, the emoticon is used in its original, typewritten form which is rotated 90 degrees.

The *:-)* smiley is used in its typewritten form on the posters and food packaging of KFC, the fast food restaurant. What is more, the slogan included an internet-specific graphostylistic technique: the word *be* was written as *B*. In the language of the internet/text messages, capital letters are often used to replace full words on the basis of homonymy. "*be smart*" (*Image 67*).

Image 67: Emoticon as slogan closing element on packaging and street poster (Budapest, Hungary 2010)

A common character saving technique in texting is to type in lower case and reserve capitals to replace word parts or complete words. In this case capitals should be read as pronounced letters. This technique is used by T-Mobile to promote its SMS service: *SMS? Ngedd L magad!* (*Ngedd L magad* < *Engedd el magad* 'take it easy') (*Image 68*). The advertisement also illustrates how digilectic forms filter into the language of advertising even in printed materials. The message in the woman's hands reads "*Rámérsz?*" ('Do you have time for me?'). This expression is very common in text messages, such as the following on taken from my text message corpus: *Szia* [nickname]*! Mikor* ***érsz rám****? Találkozzunk a Móricz Zsigmond Otéren a 6-os vili +állójánál. 6 után ráérek.Puszi:**

Image 68: Text message abbreviation in a phone advertisement (2006)

However, the influence of digilect on advertisements is not limited to smileys. In its spring 2010 and 2011 campaigns the well-known beverage company promoted

Fanta in a TV commercial with the internet-specific acronym *LOL* 'laugh out loud' (*Image 69*).

Image 69: LOL in a TV commercial and on a billboard (Budapest, Hungary 2010 and 2011)

Most probably, the intention behind using the abbreviation *fogy-6* (< *fogyhat* 'you can lose weight') in a lamp post advertisement (*Image 70*) was to attract attention rather than to save characters.

Image 70: Letter replaced with number (Kiskunhalas, Hungary 2009)

Another example for digilect's influence on the language of advertising, but this time without the intention to draw attention or enhance promotion, appears in *Image 71* which uses the textese *2hetente* instead of the standard form *kéthetente* 'every two weeks'.

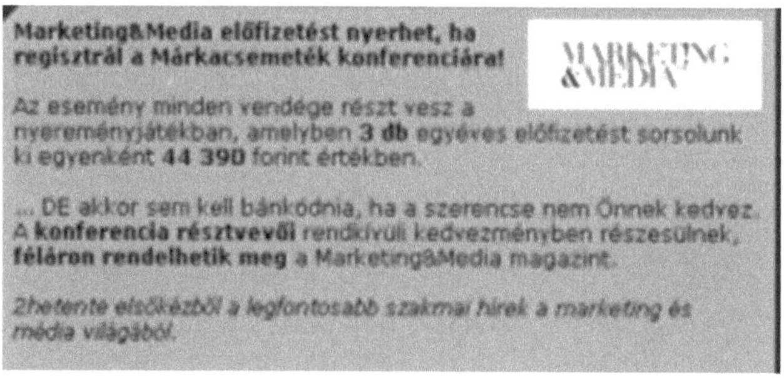

Image 71: Abbreviation with number: 2hetente – in an admail (2010)

By contrast, the slogan of the Advertising Week (*Image 72*) consciously applies the digilectic technique of replacing words with digits: *5let nélkül a reklám nem 6 – reklám7* (< *Ötlet nélkül a reklám nem hat – Reklámhét.* 'Advertisements are worth nothing without ideas – Advertising Week'). The opening page of the website features another instance of this technique: "*1 bulikonferencia a reklámról*" where the indefinite article *egy* 'a/an' is replaced with the digit *1* (also pronounced as *egy*).

Image 72: Abbreviation with number: "5let nélkül a reklám nem 6, Reklám7" – on a street poster and website (Budapest, Hungary 2010)

The indefinite article is replaced the same way in *Image 73* where the text appears as a text message on a mobile phone display (*Este 1 fröccs?* < Este egy fröccs 'A drink in the evening?').

Image 73: Abbreviation with number: "Este 1 fröccs?" – on a street poster (Budapest, Hungary 2010)

As a conclusion, the effects of digilect on the language of advertisements can be summarized as follows.. The most common elements which have made their way from the digital language variety to printed advertisements are emoticons, and to a certain extent, abbreviation techniques. This influence was notable as early as in 2004 (the year when I recorded the first example) and has only become stronger since then: more and more advertisements incorporate digilectic elements today. As my corpus of advertisements is small and not representative, it only allowed for the presentation of some examples but did not permit any general or statistically significant conclusion.

The examples, however, suggest that knowledge of the digital language variety (digilect) is increasingly important to understand certain advertisements.

6 Summary and conclusions. Directions for further research

6.1 Summary and conclusions

The book aimed to document the emergence and main features of a new language variety, the *digilect*, as well as to examine the interaction between digilectic texts and other text types. To this end, it introduced the term *digilect* following the theoretical discussion of the (linguistic) impact of devices capable of forwarding information, and the relationship between orality and literacy. In a broader sense, digilect is the characteristic, although heterogeneous language variety of computer-mediated communication.

Digilect was described from four perspectives which were separated from each other by "demarcation lines" (cf. Bańczerowski 2008: 77) exclusively for the purpose of the analysis. The four perspectives were the following: pragmatic-textological, lexical, grammatical and formal. All statements in the chapters were supported by authentic examples taken from the texts of digilect (cf. *Image 74*).

Texts analyses and surveys showed that the texts of digilect as a language variety (sociolect, mediolect or register) are in continuous interaction with other text types; they influence their vocabulary, pragmatic relationships and formal characteristics (and it is reasonable to assume that in the long run this influence will extend to grammatical characteristics as well).

As a conclusion, it may be established that internet (digital) literacy functions as a way to express group identity and also to exclude others from communication. The digital gap may not only refer to the lack of computer skills but also to not knowing the special language and signs of digital communication. Common background knowledge has a key role in every form of communication. But effective participation in internet communication absolutely requires knowledge of the components of digilect, including emoticons, special abbreviations and vocabulary.

6.2 The impact of the described digilect concept. Directions for further research

This book is the revised and updated text of my dissertation written in 2011 (and defended in 2012), which has been extended to include recent research findings. The main findings of the dissertation have been published in various studies, and the Hungarian manuscript of the book was also available for reference to other

DOI 10.1515/9783110499117-006

researchers. The following results indicate that the concept was used and further developed by others.

Image 74: The essence at a glance: characteristic features of digilect in a chain of comments on Facebook

The term "digilect" inspired several other authors (i. a. Fazakas 2015; Gonda 2015; Gyarmathy 2013; Márku 2015; Marton 2015; Petykó 2012a and 2012b; Porkoláb 2015; Simon 2014; Zimányi 2014). My research findings relating to the neologisms of digital communication (the so-called "netologisms") have been and are still used in lexicography and grammatical research focusing on the types of word formation (i. a.: Bódi 2011; Daróczi 2016; Sz. Tóth 2013; Veszelszki ed. 2012; Zachar 2013, 2014). The work also contributed to spoken language research, as well as to examinations dealing with discourse markers and grammaticalization (Dér 2013a, 2013b; Domonkosi 2012, 2013; Kugler 2015; Schirm 2010). The book may also be useful for finding out how hyperlinked texts are changing reading comprehension strategies (cf. Gonda 2015). Combining visuality, visual communication and visual learning with linguistics may also bring new and surprising results (cf. Veszelszki 2013b; Domokos–Vargha 2015; Hortoványi 2015; Glózer 2015), and may also influence other fields of study (such as folklore studies: Balázs 2007a, 2015; Domokos 2014; or marketing: Horváth 2014, 2015;

Horváth–Mitev 2016a and 2016b). The analysis of advertisements and commercials with linguistic methods may even help to build the foundations of a new discipline, the "marketolingusitics" (cf. Veszelszki 2014; Gyuricza 2016 who already use this term). Researchers of the so called "gamer speech", i. e. the language used when talking during or about video games, may also draw on the findings of this book (cf. Balogh 2012, 2014).

The examples described in the book may also form a starting point for research on the changes of spelling. It also seems useful to extend theoretical and practical text typology analyses to digilectic texts as well. The topics mentioned in the book but not discussed in detail due to thematic restrictions, are recommended for research from a semiotic, visual philosophical, and rhetorical perspective as well (cf. Series *Visual Learning*: Benedek–Nyíri eds. 2010–2015, Benedek–Veszelszki eds. 2016; Aczél 2012, 2013, 2014, 2015). A contrastive examination of digilect extended to multiple languages would help to determine the universal and language-specific features of digital communication. The findings of the book may also be relevant for speech technology and speech synthesis (e. g.: in reading aloud text messages and e-mails). Furthermore, the results may be utilized in education and education technology as well: presenting current trends of linguistic changes may serve both as syllabus and auxiliary material helping teacher-student communication (cf. digital natives and outsiders). When teaching language and communication in schools, classroom work may involve the examination of new text types and the comparison of different registers.

There is new potential in examining the pragmatic characteristics of online discourse as well. Digital pragmatic research fits into the latest wave of international internet language studies, and can be considered as a completely novel endeavour among studies dealing with online discourses in the Hungarian language; though only case studies have been published so far (e. g. on trolling by Petykó 2013; Veszelszki 2015d; and on online bereavement: Veszelszki–Parapatics 2014, 2016).

My future research plans include examination of the interferences between digilectic and literary texts (among others: the works of Hungarian poet Dániel Varró, the Dutch novelist Jessica Durlacher, the author under the pseudonym Jake Smiles, the Transsylvanian writer Noémi Jancsó, the Hungarian novelist-essayist Endre Kukorelly). From a linguistic anthropological and semiotic perspective, the language usage strategies of audiovisual communication technologies (Skype, webcam, Snapchat) and their long-term effects also provide a good area for research.

Annex

Annex 1

A day in the internet (2009) – What happens in an internet minute? (2016)

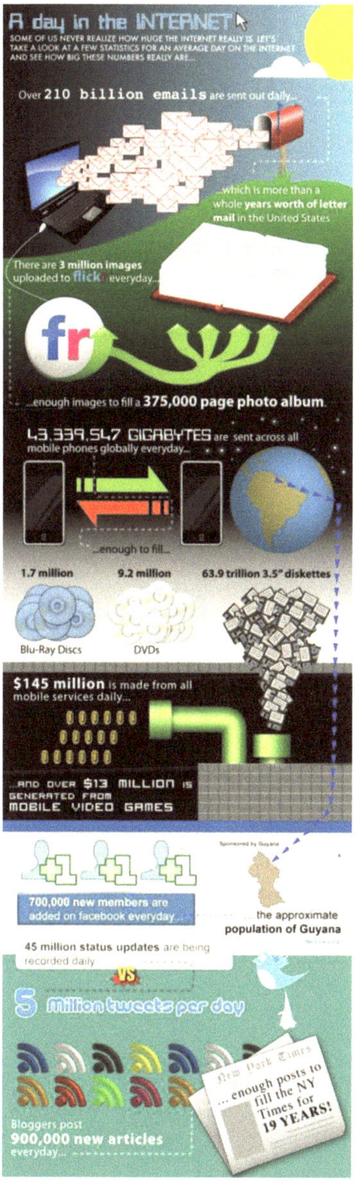

Source: http://www.onlineeducation.net/2009/12/08/a-day-in-the-internet/ [08. 12. 2009]

DOI 10.1515/9783110499117-007

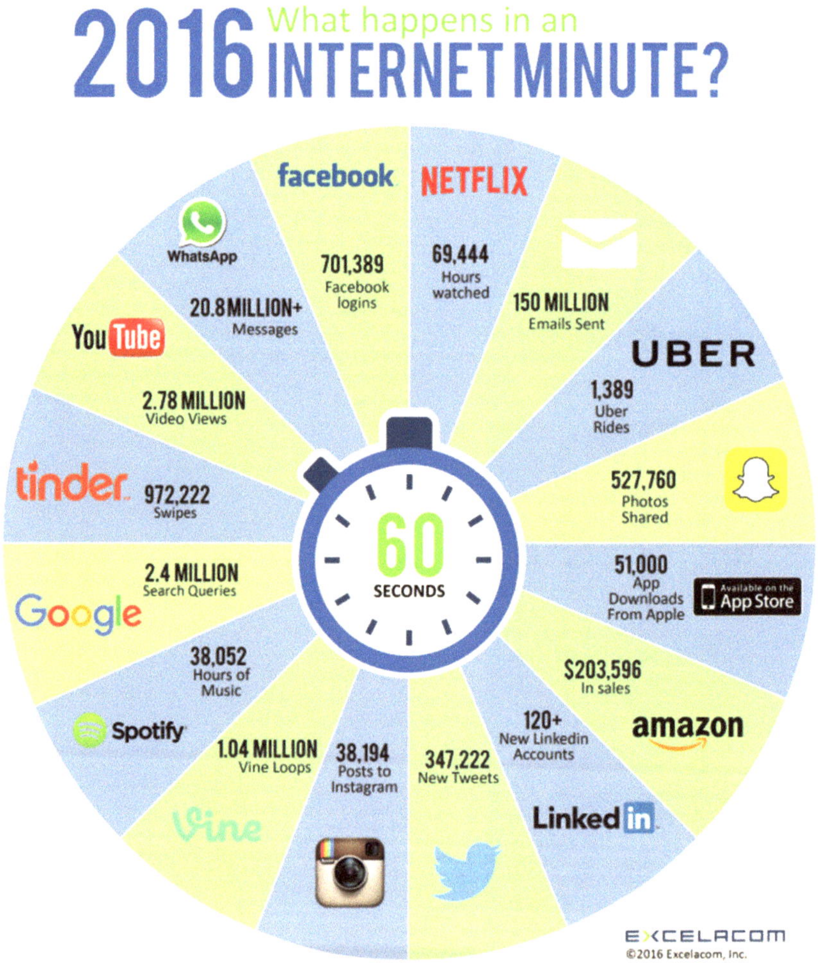

Source: http://www.visualcapitalist.com/what-happens-internet-minute-2016/ [22. 08. 2016]

Annex 2

Abbreviations: English abbreviations and their meaning

@	at	MOB	mobile
AFAIK	as far as I know	MSG	message
AFK	away from the keyboard	MYOB	mind your own business
ASL	age, sex, location	NE1	anyone
ATB	all the best	NOYB	none of your business
B	be	NO1	no one
BAK	back at the keyboard	OTOH	on the other hand
BBL	be back late(r)	OIC	oh I see
BCNU	be seeing you	PCM	please call me
BFN	bye for now	PLS	please
B4	before	PPL	people
BRB	be right back	R	are
BTW	by the way	ROTF(L)	rolling on the floor
C	see	RUOK	are you okay?
CUL8R	see you later	SIT	stay in touch
F2F	face to face	SOM1	someone
F2T	free to talk	SPK	speak
FWIW	for what it's worth	TTYL	talk to you later
FYI	for your information	TX / THX	thanks
GAL	get a life	U	you
GR8	great	WAN2	want to
HAND	have a nice day	W/	with
H8	hate	WKND	weekend
HSIK	how should I know?	WU	what's up?
HTH	hope this helps	X	kiss
IMHO	in my humble opinion	XOXOX	hugs and kisses
IMO	in my opinion	YMMV	your mileage may vary
IOW	in other words	YR	your
JIC	just in case	2	to, too
JK	just kidding	2DAY	today
KIT	keep in touch	2MORO	tomorrow
KWIM	know what I mean	2NITE	tonight
L8R	later	3SUM	threesome
LOL	lots of luck / laughing out loud	4	for

Source:
http://www.askoxford.com/betterwriting/emoticons/?view=uk [02. 01. 2006]

Further abbrevations:
Crystal 2008 (in English and eleven other languages)
Schlobinski 2009: 27–40, 66–79 (in English, German, French, Italian, Spanish, Portuguese, Swedish, Chinese and Japanese)

DOI 10.1515/9783110499117-008

Annex 3

Abbreviations: character-saving techniques in digilect.
Own collection based on the results of the 2010 questionnaire survey.

Type of abbreviation (1)	Type of abbreviation (2)	Abbreviation	Meaning	Occurrences[71]
initialism	replacing letters with numbers	g2g	got to go (English), mennem kell	1
initialism	replacing letters with numbers	j7en	jövő héten	1
initialism	replacing letters with numbers	j8	jó éjt	1
initialism	other	x	kiss (English), csók; per (olasz); ksz, sz, ssz helyett	5
initialism	omission of vowels	nM, nm	nem; nincs mit	29
initialism		azt m.	azt mondta	1
initialism		bb	bye-bye (English), viszlát	6
initialism		bby	baby, bébi	1
initialism		bbye	bye-bye (English), viszlát	1
initialism		Bp, bp	Budapest	3
initialism		D	Déli pályaudvar	1
initialism		etc.	et cetera, satöbbi	5
initialism		f5[72]	fél öt	1
initialism		f6	fél hat	1
initialism		f7	fél hét	1
initialism		fb	facebook	3
initialism		h	hogy; ha	326
initialism		h2	háromnegyed kettő	1
initialism		hf	házi feladat	3
initialism		hh	hogyhogy	8
initialism		ho	home office (English)	1
initialism		hszp	hali, szia, puszi	1
initialism		hv?	hogy vagy?	1
initialism		íj	így jártál	2

71 The respondents were prompted in an open question to list the abbreviations they regularly use in digital communication (e.g. in instant messaging or text messaging). The numbers in this column indicate how many times the given form was mentioned by the respondents.
72 All mentioned examples are listed here despite the obvious similarities and analogies with other examples.

DOI 10.1515/9783110499117-009

Type of abbreviation (1)	Type of abbreviation (2)	Abbreviation	Meaning	Occurrences[71]
initialism		IT	information technology (English)	1
initialism		JC	Jesus Christ (English)	1
initialism		K	kiss (English); Keleti pályaudvar	2
initialism		kb, kb.	körülbelül	18
initialism		KS	Kristen Stewart (filmszereplő)	1
initialism		lsz	légy szíves	1
initialism		m	mint, mert; miért	20
initialism		mh	merthogy	1
initialism		mm	mármint	7
initialism		mmofps	massively multiplayer online first person shooter (játéktípus)	1
initialism		n	nagyon; nem	3
initialism		n/a	not applicable, not available (English)	1
initialism		n3	negyed 3	1
initialism		n7	negyed 7	1
initialism		nc	no comment (English)	1
initialism		np	no problem (English)	5
initialism		nt	nem tudom	1
initialism		p	puszi	19
initialism		RP	Robert Pattinson (actor)	1
initialism		skh	semmi közöd hozzá	1
initialism		stb	satöbbi	37
initialism		sz!	szeretlek; szia	3
initialism		SzK!	szervusz Kedves	1
initialism		u.h.	úgyhogy	1
initialism		ua	ugyanaz	1
initialism		ue.	ugyanez	1
initialism		v	vagy; van; volt	33
initialism		v-v	vagy-vagy	1
initialism		v	vagy	2
initialism		WfH	working from home (English)	1
internet-specific acronym	substitution of a letter with another letter	CYA	see you (English)	3
internet-specific acronym	replacing letters with numbers	cul8r	see you later (English)	1
internet-specific acronym	initialism	br	best regards (English)	1

Type of abbreviation (1)	Type of abbreviation (2)	Abbreviation	Meaning	Occurrences[71]
internet-specific acronym	initialism	hsz	hányós szmájli	1
internet-specific acronym	initialism	mmorpg	Massively Multiplayer Online Role-Playing Game (English)	1
internet-specific acronym	initialism	mmorts	Massively Multiplayer Online Realtime Strategy Game (English)	1
internet-specific acronym	initialism	brb	be right back (English)	12
internet-specific acronym	initialism	btw	by the way (English); between (English)	8
internet-specific acronym	initialism	CTN	can't talk now (English)	1
internet-specific acronym	initialism	cu	see you soon (English)	4
internet-specific acronym	initialism	DND	do not disturb (English)	1
internet-specific acronym	initialism	frsz	fekve röhögős szmájli	1
internet-specific acronym	initialism	FYA	for your attention (English)	1
internet-specific acronym	initialism	FYI	for your information (English)	8
internet-specific acronym	initialism	FYKI	for your kind information (English)	2
internet-specific acronym	initialism	hdl	hab dich lieb (German), szeretlek	1
internet-specific acronym	initialism	IDK	I don't know (English)	1
internet-specific acronym	initialism	imho	in my humble opinion (English)	4
internet-specific acronym	initialism	imo	in my opinion (English)	1
internet-specific acronym	initialism	IRL	in real life (English)	1
internet-specific acronym	initialism	lmao	laughing my ass off (English)	6
internet-specific acronym	initialism	lmfao	laughing my fucking ass off (English, durva)	2
internet-specific acronym	initialism	lol	laughing out loud (English)	62
internet-specific acronym	initialism	omfg	oh my fucking god (English, durva)	5

Type of abbreviation (1)	Type of abbreviation (2)	Abbreviation	Meaning	Occurrences[71]
internet-specific acronym	initialism	OMFRG	oh my freaking God (English)	1
internet-specific acronym	initialism	omg, OMG	oh my god (English)	30
internet-specific acronym	initialism	rofl	rolling on the floor laughing (English)	11
internet-specific acronym	initialism	ROFLMAO	rolling on the floor laughing my ass off (English)	2
internet-specific acronym	initialism	rotfl	rolling on the floor laughing (English)	1
internet-specific acronym	initialism	SYL	see you later (English)	1
internet-specific acronym	initialism	szvsz	szerény véleményem szerint	9
internet-specific acronym	initialism	ttyl	talk to you later (English)	2
internet-specific acronym	initialism	wtf	what the fuck? (English)	16
internet-specific acronym	initialism	WTH	what the hell? (English)	1
internet-specific acronym		ACSA	akkor és csak akkor	1
internet-specific acronym		afaik	as far as I know (English)	3
internet-specific acronym		afk	away from keyboard (English)	3
internet-specific acronym		aka	also known as (English)	1
internet-specific acronym		AP	action point (English)	1
internet-specific acronym		asap	as soon as possible (English)	22
omission of vowels		hlnp	holnap	1
omission of vowels		ht	hát	1
omission of vowels		jó éjszkt	jó éjszakát	1
omission of vowels		jv	jövő	1
omission of vowels		kk	kay-kay (English), okay	2
omission of vowels		kv	kávé	1
omission of vowels		lht	lehet	1
omission of vowels		mjd	majd	1
omission of vowels		mrt	mert	1
omission of vowels		mst	most	1
omission of vowels		pls	please (English)	11
omission of vowels		plz	please (English)	3
omission of vowels		ppl	people (English)	2
omission of vowels		s	és	1

Type of abbreviation (1)	Type of abbreviation (2)	Abbreviation	Meaning	Occurrences[71]
omission of vowels		szrtlk	szeretlek	3
omission of vowels		szvl	szóval	1
omission of vowels		vlt	volt	1
omission of vowels	initialism	k	ok; késem	6
omission of vowels	initialism	L	el – away; love (English)	5
capitalising letters whose letter name is pronounced	replacing letters with numbers	Lm1ek	elmegyek	1
capitalising letters whose letter name is pronounced		fL	fel	1
capitalising letters whose letter name is pronounced		Lmegy	elmegy	1
capitalising letters whose letter name is pronounced		LmNtM	elmentem	1
capitalising letters whose letter name is pronounced		Mber	ember	1
capitalising letters whose letter name is pronounced		Ste	este	1
capitalising letters whose letter name is pronounced		veTM	vettem	1
omission of vowels	substitution of a letter with another letter	thnx	thanks (English)	1
substitution of a letter with another letter		pux	puszillak	2
substitution of a letter with another letter	replacing letters with numbers	U2	you too (English)	1
substitution of a letter with another letter	phonetic spelling	u	you (English)	5
substitution of a letter with another letter	phonetic spelling	u r	you are (English)	3
substitution of a letter with another letter	word formation	puxi	puszi	1
substitution of a letter with another letter		accem	azt hiszem	1
substitution of a letter with another letter		cia	szia	2
substitution of a letter with another letter		civecske	szívecske	1

Type of abbreviation (1)	Type of abbreviation (2)	Abbreviation	Meaning	Occurrences[71]
substitution of a letter with another letter		emléxik	emlékszik	1
substitution of a letter with another letter		ezexerint	ezek szerint	1
substitution of a letter with another letter		igyexem	igyekszem	1
substitution of a letter with another letter		jöhetx	jöhetsz	1
substitution of a letter with another letter		kex	keksz	1
substitution of a letter with another letter		lefeXem	lefekszem	1
substitution of a letter with another letter		lex	lesz	1
substitution of a letter with another letter		lexebb	legszebb	1
substitution of a letter with another letter		miJen	milyen	1
substitution of a letter with another letter		ojan	olyan	1
substitution of a letter with another letter		öxe	össze	1
substitution of a letter with another letter		ql	cool (English)	1
substitution of a letter with another letter		qtya	kutya	1
substitution of a letter with another letter		qu	quanto (Italian)	1
substitution of a letter with another letter		see u	see you (English)	4
substitution of a letter with another letter		soxeretettel	sok szeretettel	1
substitution of a letter with another letter		soxor	sokszor	3
substitution of a letter with another letter		sya	szia	1
substitution of a letter with another letter		xtem	szerintem	2
substitution of a letter with another letter		y?	why? (English)	3
substitution of a letter with another letter		yay	jaj	1
replacing letters with numbers	substitution of a letter with symbol	1x	egyszer	6

Type of abbreviation (1)	Type of abbreviation (2)	Abbreviation	Meaning	Occurrences[71]
replacing letters with numbers	substitution of a letter with symbol	2*	kétszer	1
replacing letters with numbers	substitution of a letter with symbol	2x	kétszer	4
replacing letters with numbers	substitution of a letter with symbol	6x	hatszor	1
replacing letters with numbers	phonetic spelling	2u	to you (English)	2
replacing letters with numbers	phonetic spelling	4u	for you (English)	4
replacing letters with numbers	phonetic spelling	4U2	for you too (English)	1
replacing letters with numbers	capitalising letters whose letter name is pronounced	7VG-n	hétvégén	1
replacing letters with numbers		1	egy	12
replacing letters with numbers		2	to, too, two (English)	2
replacing letters with numbers		4	four, for (English)	1
replacing letters with numbers		6	-hat; sei (Italian)	2
replacing letters with numbers		7	hét	7
replacing letters with numbers		10min.	tíz perc	1
replacing letters with numbers		10perc	tíz perc	1
replacing letters with numbers		1ben	egyben	3
replacing letters with numbers		1éb	other	1
replacing letters with numbers		1ébként	otherként	1
replacing letters with numbers		1edül	egyedül	2
replacing letters with numbers		1előre	egyelőre	2
replacing letters with numbers		1enként	egyenként	1
replacing letters with numbers		1etem	egyetem	1
replacing letters with numbers		1ik	egyik	4
replacing letters with numbers		1re megy	egyre megy	1
replacing letters with numbers		1szer	egyszer	5
replacing letters with numbers		1szerű	egyszerű	1
replacing letters with numbers		1ütt	együtt	5
replacing letters with numbers		2en	ketten	1
replacing letters with numbers		2es	kettes	1
replacing letters with numbers		2féle	kétféle	1
replacing letters with numbers		2ség	kétség	1
replacing letters with numbers		4ever	forever (English)	1
replacing letters with numbers		4kor	négykor	1

Type of abbreviation (1)	Type of abbreviation (2)	Abbreviation	Meaning	Occurrences[71]
replacing letters with numbers		5let	ötlet	6
replacing letters with numbers		6ásos	hatásos	2
replacing letters with numbers		6kor	hatkor	1
replacing letters with numbers		7en	héten	6
replacing letters with numbers		7fő	hétfő	6
replacing letters with numbers		7h	hét óra	1
replacing letters with numbers		7vége	hétvége	15
replacing letters with numbers		7végén	hétvégén	5
replacing letters with numbers		7Vge	hétvége	1
replacing letters with numbers		a 7 minden napján	a hét minden napján	1
replacing letters with numbers		e5vös	Eötvös	1
replacing letters with numbers		elhoz6om	elhozhatom	1
replacing letters with numbers		elm1ek	elmegyek	1
replacing letters with numbers		eltart6 egy darabig	eltarthat egy darabig	1
replacing letters with numbers		ezen a 7en	ezen a héten	1
replacing letters with numbers		felhív6lak	felhívhatlak	1
replacing letters with numbers		ír6sz	írhatsz	1
replacing letters with numbers		jó8	jó éjt	35
replacing letters with numbers		jó8vágyat	jó étvágyat	2
replacing letters with numbers		jövő 7	jövő hét	2
replacing letters with numbers		jövő7en	jövő héten	1
replacing letters with numbers		ki5lünk	kiötlünk	1
replacing letters with numbers		lát6	láthat	1
replacing letters with numbers		le7	lehet	1
replacing letters with numbers		m1ek	megyek	8
replacing letters with numbers		me2	me too (English)	1
replacing letters with numbers		még1szer	még egyszer	2
replacing letters with numbers		meg6ó	megható	1
replacing letters with numbers		mind1	mindegy	95
replacing letters with numbers		mind1ik	mindegyik	1
replacing letters with numbers		mind2	mindkét, mindkettő	39
replacing letters with numbers		mind3	mindhárom	2
replacing letters with numbers		mond6ó	mondható	1
replacing letters with numbers		találkoz6unk	találkozhatunk	1
substitution of a letter with symbol	replacing letters with numbers	+6ó	megható	1
substitution of a letter with symbol	replacing letters with numbers	+old6ó	megoldható	1
substitution of a letter with symbol	replacing letters with numbers	+6ározó	meghatározó	1
substitution of a letter with symbol	replacing letters with numbers	Ë8!	Jó éjt!	1

Type of abbreviation (1)	Type of abbreviation (2)	Abbreviation	Meaning	Occurrences[71]
substitution of a letter with symbol	capitalising letters whose letter name is pronounced	+kLlTNnM	meg kell tennem	1
substitution of a letter with symbol	capitalising letters whose letter name is pronounced	Ë7VGt!	Jó hétvégét!	1
substitution of a letter with symbol	clipping	+dum	megdumáltuk, megbeszéltük	1
substitution of a letter with symbol	clipping	/pill	per pillanat	1
substitution of a letter with symbol		---/--@	rózsa	1
substitution of a letter with symbol		+beszéljük	megbeszéljük	2
substitution of a letter with symbol		+főzöm	megfőzöm	1
substitution of a letter with symbol		+int	megint	1
substitution of a letter with symbol		+kaptam	megkaptam	1
substitution of a letter with symbol		+mondom	megmondom	1
substitution of a letter with symbol		+nézem	megnézem	3
substitution of a letter with symbol		+puszil	megpuszil	1
substitution of a letter with symbol		+talál	megtalál	1
substitution of a letter with symbol		+van	megvan	3
substitution of a letter with symbol		&	és	2
substitution of a letter with symbol		*@lek	szeretlek	1
substitution of a letter with symbol		/sze	persze	1
substitution of a letter with symbol		+	meg (igekötő, kötőszó)	21
substitution of a letter with symbol		+csinálom	megcsinálom	3
substitution of a letter with symbol		+fogom tenni	meg fogom tenni	1
substitution of a letter with symbol		+határozó	meghatározó	1
substitution of a letter with symbol		+veszem	megveszem	1
substitution of a letter with symbol		<3m	szívem	1
substitution of a letter with symbol		0v	ne válaszolj	1
substitution of a letter with symbol		el+yek	elmegyek	1
substitution of a letter with symbol		millióX	milliószor	1
clipping		pussz	puszillak	4
clipping	substitution of a letter with another letter	köcc	köszönöm	1
clipping	substitution of a letter with another letter	köcce	köszönöm	1
clipping	substitution of a letter with another letter	puff	puszillak	1

Type of abbreviation (1)	Type of abbreviation (2)	Abbreviation	Meaning	Occurrences[71]
clipping	phonetic spelling	téll	tényleg	1
clipping	initialism	cs	csak; csókol; csá; Counter Strike	9
clipping	initialism	hha	hogyha	2
clipping	word formation	ari	aranyos	3
clipping	word formation	bocsi	bocsánat	2
clipping	word formation	csini	csinos	1
clipping	word formation	csörizz	csörögj, telefonálj	1
clipping	word formation	figyi	figyelj	3
clipping	word formation	furi	furcsa	1
clipping	word formation	kivi	kíváncsi	4
clipping	word formation	köszcsí	köszönöm	1
clipping	word formation	köszi	köszönöm	4
clipping	word formation	mérgi	mérges	1
clipping	word formation	pihi	pihenés	1
clipping	word formation	puszat	puszi	1
clipping	word formation	puszcsí	puszi	1
clipping	word formation	pusszancs	puszillak	1
clipping	word formation	sajna	sajnos	1
clipping	word formation	szercsi	szeretlek	1
clipping	word formation	szeri	szeretlek	2
clipping	word formation	szivi	szívesen	2
clipping	word formation	szupi	szuper	2
clipping	word formation	tali	találkozó, találkozzunk, találkozni	11
clipping	word formation	talizni	találkozni	1
clipping	word formation	talizunk	találkozunk	1
clipping	word formation	talizzunk	találkozzunk	1
clipping	word formation	uncsizik	unatkozik	1
clipping	word formation	ügyi	ügyes	1
clipping	word formation	üzi	üzenet	2
clipping	word formation	vagi	vagány	1
clipping	contraction	kivi vok	kíváncsi vagyok	1
clipping		ak	akkor	1
clipping		akk	akkor	14
clipping		akko	akkor	1
clipping		ált	általában, általános	2
clipping		am	amúgy	15
clipping		ápr	április	1
clipping		asp	aspetti – wait (Italian)	1
clipping		azé	azért	1
clipping		azt	aztán	1
clipping		bekapcs	bekapcsol	1

Type of abbreviation (1)	Type of abbreviation (2)	Abbreviation	Meaning	Occurrences[71]
clipping		besz	beszélni, beszélünk, beszélek…	3
clipping		besz később	beszélünk később	1
clipping		besz még	beszélünk még	1
clipping		bocs	bocsánat	1
clipping		csin	csinálsz, csinálni, csinálok	2
clipping		csü	csütörtök	1
clipping		csüt	csütörtök	5
clipping		d	de	2
clipping		feb	február	1
clipping		figy	figyelj	5
clipping		figgy	figyelj	1
clipping		folyt. kov.	folytatás következik	1
clipping		grat	gratulálok	1
clipping		haszn	használ	1
clipping		hé	hétfő	1
clipping		holn	holnap	2
clipping		ill.	illetve	5
clipping		infó	információ	1
clipping		jan	január	1
clipping		jóétv	jó étvágyat	1
clipping		jon	jó napot	1
clipping		ke	kedd	1
clipping		képz.	képzeld	1
clipping		kira	király	2
clipping		kösz	köszönöm	1
clipping		maj	majd	1
clipping		márc	március	1
clipping		max	maximum	2
clipping		me	mert	1
clipping		memo	emlékeztető	1
clipping		mer	mert	4
clipping		micsi	mit csinálsz?	1
clipping		micsin	mit csinálsz?	1
clipping		micsina?	mit csinálsz?	1
clipping		mié	miért	1
clipping		miko	mikor	1
clipping		min	minimum; perc	2
clipping		nemtu	nem tudom	1
clipping		net	internet	3
clipping		pé	péntek	1
clipping		pént.	péntek	2
clipping		pill	pillanat, azonnal jövök	34

Type of abbreviation (1)	Type of abbreviation (2)	Abbreviation	Meaning	Occurrences[71]
clipping		pu	pusz	1
clipping		pusz	puszi	8
clipping		re	ismét itt vagyok, visszajöttem; válasz	11
clipping		rem	remélem	8
clipping		semmi kül	semmi különös	1
clipping		sik	sikerül, sikerült	1
clipping		sze	szerda	1
clipping		szer Téged	szeretlek	1
clipping		szere	szeretlek	2
clipping		sziv	szívesen	1
clipping		szo	szombat	3
clipping		szomb	szombat	1
clipping		tal	találkozó, találkozzunk, találkozni	9
clipping		tel, tel.	telefon	3
clipping		TW	Twilight	1
clipping		udv	üdvözlettel	1
clipping		úgyh	úgyhogy	4
clipping		üdv	üdvözlettel	7
clipping		vas.	vasárnap	3
contraction	substitution of a letter with another letter	wok	vagyok	1
contraction	replacing letters with numbers	m1	mindegy; megy	3
contraction	initialism	u.i.	ugyanis; utóirat	1
contraction	initialism	uh, úh	úgyhogy	8
contraction	initialism	ui	ugyanis	2
contraction	initialism	ún.	úgynevezett	2
contraction	initialism	uvé	utóvizsga	1
contraction	initialism	vh	valahogy	1
contraction		akr	akkor	1
contraction		aven	alapvetően	1
contraction		bmi	bármi	1
contraction		bmikor	bármikor	1
contraction		bnő	barátnő	4
contraction		cmq	comunque (Italian), mindenesetre	1
contraction		csk	csak	1
contraction		db	darab	2
contraction		de, de.	délelőtt	15
contraction		delőtt	délelőtt	1
contraction		du, du.	délután	18

Type of abbreviation (1)	Type of abbreviation (2)	Abbreviation	Meaning	Occurrences[71]
contraction		dután	délután	2
contraction		egs	egészség	1
contraction		fadta	feladta	1
contraction		ffi	férfi	1
contraction		ftó	Fehértó	1
contraction		gz	gratz, congratulations (English)	1
contraction		hnap	holnap	23
contraction		jök	jövök	1
contraction		jtm	Je t'aime (French), I love you	1
contraction		KePu	Keleti pályaudvar	1
contraction		Ktg; ktg.	költség	2
contraction		kvt	könyvtár	1
contraction		lszi	légy szíves	1
contraction		meek	megyek	1
contraction		mek	megyek	8
contraction		mjuk	mondjuk	1
contraction		mképp	mindenképp	1
contraction		mkettő	mindkettő	2
contraction		mki	mindenki	5
contraction		mmondom	megmondom	1
contraction		mta	mondta	2
contraction		mtam	mondtam	1
contraction		mva	múlva	1
contraction		nan	nagyon	1
contraction		naon	nagyon	3
contraction		nem tom	nem tudom	8
contraction		nemtod	nem tudod	1
contraction		nemtom	nem tudom	48
contraction		non	nagyon	1
contraction		ntom	nem tudom	2
contraction		pl	például	36
contraction		psze	persze	1
contraction		sry	sorry, bocs	1
contraction		sz.tem	szerintem	1
contraction		szal	szóval	18
contraction		szasz	szevasz	1
contraction		szgép	számítógép	2
contraction		sznap	szülinap	1
contraction		sznted	szerinted	1
contraction		szte	szerinte	1
contraction		szted	szerinted	6
contraction		sztem	szerintem	77

Type of abbreviation (1)	Type of abbreviation (2)	Abbreviation	Meaning	Occurrences[71]
contraction		sztm	szerintem	1
contraction		szül.nap	születésnap	1
contraction		szval	szóval	2
contraction		taljunk	találkozzunk	1
contraction		talkozunk	találkozunk	1
contraction		tgnap	tegnap	2
contraction		thx	thanks (English)	17
contraction		tkp	tulajdonképp, tulajdonképpen	5
contraction		tleg	tényleg	1
contraction		tma	téma	2
contraction		tnap	tegnap	12
contraction		tod	tudod	3
contraction		tok	tudok	6
contraction		tom	tudom	19
contraction		toom	tudom	1
contraction		tulképp	tulajdonképp, tulajdonképpen	5
contraction		tulképpen	tulajdonképp, tulajdonképpen	1
contraction		uannyi	ugyanannyi	1
contraction		uaz	ugyanaz	1
contraction		uott	ugyanott	1
contraction		utsó	utolsó	2
contraction		uúgy	ugyanúgy	1
contraction		valszeg	valószínűleg	29
contraction		valszín	valószínűleg	1
contraction		valszínűleg	valószínűleg	1
contraction		vaok	vagyok	2
contraction		vd	veled	1
contraction		vhogy	valahogy	8
contraction		vhol	valahol	13
contraction		vhova	valahova	2
contraction		vki	valaki	44
contraction		vkinek	valakinek	1
contraction		vkivel	valakivel	2
contraction		vlmi	valami	1
contraction		vmely	valamely	2
contraction		vmelyik	valamelyik	3
contraction		vmennyi	valamennyi	1
contraction		vmi	valami	46
contraction		vmien	valamilyen	1
contraction		vmikor	valamikor	18
contraction		vmilyen	valamilyen	3

Type of abbreviation (1)	Type of abbreviation (2)	Abbreviation	Meaning	Occurrences[71]
contraction		vmit	valamit	6
contraction		vmivel	valamivel	1
contraction		vna	volna	1
contraction		vnap	vasárnap	1
contraction		vok	vagyok	65
contraction		vsnap	vasárnap	1
contraction		vsz	valószínű, valószínűleg	1
contraction		vszeg	valószínűleg	3
contraction		vszínű	valószínű	1
contraction		vt	volt	5
contraction		vtam	voltam	4
contraction		vtok	vagytok	1
contraction		vtunk	voltunk	1
contraction		vunk	vagyunk	1
contraction		wlc	welcome (English)	1
phonetic spelling	clipping	jocc	jó éjszakát	1
phonetic spelling	contraction	aszem	azt hiszem	2
phonetic spelling	contraction	aszonta	azt mondta	1
phonetic spelling	contraction	asszed	azt hiszed	1
phonetic spelling	contraction	axem	azt hiszem	1
phonetic spelling	contraction	uccsó	utolsó	1
phonetic spelling		ajmáá	aj már, ejnye már	1
phonetic spelling		asszem	azt hiszem	124
phonetic spelling		b	be (English)	1
phonetic spelling		b/c	because (English)	1
phonetic spelling		c	see (English)	1
phonetic spelling		c u	see you (English)	1
phonetic spelling		c u soon	see you soon (English)	1
phonetic spelling		dógoz	dolgozom	1
phonetic spelling		dunno	don't know (English)	1
phonetic spelling		elmonta	elmondta	1
phonetic spelling		how r u	how are you? (English)	3
phonetic spelling		joccakát	jó éjszakát	3
phonetic spelling		kék	kell	1
phonetic spelling		kúl	cool (English)	1
phonetic spelling		lájk	kedvelem	3
phonetic spelling		láwcsi	szeretet	1
phonetic spelling		lécc	légy szíves	1
phonetic spelling		lécci	légy szíves	16
phonetic spelling		léccy	légy szíves	1
phonetic spelling		légyszi	légy szíves	2
phonetic spelling		luv ya	I love you (English)	1
phonetic spelling		luw	love (English)	1
phonetic spelling		má	már	1

Type of abbreviation (1)	Type of abbreviation (2)	Abbreviation	Meaning	Occurrences[71]
phonetic spelling		mé	miért	2
phonetic spelling		mék	megyek	1
phonetic spelling		mingyá	mindjárt	1
phonetic spelling		minnyá	mindjárt	2
phonetic spelling		mittomén	mit tudom én	1
phonetic spelling		mongyuk	mondjuk	2
phonetic spelling		monnyuk	mondjuk	3
phonetic spelling		monyuk	mondjuk	1
phonetic spelling		műxik	működik	1
phonetic spelling		nájsz	nice (English)	1
phonetic spelling		okés	rendben	1
phonetic spelling		péz	pénz	1
phonetic spelling		plíz	please (English)	1
phonetic spelling		teccik	tetszik	4
phonetic spelling		télleg	tényleg	9
phonetic spelling		töknyóc	tök nyolc, teljesen mindegy	1
phonetic spelling		ucca	utca	1
phonetic spelling		ződ	zöld	1
euphemistic abbreviation of swear words	initialism	ft	f*sz tudja	1
euphemistic abbreviation of swear words		****	(káromkodás helyett)	1
euphemistic abbreviation of swear words		....		1
euphemistic abbreviation of swear words		b...meg	b*szd meg	1
euphemistic abbreviation of swear words		bazzeg	b*szd meg	1
euphemistic abbreviation of swear words		bhäázz	(káromkodás)	1
euphemistic abbreviation of swear words		bmeg	b*szd meg	1
euphemistic abbreviation of swear words		k.jó	k*rva jó	1
euphemistic abbreviation of swear words		k/qrva	k*rva	1
euphemistic abbreviation of swear words		kva	k*rva	1
other	phonetic spelling	nimmit	nincs mit	1
other		!!!!!	(fontos)	1
other		(K)	kiss, csók	1
other		(L)	love, szeretlek	2

Type of abbreviation (1)	Type of abbreviation (2)	Abbreviation	Meaning	Occurrences[71]
other		?	nem értem a kérdést, fejtse ki bővebben (visszakérdezés); miért/hogy/mi/ki...	4
other		???	mi van?	2
other		^^^^	mountains	1
other		a-aa		1
other		amúgymeg	amúgy meg	1
other		azabaj	az a baj	1
other		csá	(köszönés)	1
other		csácsi	(köszönés)	1
other		cső	cső (köszönés)	2
other		énis	én is	1
other		eNyi	ennyi	1
other		estét	Jó estét!	1
other		ezaz	ez az	1
other		fél9ig	fél kilencig	1
other		hali	(köszönés)	2
other		haller	(köszönés)	1
other		hi	(köszönés)	8
other		hy	(köszönés)	3
other		ien	ilyen	1
other		jah	ja, igen	1
other		jo7++	jó éjt (programozós-zleng)	1
other		jólvan	jól van	1
other		mi a bre?	mi van?	1
other		mien	milyen	1
other		mivan	mi van?	1
other		mizu	mi a helyzet? mizujs? mi újság?	11
other		mizujs	mi a helyzet? mizujs? mi újság?	3
other		nagyonjó	nagyon jó	1
other		'napot	jó napot	2
other		nemáá	ne már	1
other		nemakarom	nem akarom	1
other		nembaj	nem baj	1
other		nemfog	nem fog	1
other		nemtudom	nem tudom	2
other		ok	rendben	21
other		okok	oké, oké	1
other		oks	rendben	1
other		oksi	rendben	1

Type of abbreviation (1)	Type of abbreviation (2)	Abbreviation	Meaning	Occurrences[71]
other		ottvan	ott van	1
other		pá	(köszönés)	1
other		pff	(nemtetszés)	2
other		ptz	puszi /cupp	1
other		'reggelt,	jó reggelt	1
other		sorry	bocsánat	1
other		sos	segítség	2
other		teis	te is	1
other		tökjó	tök jó	1
other		X&O	csókol és ölel	1
other		xd, XD	nevet	4
other		xké	perché (olasz)	1
other		xoxo	csókok	1
other		:re	visszatértem	1
Undefined[73]		asd		1
undefined		BC		1
undefined		dp		1
undefined		eg.		1
undefined		EOB		1
undefined		g9		1
undefined		gf		1
undefined		gg		1
undefined		gl		2
undefined		go		1
undefined		hvg		1
undefined		ID		1
undefined		j		1
undefined		jk		1
undefined		khm		1
undefined		LBTD		1
undefined		lkj.		1
undefined		loál		1
undefined		mhm		1
undefined		mi		1
undefined		omw		1
undefined		pada		1
undefined		PM		1
undefined		slt		1
undefined		so		1

73 The respondent failed to define the meaning and usage of the form, and no clear explanations were found in the internet.

Type of abbreviation (1)	Type of abbreviation (2)	Abbreviation	Meaning	Occurrences[71]
undefined		szi		1
undefined		szt		1
undefined		ti.		1
undefined		vs		1
undefined		w		1
undefined		w/		2
undefined		w00t		1
undefined		wow		2
undefined		xxx		1

Annex 4

Using emoticons on the computer

Iwiw profile page (2009)

Nem	Nő
Családi állapot	egyedülálló
Életkor	18
Magamról	hiiiii(v.v) nagyon unatkozhatsz, ha ezt elolvasod "magamról": túl érintett vagyok ebben a kérdésben, talán egy kívülállót kérdezzzzz (^^,). Amúgy meg: Tavaly csak bohó voltam, idén már balga vagyok. seggigérő hosszú fekete egyenes hajat szeretnék (^^,); nem akarok megöregedni ; hosszú szempillákat szeretnék; kókuszillatú bőrt szeretnék xD; most ebben a pillanatban egy csokis shake igazán jól esne (^^,); imádom a csillagokat nézni , de a legjobban azt szeretem, ha látom is őket (ami a fényszennyezés miatt néhol lehetetlen..,:'(); (L) sééééta jól esik bármikor ; imáááfok túlelemezgetni-analizálgatni {u.u}; sütnifőznienni ; nincs névnapom, sosem tartottam...: ha vki egyszer tart nekem hozzámegyek xD; tudtad, hogy szeretek utazni???^^,, nahm mostmár ezt is tudod =); továbbá önmagam szórakoztatása céljából hajszeszt és mindenféle kenceficét pakolok a hajamra, hogy ezzel is támogassam a kenceficepart. OH. és talán az egyik legfontosabb dolog, amit tudni illik rólam, hogy imádom a fagyit bármikor,bárhol,bárkivel-kategóriás étel . és i love szőlő . Jahm és 37-es a lábam, de néha 38-as. Tágulékony a drága . Az éreccséégi káros az egészségre, ezt egyéni tapasztalatokból állapítottam meg, nem hivatalos felmérés alapján. Ennyit "magamról". Veled mű[illegible]ság??? {c.c} **
Anyanyelv	kínai
Beszélt nyelvek	kínai, magyar, angol, francia
Születésnap	1990. aug. 31.
Állatok	nincs fűtestvérem, akit állatnak nevezhetnék. úgyhogy nincs :D
Ezt csinálom, mikor dolgozom	megpróbálok nem megbukni & tolmácsolok
Ezt csinálom, mikor nem dolgozom	biztos nem azt, amit kéne (a)

DOI 10.1515/9783110499117-010

Facebook profile page (2010)

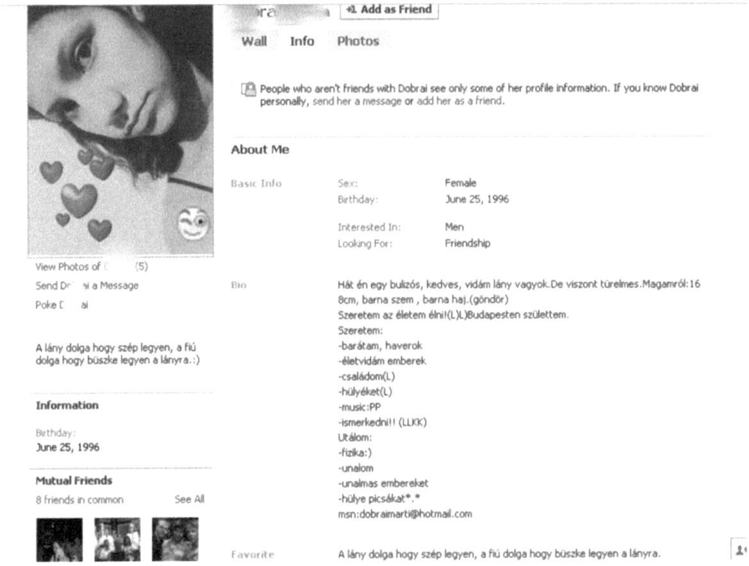

Facebook comments (2010)

Facebook comments (2010): *Rajság van*

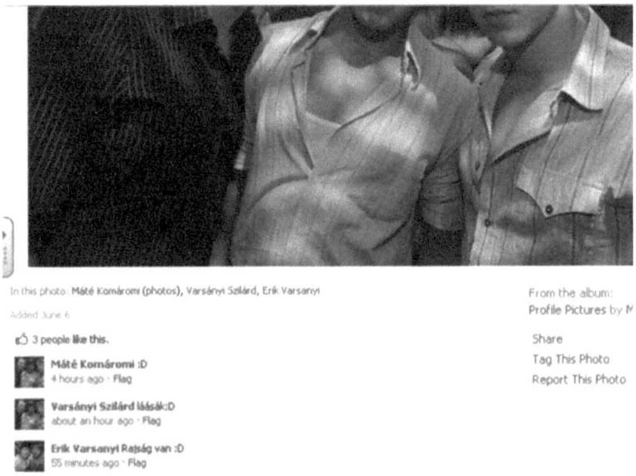

Connecting visual and verbal communucation

Facebook profile page (2010)

Bio

,,Figyelj rám, és egy vidám élettel hálálom meg!"
Hogy is jellemezném magam.?..egy jó szándékú dumagép,ez a megfelelő :P εïз Jó néha tényleg túlzáásba viszem,na...bocsi!deháát igy szerettek!(L)Igényem van a társaságra, a közönségre!Színt viszek az emberek életében :D szeretem az élénk árnyalatokat,a kedélyes hangulatot sugárzó elemeket,Ez asszem a szobámon is +látszik :$ Általában véve változatosabban öltözködöm,mint az átlag,nekem ez tetszik (: ..szeretekEzt-azt csinálni,de valamit mindig muszály csinálnom,az élénk és mozgékony talán enyhe kifejezés rám:$ bárkivel könnyen teremtek kapcsolatot. Részben azért,mert bármikor szívesen kezdeményezek beszélgetést,mert kiváncsiságom folytán minden érdekel :$ Nagy szemekkel figyelem,akármiről mesél bárki,hogy aztán hozzátehessem,amit én hallottam minderől.Na jó tegyük hozzá: azért bármi nem érdekel. Száraz tényekkel,adatokkal nem lehet megfogni a fantáziám.Inkább az emberek érdekelnek,események,sztorik és pletykák. A legjobb pedig,ha mindezt minél színesebb,viccesebb körítéssel adják elő nekem :P „Az élet nagy kaland,átírható színdarab.És hidd el, csak rajtad áll,hogy jó vagy rossz, mi Rád vár..."..horoszkópom alapján: "a Rákokat elég nehéz megérteni, mert érzelmeik hullámzásával, a viselkedésük, gondolkodásmódjuk is hullámzik.Ők a tipikus hangulatemberek, akik egyik percben még a "szakadék szélén" érzik magukat, másik pillanatban, meg már a világot szeretnék megváltani." VigyáZz velem mert lesem tagadhatom ezeket szeretem a LiLát & ♥мιcђaєł Jacksoŋ
Nem vagyok tökéletes lány.A hajam sem áll mindig jól és gyakran elbukom.Ügyetlen vagyok,és néha összetörik a szívem.Van hogy veszekszem a barátaimmal és vannak napok,amikor semmi sem sikerül.De ha jobban belegondolok és egyet hátra lépek akkor úgy érzem,hogy ez különleges élet a valóság és lehet,de csak lehet,hogy szeretek tökéletlen lenni.

Favorite Quotations

Hinnünk kell, hogy tehetségesek vagyunk valamiben és ezt a valamit bármi áron is,el kell érnünk..!

„Figyelj rám, és egy vidám élettel hálálom meg!"
Hogy is jellemezném magam.?..egy jó szándékú dumagép,ez a megfelelő:P εïз Jó néha tényleg túlzáásba viszem,na...bocsi!deháát igy szerettek!(L)Igényem van a társaságra, a közönségre!Színt viszek az emberek életében :D szeretem az élénk árnyalatokat,a kedélyes hangulatot sugárzó elemeket,Ez asszem a szobámon is +látszik:$ Általában véve változatosabban öltözködöm,mint az átlag,nekem ez tetszik (: ..szeretekEzt-azt csinálni,de valamit mindig muszály csinálnom,az élénk és mozgékony talán enyhe kifejezés rám :$ bárkivel könnyen teremtek kapcsolatot. Részben azért,mert bármikor szívesen kezdeményezek beszélgetést,mert kíváncsiságom folytán minden érdekel:$ Nagy szemekkel figyelem,akármiről mesél bárki,hogy aztán hozzátehessem,amit én hallottam minderől.Na jó tegyük hozzá: azért bármi nem érdekel. Száraz tényekkel,adatokkal nem lehet megfogni a fantáziám.Inkább az emberek érdekelnek,események,sztorik és pletykák. A legjobb pedig,ha mindezt minél színesebb,viccesebb körítéssel adják elő nekem :P „Az élet nagy kaland,átírható színdarab.És hidd el, csak rajtad áll,hogy jó vagy rossz, mi Rád vár..."..horoszkópom alapján: "a Rákokat elég nehéz megérteni, mert érzelmeik hullámzásával, a viselkedésük, gondolkodásmódjuk is hullámzik.Ők a tipikus hangulatemberek, akik egyik percben még a "szakadék szélén" érzik magukat, másik pillanatban, meg már a világot szeretnék megváltani." VigyáZz

velem mert lesem tagadhatom ezeket szeretem a LiLát & ♥мιcђaɛl Jacкsoɳ ıll.ıl ıı.ıll.ılıı.ıll.ılıı.ıll.ılıı.ıll.ılıll.ılıı.ıll.ılıı.ıll.ılıı.ıll.ılıı.ıll.ılıll.ılıı.ıll.ılıı.ıll.ılıı.ıll.ılıı.ıll.ıl Nem vagyok tökéletes lány.A hajam sem áll mindig jól és gyakran elbukom.Ügyetlen vagyok,és néha összetörik a szívem.Van hogy veszekszem a barátaimmal és vannak napok,amikor semmi sem sikerül.De ha jobban belegondolok és egyet hátra lépek akkor úgy érzem,hogy ez különleges élet a valóság és lehet,de csak lehet,hogy szeretek tökéletlen lenni.

Annex 5

Emoticons in the survey (2010)

Emoticon	Meaning	Occurrences
%)		
0	ölelés 'hug'	
0:-)	mosolygós (szent) fej 'smiling holy head'	
(*)		
(*_*)		
('.')		
(:		5
(:)		
(\('*'/)		
(|)		
(=	bal kezes szmájli 'left-handed smiley'	3
(>_<)		
(6)		
(8		
(A)		3
(finger)		
(flag:hu)		
(flag:XX)		
(flower)		
(h)		
(H)		2
(h5)		
(hug)	ha a másik felet vigasztalni kell, ölelés 'hug'	2
(K)	elköszönés 'saying good-bye'	10
(l)		
(L)	szeretet kifejezésekor, barátnőtől elköszönés, szív 'love, saying good-bye to a good friend, heart'	21
(mooning)		
(N)		
(o_O)		
(rofl)		2
(S)		
(swear)		
(toivo)		
(Y)		2
):		2
*-)		
**		
-		3
,		2
.		

DOI 10.1515/9783110499117-011

Emoticon	Meaning	Occurrences
_		
o		
prrrr-prrrr-prrr		
u		
,!,,		
-.-'		3
-.-	nem érdekes, frusztráltság, unottság 'not interesting, frustrated, bored'	17
-.-"		7
•__•		
/bow		
/hug		
/kiss		
/o\		
/shy		
: /		
: \		
: D		
:#		
:-#		2
:$	pirulás, zavarban van, szégyenlős 'embarrassment'	27
:"(		
:'-(		
:'(	sírás, komolytalan hiszti 'crying, not so serious problem'	43
:-(	szomorú 'sad'	40
:(	szomorúság, rossz hír 'sadness, bad news'	256
:((		6
:-((		
:-(((		
:(((	sajnálom 'I am sorry'	2
:((((		2
:"""")		
:')		3
:)	ha a másik fél valami kedveset mond; mosoly, nem bántásból írta, olvasta az üzenetet, jókedv, vicces, örülök, kedves, vidám 'happy, nice, good mood, I am glad, funny'	339
:-)	vidám 'happy'	69
:))		13
:-))		3
:)))		5
:-))))		3
:))))		4
:)))))))))		
:-*	csók, csókot dob 'kisses'	6
:*	puszi 'kiss'	16

Emoticon	Meaning	Occurrences
:,(		
:,-)		
:,)		
:-/	el van kenődve, savanyú, szájhúzás 'skeptical, perplexed'	14
:/	eltűnődés 'thinking'	21
::::)	vízipók 'water spider'	
:@	mérges 'angry'	41
:-\		
:\		
:]	vigyor 'smile'	6
:|		8
:-|	rezignált beletörődés 'resignation'	5
:}		
:<		
:-<*		
:>		
:×		
:-¬)		
:0		
:-1		
:ASD		
:b		
:-b		
:d		
:Ð		
:D	vicces, nevet 'funny, laugh'	235
:'D		2
:-D	vicces, vigyor, poénos szituáció, kacag 'laugh, funny'	28
-:D		
:D:D		
:D:D:D		2
:-DD		
:DD		3
:DDD		
:DDDD		
:DDDDDD		
:E		
:É	oldalra hajtott fejű mosolygós 'smiling'	
:G		
:H	a nagy, imádattól csillogó szemű szmájli 'spark-eyed from love'	2
:l		3
:-l		2
:-o		6
:-O		8

Emoticon	Meaning	Occurrences
:o	csodálkozás, meglepetés 'surprise, wondering'	8
:O	csodálkozás, meglepődés, nagy szopás, megdöbbent 'surprise, wondering, sucking'	57
:o)		3
:O)		
:p		7
:P	poénkodás, nyelvkiöltés, nincs mit hozzászólni 'no comment, funny, tongue'	113
:-p		2
:-P	nyelvkiöltés, gusztustalanság 'tongue, disgusting'	12
:PP		
:-s		
:-S	kényelmetlen szituáció 'embarrassment'	5
:S	furcsa, rossz dologról van szó, jaj, nem jó, rosszullét, zavarba hoz 'strange situation, not good, embarrassment, discomfort'	95
:s		2
:T		
:X		
:-X		2
:x	puszi 'kiss'	2
;(		11
;-(	sír 'crying'	
;)	kacsintás 'wink'	97
;-)		20
;-))		2
;[		
;\		
;]		
;<		
;>		
;-D		
;D		4
;P		6
;-P	huncut nyelvkiöltés 'mischievous'	
@.@		
\\o o//		
\^o^/		
\m/		
\o o/		
\o/		
^		
^.^		
^^'		

<table>
<tr><th>Emoticon</th><th>Meaning</th><th>Occurrences</th></tr>
<tr><td>^^</td><td>vidám szemek, gúny, szarkazmus ‘happy eyes, irony, sarcasm’</td><td>26</td></tr>
<tr><td>^-^</td><td></td><td></td></tr>
<tr><td>‘^^</td><td></td><td></td></tr>
<tr><td>^^"</td><td></td><td>2</td></tr>
<tr><td>^-^"</td><td></td><td></td></tr>
<tr><td>^_^</td><td></td><td></td></tr>
<tr><td>^___^</td><td></td><td></td></tr>
<tr><td>^o)</td><td>nem kellemes meglepetés ‘awkward surprise’</td><td></td></tr>
<tr><td>-_-.</td><td></td><td></td></tr>
<tr><td>_@_'/'</td><td>csiga ‘snail’</td><td></td></tr>
<tr><td>-__-</td><td></td><td></td></tr>
<tr><td>-{@</td><td>rózsa ‘rose’</td><td></td></tr>
<tr><td>|:</td><td></td><td></td></tr>
<tr><td>~~~</td><td></td><td></td></tr>
<tr><td>˘˘</td><td></td><td></td></tr>
<tr><td>‹3</td><td></td><td></td></tr>
<tr><td>++</td><td></td><td></td></tr>
<tr><td><3</td><td>szív ‘heart’</td><td>26</td></tr>
<tr><td><8-)</td><td></td><td></td></tr>
<tr><td>=(</td><td></td><td>2</td></tr>
<tr><td>=)</td><td></td><td>17</td></tr>
<tr><td>=))</td><td></td><td>2</td></tr>
<tr><td>=)))</td><td></td><td></td></tr>
<tr><td>=^^=</td><td></td><td></td></tr>
<tr><td>=^o^=</td><td></td><td></td></tr>
<tr><td>=D</td><td></td><td>4</td></tr>
<tr><td>=O</td><td></td><td></td></tr>
<tr><td>=P</td><td>pimasz, ironikus megjegyzés után ‘irony’</td><td>2</td></tr>
<tr><td>--> <--</td><td></td><td></td></tr>
<tr><td>>.<</td><td></td><td>2</td></tr>
<tr><td>>:(</td><td></td><td>2</td></tr>
<tr><td>>:|</td><td></td><td></td></tr>
<tr><td>>:-|</td><td></td><td>2</td></tr>
<tr><td>>:-<</td><td></td><td></td></tr>
<tr><td>><</td><td></td><td></td></tr>
<tr><td>°_O</td><td></td><td></td></tr>
<tr><td>00</td><td></td><td></td></tr>
<tr><td>8-)</td><td></td><td>2</td></tr>
<tr><td>8)</td><td></td><td>5</td></tr>
<tr><td>8))</td><td></td><td></td></tr>
<tr><td>8/</td><td></td><td></td></tr>
<tr><td>8D</td><td></td><td>2</td></tr>
<tr><td>8Đ</td><td></td><td></td></tr>
<tr><td>8-O</td><td>kigúvadt szemmel csodálkozó ‘surprise’</td><td></td></tr>
</table>

Emoticon	Meaning	Occurrences
ASD		
B-)		
B)		
C:		
D)		
D:		
d:		
D:)		
hihi		
ikszdé		2
LoL		
lol		3
LOL		4
n__n		
O.O		
o.O'		
O.o	meglepődés 'surprise'	4
o.O	csodálkozó szemek 'wondering, surprise'	8
O.o"		
O:		
o^^o		
O_o		
o_O		2
O_O		
o_o		
o>-<|:		
OMG		2
Oo		
oO		2
-o-o-	bikinifelső 'bikini top'	
OwO		
q:-)		
T.T		2
T__T		2
T_T		
ToT		
WTF		
x	csók 'kiss'	
X!	csók 'kiss'	
X"D		2
X"DDDD		
X')		
X)		2
x)		5
x_x		

Emoticon	Meaning	Occurrences
X'D		
xd		5
xD		40
XD	tetszik, kinevet 'like, laugh out'	60
XDD		
xo		
xoxo		2
xP		3
XP	fárasztó humor után 'tiring humour'	
animált, beépített		38
saját hangulatjelek		2

Annex 6

Questionnaires for elementary school students (examples)

A számítógépet azért szidják, mert Sokan függők lesznek legalábbis azok akik nagyon sokat gépeznek

Az e-mail abban különbözik a csevegéstől, hogy nem rögtön válaszol rá az illető hanem mondjuk 1-2 nap múlva

Az sms abban különbözik a csevegéstől, hogy telefonon lehet küldeni a másikat meg számítógépen.

Az sms abban különbözik a csevegéstől, hogy romantikusabb ;)

Erről a témáról ez jutott még eszembe: (Isten áldja a webkamerát!)

Ha nem lenne mobilom, meghalnék ∇ xD

Amióta mobilom van, könnyebb kommunikálni ∇

A számítógép hasznos, mert msn-ezni és, MyVipet lehet használni ∇ !! (és vidéki emberekkel)

A számítógépet azért szidják, mert nem értenek hozzá és irigykednek ∇

Az e-mail abban különbözik a csevegéstől, hogy sokkal lassabb és unalmas ∇ :D

Az sms abban különbözik a csevegéstől, hogy pénzbe kerül ∇ :P

Erről a témáról ez jutott még eszembe: szeretek msn-ezni, de nem viszem túlzásba, van másra is használható a számítógép ∇.

Erről a témáról ez jutott még eszembe: most inkább a számítógép előtt ülnék, és myvipen nézegetném a kedvenc képeim róla :) (L) + msn (:)) !!

Ha nem lenne mobilom, nem lenne teljes az életem xD

Ha nem lenne mobilom, hát olyan nem lenne xD

Amióta mobilom van, hamarabb el tudom érni a barátaimat :)

A számítógép hasznos, mert a távoli barátaimmal is tudom tartani a kapcsolatot :)

A számítógépet azért szidják, mert sokat ülök előtte xD

Az e-mail abban különbözik a csevegéstől, hogy rosszabb és lassabb és nem online :/

Az sms abban különbözik a csevegéstől, hogy nem hallom a partner hangját

Erről a témáról ez jutott még eszembe: szeretek msnen olyanokkal camozni és mikrofonozni, akik messze vannak tőlem :)

Ha nem lenne mobilom, nem tudnék bárhonnan üzenni másoknak

Amióta mobilom van, boldogabb vok (vagyok xD)

A számítógép hasznos, mert szinte bármit lehet rajta csinálni

A számítógépet azért szidják, mert gonoszak a felnőttek xD

Az e-mail abban különbözik a csevegéstől, hogy nem ~~tol~~ látjuk, mit írtunk előbb

Az sms abban különbözik a csevegéstől, hogy drága

Erről a témáról ez jutott még eszembe: semmi! xD

Erről a témáról ez jutott még eszembe: XD LOL <3 szeretem a smiley-kat! :P

DOI 10.1515/9783110499117-012

Annex 7

Excerpts from test papers written in class (secondary and higher education) and at a secondary school final examination (from 14–20 year old students; examples)

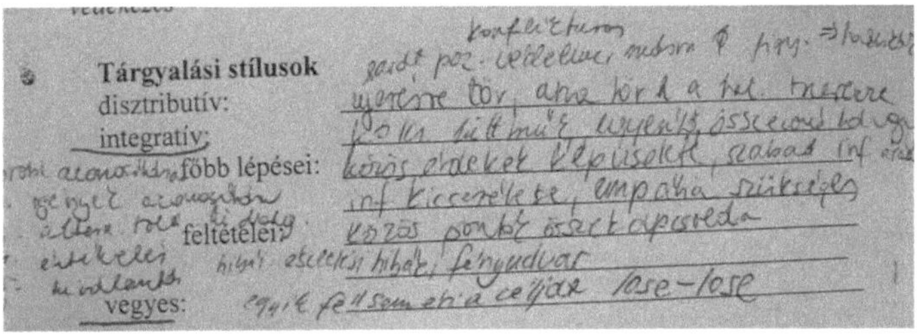

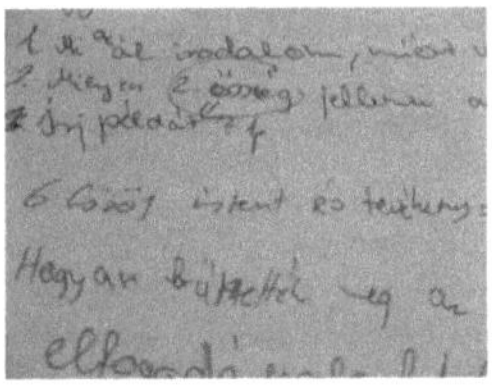

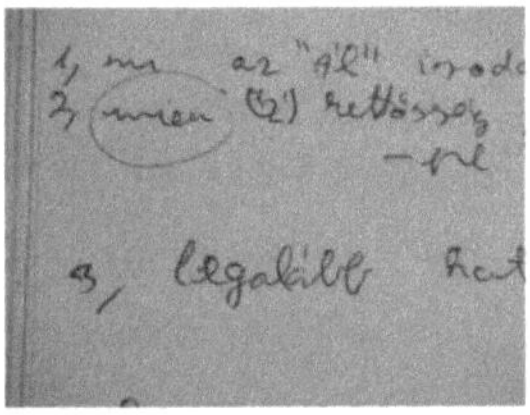

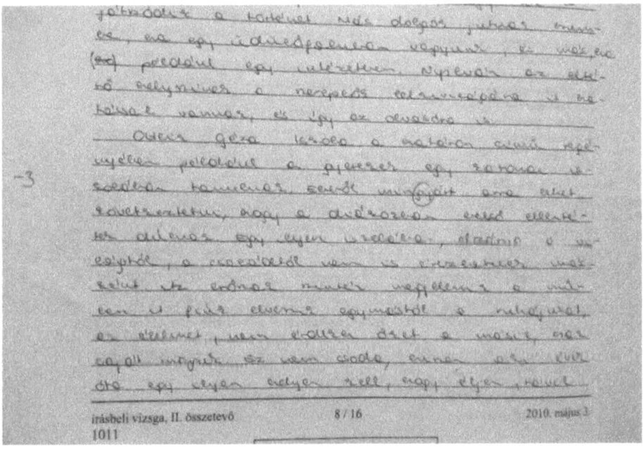

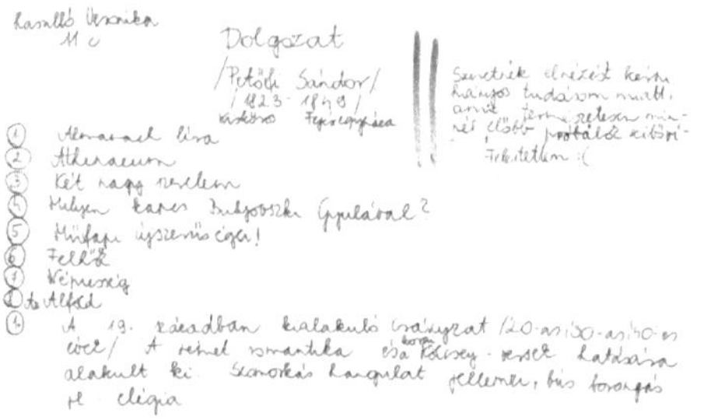

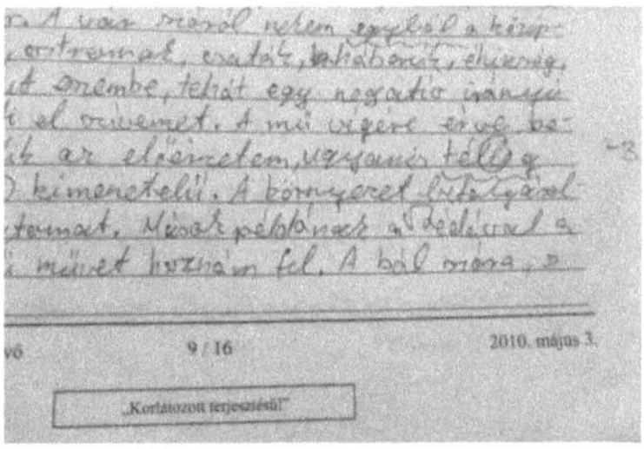

DOI 10.1515/9783110499117-013

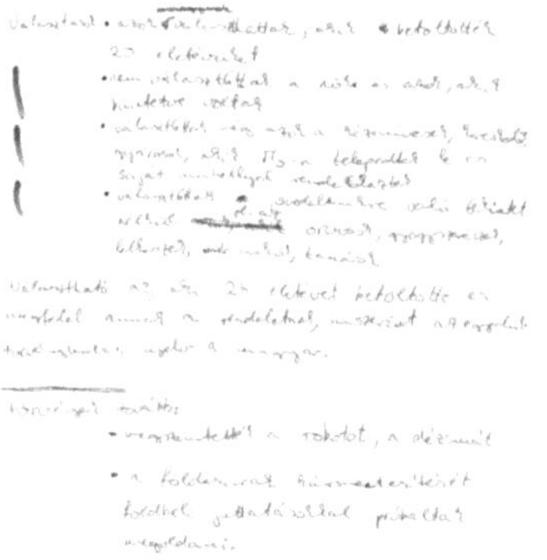

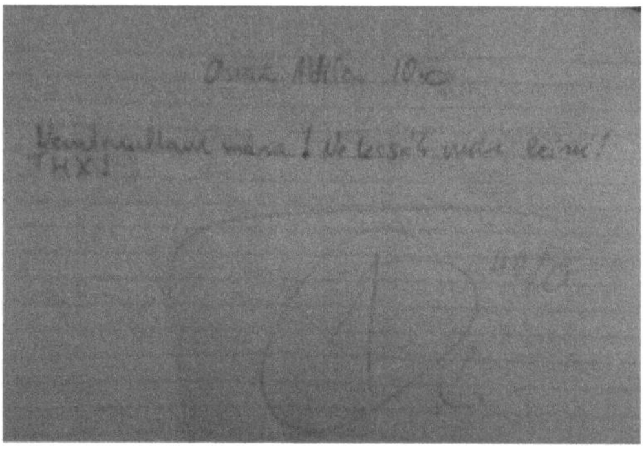

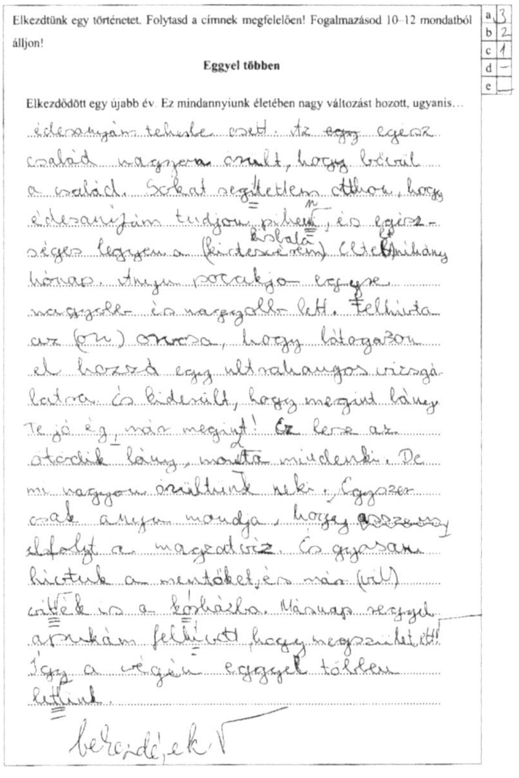

8. évfolyam — AMNy1 feladatlap / 8

10. Elkezdtünk egy történetet. Folytasd a címnek megfelelően! Fogalmazásod 10–12 mondatból álljon!

Eggyel többen

Elkezdődött egy újabb év. Ez mindannyiunk életében nagy változást hozott, ugyanis… édesanyám terhes lett. Ez egy egész család nagyon örült, hogy bővül a család. Sokat segítettem otthon, hogy édesanyám tudjon pihenni, és egészséges legyen a (kistestvérem) kisbaba. Eltelt néhány hónap. Anyu pocakja egyre nagyobb és nagyobb lett. Felhívta az (őt) orvosa, hogy látogasson el hozzá egy ultrahangos vizsgálatra. És kiderült, hogy megint lány. Te jó ég, már megint! Ez lesz az ötödik lány, mondta mindenki. De mi nagyon örültünk neki. Egyszer csak anyu mondja, hogy elfolyt a magzatvíz. És gyorsan hívtuk a mentőket, és már (vitt) vitték is a kórházba. Másnap reggel apukám felhívott hogy megszületett! Így a végén eggyel többen lettünk.

bekezdések √

a	3
b	2
c	1
d	–
e	–

2010. január 23.

Annex 8

Classroom notes taken by secondary school and university students aged 14 to 20 (examples)

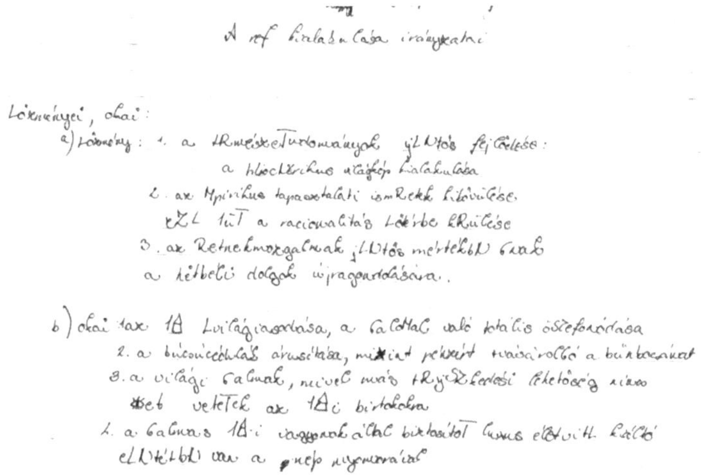

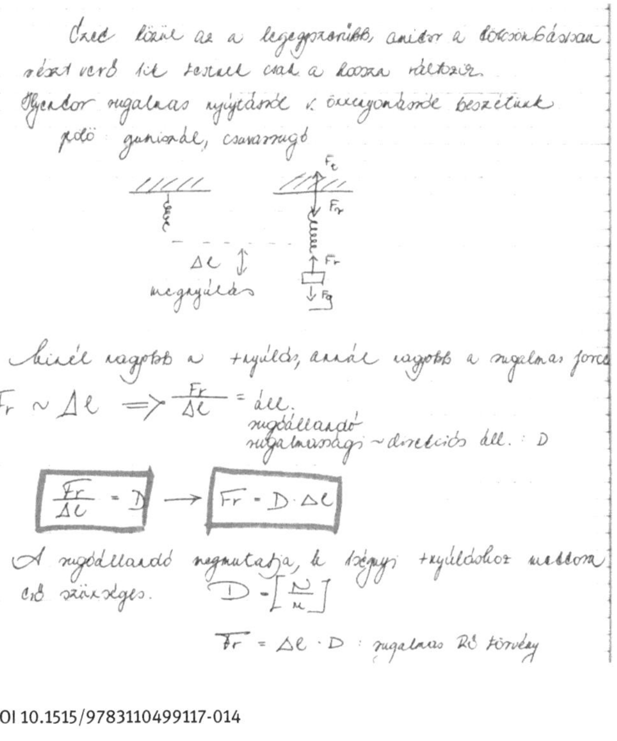

DOI 10.1515/9783110499117-014

Annex 9

Dialogue letters: notes passed in class by students aged 14 to 20 (examples)

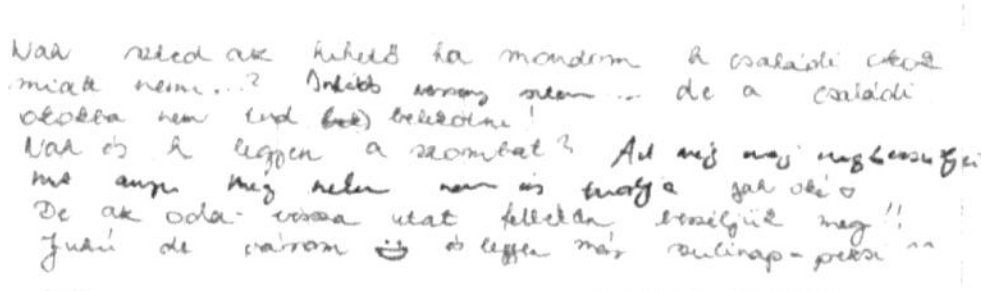

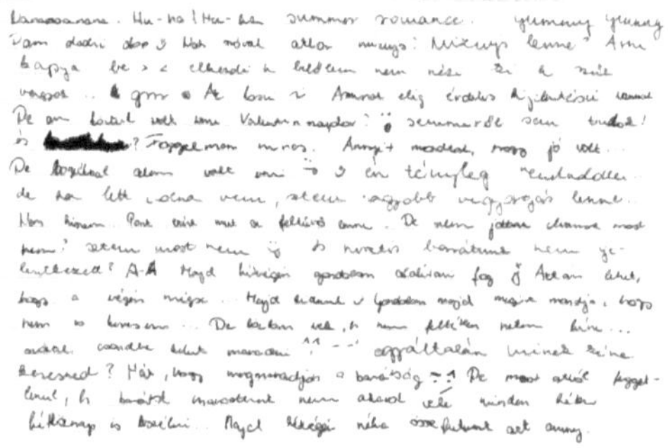

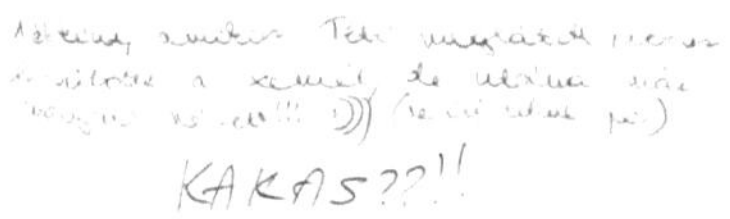

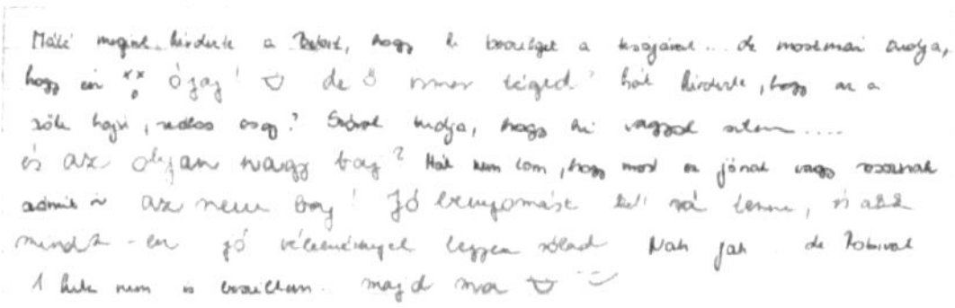

DOI 10.1515/9783110499117-015

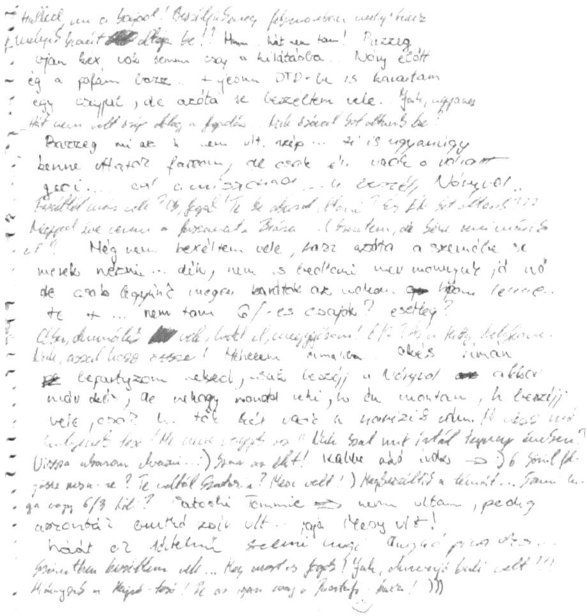

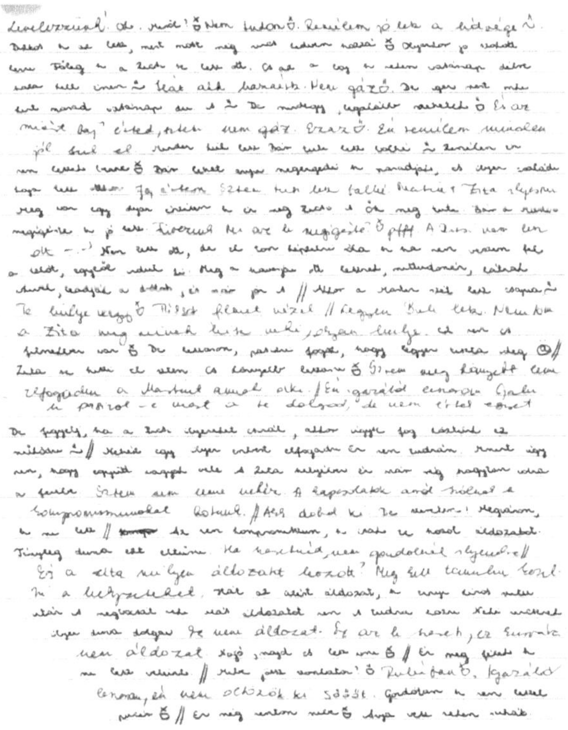

Annex 10

Digilect in orality. Spontaneous speech corpus

The collection is designed as a linguistic diary on the basis of the idea and guidance of Attila Benő and István Lanstyák.

In addition to my own observations, I also used the relevant data from the linguistic diaries of students of ELTE Faculty of Humanities, subject to their consent and with the indication of the author's monogram. Authors of the linguistic diaries used: Dorottya Hlatki, Dániel Konkoly, Szandra Nyárondi, Róbert Pusztai, Roland Szabó, Alexandra Tóth. Further contributors: Bianka Duka, Tamás Kruzslicz, Nóra Kugler and Ildikó Veszelszkiné Huszárik.

AFK

I. Temporal and spatial limits of the speech event

1. 2011. május

2. Kiskunhalas, középiskola

3. **a)** aktív résztvevők: 9. évfolyamos osztály (15 éves diákok, fiúk-lányok), tanár (VHI)

b) passzív résztvevő: -

c) viszony: diák-diák és tanár-diák viszony

II. Description of the speech event

1. Diákok: „Az *áefká* azt jelenti, hogy valaki nincsen ott, a kérdezett helyen."

2. az *AFK* internet-specific acronym használata élőbeszédben

III. Linguistic analysis of the speech event

1. tényforma: *AFK*; célforma: „nincs ott"

2. Az *AFK* akronima jelentése számítógépes kommunikációban: „away from keyboard", azaz 'nincs a gépnél'. Szóbeli használata során azt jelöli, hogy a kérdezett személy nincs a kérdéses helyen.

ASD (1)

I. Temporal and spatial limits of the speech event

1. 2009. április

2. Budapest

3. **a)** aktív résztvevők: 16–17 év körüli fiúk

b) passzív résztvevő: megfigyelő (KN)

c) viszony: egyikőjük a megfigyelő fia, egymásnak osztálytársai

DOI 10.1515/9783110499117-016

II. Description of the speech event

1. „A fiam sokat csetel a számítógépen. Amikor éppen nem tud semmit mondani a partnerének, leüti az ASD billentyűket – ezek egymás mellett helyezkednek el a billentyűzeten. Ez jelzi azt, hogy ott van, nem ment el a géptől, de éppen nincs mondandója. Beszédben is használják, kb. 'izé' jelentése van."

2. az *ASD* internet-specific acronym használata élőbeszédben

III. Linguistic analysis of the speech event

1. tényforma: *ASD*; célforma: „itt vagyok", „izé"

2. Az egymás mellett elhelyezkedő billentyűk leütése számítógépes kommunikációban fatikus jelzés, azt jelöli, hogy a partner a gépnél van, de most éppen nem tud mit mondani. Szóban 'izé' jelentése, töltelékszó-funkciója van.

ASD (2)

I. Temporal and spatial limits of the speech event

1. 2009. szeptember

2. Budapest, számítógép előtt ülve

3. **a)** aktív résztvevők: két 17 éves diák

b) passzív résztvevő: megfigyelő (SZR)

II. Description of the speech event

1. A Ventrillo nevű számítógépes szoftvert mutatta be a számítógépfüggő fiú. A program a Skype-hoz hasonlít, ám ezt játék közben használják. Bejelentkezéskor az első megjelenő név az Asdor volt. A becenév előállítása (Andor > *Asd*or) annyira lenyűgözte a társaságot, hogy azóta is előszeretettel használják az ASD kifejezést, gyakorlatilag bármire.

2. számítógépes kommunikációból származó kifejezés

III. Linguistic analysis of the speech event

1. –

2. A három betűből álló „szó" funkcióját tekintve hasonlít a *para, flash* szavakhoz. A huszonévesek gyakran használják olyan szituációkban, amikor valami nem átlagos dologgal találkoznak, és próbálják kifejezni ezt az érzést. Az asd a számítógép billentyűzetén három egymást követő betű kombinálásából jött létre, ezért találkozott vele az adatközlő először informatikai, számítástechnikai közegben. Ez a három, egymás mellett álló betű három különböző szmájlit is jelenthet::A;:D,:S. E három teljesen különböző jelentésű emotikon felől is megközelíthető az *asd* szó jelentése az egyetemista adatközlő szerint.

ASD (3)

I. Temporal and spatial limits of the speech event

1. 2011. május

2. Kiskunhalas, középiskola

3. **a)** aktív résztvevők: 9. évfolyamos osztály (15 éves diákok, fiúk-lányok), tanár (VHI)
b) passzív résztvevő: -
c) viszony: diák-diák és tanár-diák viszony

II. Description of the speech event

1. Diákok: „Unalom kifejezésére használjuk az áesdé kifejezést."

2. az *ASD* internet-specific acronym használata élőbeszédben

III. Linguistic analysis of the speech event

1. tényforma: *ASD*; célforma: „itt vagyok", „izé"

2. Az egymás mellett elhelyezkedő billentyűk leütése számítógépes kommunikációban fatikus jelzés. Szóban unalmat jelző funkciója van.

ASL

I. Temporal and spatial limits of the speech event

1. 2010. május

2. Kiskunhalas, középiskola

3. **a)** aktív résztvevők: 16 éves fiú, megfigyelő (VHI)
b) passzív résztvevő: -
c) viszony: tanár-diák viszony

II. Description of the speech event

1. „Ha ismeretlenekkel csetelek, a beszélgetés az ASL rövidítéssel indul. Ez az „age sex location", vagyis életkor – nem – tartózkodási hely. Buliban is ismerkedtem már így – persze az S látható volt, csak poénból mondtam ezt."

2. a *ASL* internet-specific acronym használata élőbeszédben, tudatos nyelvi önmegfigyelés explicitté tétele

III. Linguistic analysis of the speech event

1. tényforma: *ASL*; célforma: *kor, nem, hely*

2. Az ASL angol eredetű rövidítés, jelentése: age, sex, location, azaz kor, nem, tartózkodási hely. Idegenekkel való csetes beszélgetés jellegzetes bevezető formulája.

Átácsol

I. Temporal and spatial limits of the speech event

1. 2011. február 28. hétfő

2. ELTE BTK, Budapest

3. **a)** aktív résztvevő: meghívott előadó (S. István), 60 év körüli, Kossuth Rádió szerkesztője; 20 év körüli diáklány, megfigyelő (NYSZ)

b) passzív résztvevők: a Rádióműsorok szeminárium hallgatói

c) viszony: oktató-hallgató

II. Description of the speech event

1. A szemináriumi követelményeket beszélte meg a csoport az oktatóval, például az olvasmánylistát. Az oktató erre közölte, hogy majd küld egy emailt, és abban *átácsolja* a listát. Senki sem értette, hogy mi az *átácsolás*, így az egyik hallgató megkérdezte az oktatót, aki mosolyogva közölte, hogy az *átácsolás* az, amikor valaki csatol egy fájt az e-mailhez. Tehát az *átácsolás* 'fájlcsatolást' jelent.

2. szokatlan szó használata

III. Linguistic analysis of the speech event

1. tényforma: *átácsol*; célforma: *attach*

2. Nehezítette a megértést a kifejezés szokatlansága élőbeszédben, másrészt az angol kifejezés magyarosítása. A beszélő az angol *attach* igét akarta alkalmazni; 'csatol' értelemben. Tehát az olvasmánylistát át akarta küldeni egy csatolt dokumentumban e-mailben. A szó ejtésén könnyítve magyarosan mondta ki, magyar morfémákhoz hasonlóan (*át*, *ács*).

BB

I. Temporal and spatial limits of the speech event

1. 2011. május

2. Kiskunhalas, középiskola

3. **a)** aktív résztvevők: 9. évfolyamos osztály (15 éves diákok, fiúk-lányok), tanár (VHI)

b) passzív résztvevő: -

c) viszony: diák-diák és tanár-diák viszony

II. Description of the speech event

1. Diákok: „Elköszönéskor bébét mondunk."

2. a *BB* internet-specific acronym használata élőbeszédben

III. Linguistic analysis of the speech event

1. tényforma: *BB*; célforma: „szia(sztok)"

2. A *BB* internet-specific acronym jelentése digitális kommunikációban: „bye-bye", azaz 'viszlát'. Ugyanabban a funkcióban, elköszönésként, viszont betűnként magyar kiejtés szerint ejtve *bébé*ként használja a tizenéves korosztály.

BB THE Q

I. Temporal and spatial limits of the speech event

1. 2011. május

2. Kiskunhalas, középiskola

3. **a)** aktív résztvevők: 9. évfolyamos osztály (15 éves diákok, fiúk-lányok), tanár (VHI)

b) passzív résztvevő: -

c) viszony: diák-diák és tanár-diák viszony

II. Description of the speech event

1. Diákok: „A nap végén elköszönéskor használt forma: bébé tö kú."

2. a *BB* internet-specific acronym használata élőbeszédben

III. Linguistic analysis of the speech event

1. tényforma: *BB the Q*; célforma: „sziasztok, kockák"

2. A *BB* internet-specific acronym jelentése digitális kommunikációban: „bye-bye", azaz 'viszlát'. Ugyanabban a funkcióban, elköszönésként, viszont betűnként magyar kiejtés szerint ejtve *bébé*ként használja a tizenéves korosztály. Az elköszönés kiegészül egy megszólítási formával, a *Q*-val, amely 'kockák'-ra utal. A *BB*-t és a *Q*-t magyaros, a *the* névelőt viszont angolos ejtés szerint mondják.

FACEBOOK-VICC: FÉSZEK

I. Temporal and spatial limits of the speech event

1. 2011. március 16,

2. budapesti buszjárat

3. **a)** aktív résztvevők: 14–16 év körüli fiú (zsebében csúcskategóriás telefon, feltehetőleg ért a technikai újdonságokhoz, rendszeres internetező); 13–15 körüli fiú

b) passzív résztvevők: a busz utasai (felnőttek, megfigyelő: NYSZ)

c) szereplők kapcsolata: a fiúk baráti viszonyban, a busz utasainak ismeretlenek

II. Description of the speech event

1. Az 1. fiú a másikkal beszélget, miközben a mobilját nyomkodja. Társának kommentálja az eseményeket, éppen facebookozik. Netezésről, játékokról beszélgetnek, amikor az 1. fiú a következő viccet mondja: „– Tudod, mi a közös a Twitterben és a Facebookban? –??? – Hát a *fészek.*"

2. szójáték

III. Linguistic analysis of the speech event

1. Tényforma: *fészek*, célforma: *face-ek, fészek*

2. Játék a digilektus kifejezéseivel. A *fészek* jelentése kettős: a facebook angol szó első elemének többes száma (face-ek: 'arcok'), illetve a 'madárfészek', amely a Twitter jelképére, a kék madárra utal. A szóviccet csak az értheti, aki ismeri az angol és magyar nyelvet, illetve tájékozott az internetes világban. A buszon utazó többi ember fel se kapta fejét e nyelvi leleményre.

FLAME

I. Temporal and spatial limits of the speech event

1. 2011. április 5.

2. egyetemi kollégiumi levelezőlista

3. **a)** aktív résztvevők: 21 éves kollégista; válaszolók 20 év körüli kollégisták

b) passzív résztvevők: a levelezés olvasói, megfigyelő: NYSZ

II. Description of the speech event

1. A kollégiumi levelezőlistán egyre gyakoribbak a spam- vagy személyes üzenetek. Megindult az egyre hosszabb és dühödtebb válaszüzenetek sorozata. Az üzenetküldő egy közérdekű, ám nem elsődleges prioritású levelet küldött, majd az utóiratban ezt írta: „Nem akartam trollkodni, bocsi, akit zavart a hirdetés, de kérlek, ne *flameljetek*!!!"

2. neologizmus

III. Linguistic analysis of the speech event

1. tényforma: *flamel* [flém]; célforma: ang. *flame* 'durva válaszüzenet'

2. Az angol kifejezés jelentése 'láng, fénycsóva'. Az adott szituációban azonban nem ez az elsődleges jelentése, hanem a 'durva válaszüzenet, szidalmazás'.

GG

I. Temporal and spatial limits of the speech event

1. 2011. május

2. Kiskunhalas, középiskola

3. **a)** aktív résztvevők: 9. évfolyamos osztály (15 éves diákok, fiúk-lányok), tanár (megfigyelő, VHI)

b) passzív résztvevő: -

c) viszony: diák-diák és tanár-diák viszony

II. Description of the speech event

1. Diákok: „A GG azt jelenti: gratulálok"

2. a *GG* internet-specific acronym használata élőbeszédben

III. Linguistic analysis of the speech event

1. tényforma: *GG*; célforma: „gratulálok"

2. A *GG* internet-specific acronym jelentése digitális kommunikációban: 'gratulálok'. Ugyanabban a funkcióban, betűnként magyar kiejtés szerint ejtve *gégé*ként használja a tizenéves korosztály.

GL

I. Temporal and spatial limits of the speech event

1. 2011. május

2. Kiskunhalas, középiskola

3. a) aktív résztvevők: 9. évfolyamos osztály (15 éves diákok, fiúk-lányok), tanár (VHI)

b) passzív résztvevő: -

c) viszony: diák-diák és tanár-diák viszony

II. Description of the speech event

1. Diákok: „Ha szerencsét kívánunk a másiknak, akkor *géel*t mondunk."

2. a *GL* internet-specific acronym használata élőbeszédben

III. Linguistic analysis of the speech event

1. tényforma: *GL*; célforma: „sok szerencsét"

2. A *GL* internet-specific acronym jelentése digitális kommunikációban: „good luck", azaz 'sok szerencsét'. Ugyanabban a funkcióban, elköszönésként, viszont betűnként magyar kiejtés szerint ejtve *géel*ként használja a tizenéves korosztály.

IDK

I. Temporal and spatial limits of the speech event

1. 2011. május

2. Kiskunhalas, középiskola

3. a) aktív résztvevők: 9. évfolyamos osztály (15 éves diákok, fiúk-lányok), tanár (VHI)

b) passzív résztvevő: -

c) viszony: diák-diák és tanár-diák viszony

II. Description of the speech event

1. Diákok: „Ha valamit nem tudok, azt mondom: *idéká*."

2. az *IDK* internet-specific acronym használata élőbeszédben

III. Linguistic analysis of the speech event

1. tényforma: *IDK* célforma: „nem tudom"

2. Az *IDK* internet-specific acronym jelentése digitális kommunikációban: „I don't know", vagyis ‚nem tudom'. Ugyanabban a funkcióban, betűnként magyar kiejtés szerint ejtve *idéká*ként használja a tizenéves korosztály.

IKSZDÉZIK

I. Temporal and spatial limits of the speech event

1. 2010. november

2. ELTE BTK, szemináriumi terem

3. **a)** aktív résztvevők: hallgatók (DB), tanár
c) viszony: tanár-diák

II. Description of the speech event

1. „Nálunk, Nyíregyháza környékén már használják az emotikonokat igeként is. Az ikszdézik jelentése: 'nagyon nevet', akárcsak a XD szmájlié."

2. neologizmus

III. Linguistic analysis of the speech event

1. a tényforma megegyezik a célformával: *ikszdézik*

2. A XD (iksz és dé) emotikon jelentése, az ikonikus viszony alapján, 'nagyon nevet'. Ebből -z igeképzővel új ige alakult ki.

KÉPERNYŐRE VESZ

I. Temporal and spatial limits of the speech event

1. 2011. február 21.

2. ELTE BTK kampusza, előadóterem

3. **a)** résztvevők: az előadást tartó tanár, hetven hallgató (megfigyelő: HD)
b) viszony: tanár-diák

II. Description of the speech event

1. A tanár felhívta a hallgatók figyelmét a tárgyalt irodalmi mű (amelyet már korábban feltöltött az e-learning oldalra) egy sajátosságára, mely *„azonnal szemet szúr, ha a folyóiratot képernyőre vesszük"*.

2. neologizmus

III. Linguistic analysis of the speech event

1. a tényforma megegyezik a célformával: *képernyőre vesszük*

2. A *képernyőre vesz* kifejezésnek a *kézbe vesz* szolgált alapjául. A kifejezés szokatlanságát az adathordozók és ezáltal az olvasási szokások megváltozására való szándékos figyelemfelhívás okozta. Spontán beszédben a képernyőről olvasásra is a könyv formára utaló kifejezéseket használjuk. A különbségek jelzésére a beszélő itt neologizmust alkalmazott.

Kettőspontdézik

I. Temporal and spatial limits of the speech event

1. 2010. november

2. ELTE BTK, szemináriumi terem

3. **a)** aktív résztvevők: hallgatók (DB), tanár

c) viszony: tanár-diák

II. Description of the speech event

1. „Nálunk, Nyíregyháza környékén már használják az emotikonokat igeként is. A kettőspont-dézik jelentése: 'nagyon nevet', akárcsak a:D szmájlié."

2. neologizmus

III. Linguistic analysis of the speech event

1. a tényforma megegyezik a célformával: *kettőspontdézik*

2. A:D (kettőspont és dé) emotikon jelentése, az ikonikus viszony alapján, 'nagyon nevet'. Ebből -z igeképzővel új ige alakult ki.

Lájkol

I. Temporal and spatial limits of the speech event

1. 2010. május

2. Kiskunhalas

3. **a)** aktív résztvevők: 1. 53 éves értelmiségi nő; 2. lejegyző (VÁ)

b) passzív résztvevő: -

c) viszony: gyerek-szülő viszony

II. Description of the speech event

1. Az egyik eseményről a képek Facebookra való feltöltése után így hangzott a kérdés: „*Lájkoltad már a képemet a Facebookon?*"

2. angol szó (*like*) átvétele, magyar képzővel való ellátása

III. Linguistic analysis of the speech event

1. tényforma: *lájkol*

2. lásd a 3.2.2. fejezetet

Lekattintottam

I. Temporal and spatial limits of the speech event

1. 2011. április 21.

2. kerthelyiség

3. **a)** aktív résztvevők: bölcsész-informatikus (21 éves, férfi), húszéves bölcsész

b) passzív résztvevők: 20 éves bölcsész hallgató, megfigyelő (HD)

c) viszony: baráti

II. Description of the speech event

1. A bölcsészinformatikus hallgató egy érdekes internetes írásról mondta: *„lekattintottam a netről*". Mire az egyik bölcsész visszakérdezett: *„Hogy mit csináltál?!*" Erre az első: *„Lekattintottam, ne már, hogy ez nincs meg!*" Majd körbenézett, a többiektől megerősítést várva.

2. neologizmus

III. Linguistic analysis of the speech event

1. tényforma egybeesik a célformával: *lekattintottam*

2. A *lekattintottam* informatikai zsargon, feltehetőleg a *letölteni le-* igekötőjéből és a *rákattintás* eseményéből nyeri szemantikai tartalmát. Az informatikai nyelvhasználat szinte napi gyorsaságú változásából és a számítástechnikai eszközöket alkalmazók és nem használók különbözőségéből fakadóan nem meglepő, hogy még azonos korúak között is fölmerülhet szóválasztásból adódó kommunikációs zavar.

Lenerfel

I. Temporal and spatial limits of the speech event

1. 2011. tavasz

2. Budapest, baráti társaságban

3. **a)** aktív résztvevők: huszonéves férfi, huszonéves nő (megfigyelő, TA)

b) passzív résztvevő: a férfi barátnője

II. Description of the speech event

1. „Ismerek egy fiút, aki régebben rengeteget játszott különböző online szerepjátékokkal. Olyannyira hozzászokott az ottani csevegőprogramokban használt szlenghez, hogy átvette a köznapi

életben is. Mikor a barátnője összeszidta előttem a múltkor, odasúgta nekem: »Le kellene már *nerfelni* az asszonyt.«"
2. számítógépes játékból származó kifejezés, amely beépült a mindennapi szóhasználatba

III. Linguistic analysis of the speech event
1. tényforma: *lenerfel*, célforma: *legyengít, leállít*
2. A *nerf* szó jelentése 'gyengít', számítógépes játékokban használt, angol eredetű kifejezés.

LOL

I. Temporal and spatial limits of the speech event
1. 2009. november
2. Budapest, busz
3. **a)** aktív résztvevők: két 16–17 év körüli fiú
b) passzív résztvevő: megfigyelő (VÁ)
c) viszony: a megfigyelő számára ismeretlenek, egymás osztálytársai

II. Description of the speech event
1. Egy budapesti buszon hallott párbeszédben az egyik tizenéves fiú viccet mesélt a másiknak, és az erre nem nevetett, hanem lakonikusan ennyit jegyzett meg: *LOL*.
2. a *LOL* internet-specific acronym használata élőbeszédben

III. Linguistic analysis of the speech event
1. tényforma: LOL; célforma: nevetés, nonverbális jelek
2. A LOL rövidítést főként cseten használják nevetés, jókedv kifejezésére. Jelentése: laugh out loud. Hangosan kinevet, felnevet.

Nem találkoztunk össze. Mínusz össze.

I. Temporal and spatial limits of the speech event
1. 2011. március 17.
2. egy kávézóban Füleken
3. **a)** aktív résztvevők: a megfigyelő volt osztálytársa 21 éves egyetemista, nő, többször előfordul, hogy kijavítja magát beszéd közben; megfigyelő: PR
b) passzív résztvevők: a többi vendég
c) viszony: baráti

II. Description of the speech event

1. Baráti beszélgetés során egy közös ismerősről a nő megállapította, hogy: *Nem találkoztunk össze.* Ő is észrevette, hogy ez a mondat nem egészen megfelelő, ezért mosolyogva hozzátette: *Mínusz össze.*

2. szokatlan fogalmazásmód

III. Linguistic analysis of the speech event

1. tényforma: *nem találkoztunk össze,* célforma: *nem találkoztunk*

2. A beszélő tudta, hogy a kifejezés, amit használ, nem helyes, ezért kijavította magát, olya módon, mintha MSN-en vagy Skype-on gépelt volna el valamit. Ilyenkor szokás mínuszolni vagy pluszolni a szavakat, betűket. Érdekes volt hallani, hogy élőben is így javítja ki magát.

NM

I. Temporal and spatial limits of the speech event

1. 2011. május

2. Kiskunhalas, középiskola

3. **a)** aktív résztvevők: 9. évfolyamos osztály (15 éves diákok, fiúk-lányok), tanár (VHI)
b) passzív résztvevő: -
c) viszony: diák-diák és tanár-diák viszony

II. Description of the speech event

1. Diákok: „Ha a másik megköszönöm valamit, arra *enem*mel illik felelni."

2. a *NM* internet-specific acronym használata élőbeszédben

III. Linguistic analysis of the speech event

1. tényforma: *NM*; célforma: „nincs mit"

2. Az *NM* internet-specific acronym jelentése digitális kommunikációban: 'nincs mit'. Ugyanabban a funkcióban, betűnként magyar kiejtés szerint ejtve *enem*ként használja a tizenéves korosztály.

NP

I. Temporal and spatial limits of the speech event

1. 2011. május

2. Kiskunhalas, középiskola

3. **a)** aktív résztvevők: 9. évfolyamos osztály (15 éves diákok, fiúk-lányok), tanár (VHI)
b) passzív résztvevő: -
c) viszony: diák-diák és tanár-diák viszony

II. Description of the speech event

1. Diákok: „A nem probléma, no problem kifejezést enpének szoktuk mondani."

2. a *NP* internet-specific acronym használata élőbeszédben

III. Linguistic analysis of the speech event

1. tényforma: *NP*; célforma: „nem probléma"

2. Az *NP* internet-specific acronym jelentése digitális kommunikációban: „no problem"; 'nem probléma, nem gond'. Ugyanabban a funkcióban, betűnként magyar kiejtés szerint ejtve *enpé*ként használja a tizenéves korosztály.

OFFON VAN

I. Temporal and spatial limits of the speech event

1. 2011. május

2. Budapest, kávézó

3. **a)** aktív résztvevők: huszonéves főiskolások, megfigyelő (VÁ)
   b) passzív résztvevők: a többi vendég
   c) viszony: baráti

II. Description of the speech event

1. Baráti beszélgetésben hangzott el a következő kifejezés: *XY is észrevehetné már, hogy offon vagyok, egyszerűen hagyjon békén.*

2. informatikai kifejezés használata köznapi társalgásban, átvitt értelemben

III. Linguistic analysis of the speech event

1. tényforma: *offon vagyok*, célforma: *nem vagyok elérhető*

2. Az on a technikai eszköz bekapcsolt, az off pedig a kikapcsolt állapotát jelzi. Ebből ered a 'ki van kapcsolva, valaki számára nem elérhető' jelentésű *offon van* kifejezés. Általában a kapcsolatfelvételt szándékosan kerülő magatartásra utal.

OMG

I. Temporal and spatial limits of the speech event

1. 2011. május

2. Kiskunhalas, középiskola

3. **a)** aktív résztvevők: 9. évfolyamos osztály (15 éves diákok, fiúk-lányok), tanár (VHI)
   b) passzív résztvevő: -
   c) viszony: diák-diák és tanár-diák viszony

II. Description of the speech event

1. Diák: „Ha meglepődöm, azt mondom: óemgé." Másik diák: „Vagy óemdzsí."

2. az *OMG* internet-specific acronym használata élőbeszédben

III. Linguistic analysis of the speech event

1. tényforma: *OMG*; célforma: „istenem"

2. Az *OMG* internet-specific acronym jelentése digitális kommunikációban: „oh my God", 'ó, istenem'. Ugyanabban a funkcióban, betűnként magyar vagy angol kiejtés szerint ejtve *óemgé*ként vagy *óemdzsí*ként használja a tizenéves korosztály.

Smilelike

I. Temporal and spatial limits of the speech event

1. 2011. március 24 este

2. kollégiumi büfé

3. **a)** aktív résztvevők: 22 év körüli kollégista fiú, 20 év körüli kollégista lány

b) passzív résztvevők: a büfében tartózkodó emberek, megfigyelő (NYSZ)

c) a fiú és a lány barátok

II. Description of the speech event

1. A fiú és a lány az előző esti buliról és az azon az estén készült képekről beszélgetett. A fiú csak annyit mondott a lánynak, hogy neki nagyon tetszettek a tegnap esti buli fotók és, hogy nyomott is rá egy *smilelike*-ot.[szmájlájk]

2. neologizmus

III. Linguistic analysis of the speech event

1. tényforma: *smilelike*; célforma: smiley és lájk

2. A kifejezés pontos jelentése nem egyértelmű. Jelentése például: *like-oltam a képet és kommentbe írtam egy smiley-t is.* Másik feltételezett jelentése: 'felfelé mutató hüvelykujjú emotikon a Facebookon'.

STFU

I. Temporal and spatial limits of the speech event

1. 2011. május

2. Kiskunhalas, középiskola

3. **a)** aktív résztvevők: 9. évfolyamos osztály (15 éves diákok, fiúk-lányok), tanár (VHI)

b) passzív résztvevő: -

c) viszony: diák-diák és tanár-diák viszony

II. Description of the speech event

1. Diákok: „Ha valakinek el kellene már hallgatnia, azt mondom neki: estéefú."

2. az *STFU* internet-specific acronym használata élőbeszédben

III. Linguistic analysis of the speech event

1. tényforma: *STFU*; célforma: „kuss"

2. Az *STFU* internet-specific acronym jelentése digitális kommunikációban: „shut the fuck up", azaz 'kussoljál már'. Ugyanabban a funkcióban, betűnként magyar kiejtés szerint ejtve *estéefú*ként használja a tizenéves korosztály.

THX (1)

I. Temporal and spatial limits of the speech event

1. 2010. május

2. Budapest, egyetem

3. **a)** aktív résztvevők: 19 éves egyetemista férfi, megfigyelő (VÁ)
b) passzív résztvevő: -
c) viszony: tanár-diák viszony

II. Description of the speech event

1. „'Köszi' jelentésben használom a THX-et szóban: [téháiksz]-ként kiejtve."

2. a *THX* internet-specific acronym használata élőbeszédben, tudatos nyelvi önmegfigyelés explicitté tétele

III. Linguistic analysis of the speech event

1. tényforma: *THX*; célforma: *köszönöm, köszi*

2. A *THX* angol eredetű rövidítés, contraction. Eredeti formájában: „thanks". Az alak az angol nyelvű digitális nyelvhasználatból került a magyar digilektusba, innen pedig betűnként kiejtve a magyar társalgási nyelvbe.

THX (2)

I. Temporal and spatial limits of the speech event

1. 2011. május

2. Kiskunhalas, középiskola

3. **a)** aktív résztvevők: két 15 éves diák
b) passzív résztvevő: osztálytársak, megfigyelő (tanár, VHI)
c) viszony: diák-diák és tanár-diák viszony

II. Description of the speech event

1. „Az egyik diák kért egy tollat a másiktól, és amikor megkapta, téháiksz kifejezéssel köszönte meg. Ekkor további, általuk szóban használt rövidítésekre is rákérdeztem."

2. a *THX* internet-specific acronym használata élőbeszédben, nyelvi megfigyelés

III. Linguistic analysis of the speech event

1. tényforma: *THX*; célforma: *köszönöm, köszi*

2. 1. THX (1)

WTF (1)

I. Temporal and spatial limits of the speech event

1. 2010. május

2. Budapest, egyetem

3. **a)** aktív résztvevők: 22 éves egyetemista férfi, megfigyelő (VÁ)

b) passzív résztvevő: -

c) viszony: tanár-diák viszony

II. Description of the speech event

1. „A WTF – vétéefként kiejtve káromkodás. Jelentése angolul: what the fuck. Magyarra miafasz-ként lehetne lefordítani."

2. a *WTF* internet-specific acronym használata élőbeszédben, tudatos nyelvi önmegfigyelés explicitté tétele

III. Linguistic analysis of the speech event

1. tényforma: *WTF*; célforma: *mivanmár, miafaszvan?*

2. A *WTF* angol eredetű rövidítés, internet-specific acronym. Feloldása: „what the fuck". Az alak az angol nyelvű digitális nyelvhasználatból került a magyar digilektusba, innen pedig betűnként kiejtve a magyar társalgási nyelvbe.

WTF (2)

I. Temporal and spatial limits of the speech event

1. 2011. május

2. Budapest, egyetem

3. **a)** aktív résztvevők: 18 éves egyetemista férfi, egyetemi oktató

b) passzív résztvevő: irodalmi szemináriumi csoport, megfigyelő (TA)

c) viszony: tanár-diák viszony

II. Description of the speech event

1. Tanár: „Mi a véleményük a mára feladott olvasmányról?" Hallgató: „Ez egy nagy WTF."

2. a *WTF* internet-specific acronym használata élőbeszédben

III. Linguistic analysis of the speech event

1. tényforma: *WTF*; célforma: *mivanmár, miafaszvan?*

2. 1. WTF (1)

WTF (3)

I. Temporal and spatial limits of the speech event

1. 2011. május

2. Kiskunhalas, középiskola

3. **a)** aktív résztvevők: 9. évfolyamos osztály (15 éves diákok, fiúk-lányok), tanár (VHI)

b) passzív résztvevő: -

c) viszony: diák-diák és tanár-diák viszony

II. Description of the speech event

1. Diákok: „Türelmetlenség kifejezésére használjuk a vétéef kifejezést."

2. a *WTF* internet-specific acronym használata élőbeszédben

III. Linguistic analysis of the speech event (WTF 1)

Annex 11

Summary of the survey 2010

Forms of digilect used in oral communication (opinion of the participants)

Típus	Példa	Említés	Megjegyzés
internet slang, form used in digilect	bug	1	
	dunno	1	I dont know
	lájk	3	
	lájkol	8	
	láv van	1	
	lávol (lávollak)	1	
	luv ya	1	I love you
	meglájkol	1	
	műxik	1	
	ráírni valakire	1	
	rákeresni valamire	1	
internet-specific acronym	asap	13	
	brb	1	
	btw	1	
	lol, LoooL	89	
	ROFLMAO	1	
	rtfm	1	
	omfg	2	
	omg	29	Oh, my God..., óemgé, óemdzsí
	szvsz	1	
	wtf, vétéef	15	
abbreviation creatied in digilect	[példa nélkül]	16	
	am	1	
	vok	1	
	köcc, köcce, köce	3	
	majd besz	1	
	mek	1	
	micsi	1	
	naon	1	
	nemtu	1	
	nömö = nm	1	nincs mit
	pill	4	
	szal	4	
	sztem	3	
	thx	1	
	tod, tom	2	spoken language
	tulképp	1	
	valszeg	10	
	vok	5	

DOI 10.1515/9783110499117-017

Típus	Példa	Említés	Megjegyzés
emoticon written as pronounced	[példa nélkül]	1	
	ikszdé (xD)	7	
	XD	12	
	kettőspontdé	1	
word formations with the suffixes -i or -csi, abbreviations with suffixes	figyi, figyike	3	„becézgetések”
	figyu, figyuka	2	
	kivi	1	
	köszi, köszcsi	2	
	puszcsi	2	
	szercsi, szerizlek	2	
	tali, talizni	4	
	telcsi	1	
„saying hello/ goodby with abbreviations”	üdv	4	
	chiao	1	cf. Italian ciao
	cupp	1	
	háj, hi	3	
	hali	2	
	pá	1	
	pusz, puszcsi, pussz	5	
slang	adom	1	
	csina	1	
	dettó	2	
	fuck	1	
	kira	1	
	luwcsiznám a sunádat	1	
	nice	1	
	plíz	1	please
	zsír	2	
forms originally created in spoken language	aggyá	1	
	aha		
	asszem	48	
	azta	2	
	bazz, bazze	2	
	haggyámá	1	
	jaja	1	
	jóvan	1	
	lécci, légyszi	6	
	mér	2	
	minnyá	1	
	mizu	1	
	monnyuk	1	
	nemááá	1	
	nemtom	21	
	sajna	1	
	sziv	1	szívesen

Típus	Példa	Említés	Megjegyzés
traditional (codified) abbreviation written as pronounced	stb.	2	
	IT	1	
	kábé	2	
	max	2	
	stb	1	
	ok	6	
uncategorized	eeeeee	1	
	kom?	1	
	mhm	1	
	pfff	1	
	wowm, vov	10	[váű] or [vov]
	y mint váj	1	
	yap	1	
no answer, or answer 'none'		377	

Do you think WRITING is being changed by internet and mobile communication?

Igen. = yes	220
Nem. = no	21
4sure;)	1
a felnövő nemzedék nem tud helyesen, stílusosan fogalmazni, a számítógép kijavítja a helyesírást, semmi erőfeszítésre nincs szükségük	1
A fiatal generáció számára már változott, azonban esélyesnek tartom, hogy tágabb körben fog változni.	1
A fiatalabb korosztályban jellemző, hogy hatással van az írásukra.	1
a fiatalok írását látva azt gondolom, hogy igen	1
a fiatalok írásban átveszik az az internetes beszélgetések nyelvezetét, egyre helytelenebbül fognak írni azok is, akik már nem általános iskolába/gimnáziumba járnak szerintem.	1
a fiatalok körében, akik ebben nőnek fel, IGEN.	1
A fiatalok között biztosan (10–20 éves korosztály)	1
a hagyomásnyos kézírás, vagy nyomtatott írás nem, viszont az interneten való kommunikáció teljesen más	
A használt nyelvezet igen! Van egy másodéves jogászhallgató barátnőm, aki világéletében eminens tanuló volt, de az utóbbi időben nagyon sokat csetel. Postai úton is szoktunk levelezni, és a levél csak úgy hemzseg a csetes rövidítésektől és szmájliktól.	
a helyesírás mindenképp, de születnek új szavak is néha, ami nem feltétlen jó.	1
A helyesírásunk fokozatosan romlik	1

A hivatalos biztos nem, a baráti igen. 1
A hivatalos írások biztosan nem. A magánlevelek, feljegyzések ugyanúgy egyszerűsödnek. 1
a hivatalos iratoknál nem hiszem, hogy változást okozna, mivel ott van egy hivatalos sablon, a privát jellegű beszélgetésekben meg csak olyan szempontból számít a változás, hogy ha a fiatalabb generációk hamarabb találkoznak a rövidített írásos formákkal mint a normálissal, valószínűleg rosszabb lesz a helyesírásuk 1
a kézírás háttérbe szorul, talán a kifejezések is változnak 1
a magunknak való írásban rengeteg a rövidítés: jegyzetek, napló, levelezések... 1
A mindennapi irományokban már most is megjelenik fiataloknál egymás között, kézzel írt formában, jegyzetekben. Nem tudom ezt megjósolni, előbb ezt a köznapi beszédben való elterjedése előzné meg szerintem. Az mondjuk biztos, hogy a hangulatjelek nem hivatalos iratokban sűrűn előfordulnak.
a minősége mindenképpen romlik, különösen a fiatal generáció körében 1
A nagy arányú számítógép használat miatt az ember lassacskán elfelejt kézzel írni.
A nyelvhasználat szempontjából is lehetséges, bár a nyelvtan tanárok dolga, hogy ne romboljuk szét a magyar nyelvet
A sok alulképzett tinédzserre gondolsz? Igen, már csak makogni tudnak. írni meg alig. 1
a sok rövidítés lehetséges, hogy valahogyan átszivárog 1
a számítógép mellett szocializálódó fiatalság helyesírását bizonyára nem befolyásolja pozitívan sem az internetes kommunikáció, sem az SMS írás 1
Abszolút igen, egyre kevesebben tudnak helyesen írni. 1
Aki direkt figyel ilyen szempontból, annál nem. Aki viszont eleve nem tud helyesen írni, annál rontja a helyzetet, hogy direkt nem kell figyelnie erre, elbújhat az internetes "szabály" mögött. 1
aki nem tud irni, gépelni az is tud. Ez követhető az internetes kommunikációban. 1
alapvetően nem, csak bővül új kifejezésekkel 1
Amiatt, hogy sietek (pl. sms-nél), vagy mert nem mindig jelennek meg jól a hosszú mgh-k az emailekben, szándékosan hibásan írok szavakat. Néha félek, hogy ez így bennem vagy másokban berögzül és elveszítjük a helyesírás képességét. Ezért néha ki is javítom magam, még ha így lassabb is a dolog. 1
Annak következtében, hogy meggyorsítsák az internetes kommunikációt, rengeteg rövidítés stb keletkezett, ez azonban ahhoz vezetett, hogy nagyon sok embernek romlott – az amúgy sem túl jó – helyesírása. 1
Annyiban, hogy lényegesen ritkábbá válnak a kézzel írt dolgok. Mivel mindennapjainknak nem része (miután a jegyzetelős diákévek elmúlnak), ezért külön súlyt kap, ha valamit kézzel írunk meg. Mondhatnám, ünnepivé teszi. Egy képeslapot pl. továbbra is kézzel írunk meg, hiába küldhetnénk internetes képeslapot pár kattintással. Pont azért, hogy kifejezze, hogy fontos annyira a másik, hogy rászántam papírt, időt, bélyeget, fáradságot. 1
Attol felek, ismet kicsit nehezen behatarolhato a kerdes. Ha az irasbeliseg a nyomtatott mediara vonatkozik, annak befolyasolasa szerintem nem indult meg meg (kivetel persze az

un. lektur kategoriacimkevel ellatott ag, amely mindent befogad). A keziras gyakorlata egyre jobban visszaszorul, de magaban nem valtozik. Tehat a befolyas az iras kivitelezesenek gyakorisagara, ezaltal gyakran a helyessegere iranyul, de magat az irast tartalmilag nem valtoztatja. A tavolabbi jovoben valoszinuleg a tartalmi befolyasolas is be fog kovetkezni, bar szerintem ez eleg sajnalatos. 1

Az átlagember írása igen, a művészi értékű írás lényegileg nem. 1

Az ékezetek elhagyása felé mutat. 1

Az emberek többet írnak. Az energiatakarékosság miatt egyes szavak rövidülnek, jelekké válnak

Az én írásom szerintem nem sokat változott, a szmájlikat kivéve. Nagy átlagban viszont igen, sokkal igénytelenebbek az emberek főleg helyesírásilag, központozásilag, de szóhasználatilag is, és túl sok a rövidítés, ami sms esetén érthető, de otherként nem szeretem. Aztán lehet, hogy régebben sem tudtak helyesen írni az emberek, csak ez nem volt olyan nyilvános. 1

Az én példámból kiindulva nem, de találkoztam már ellenpéldával is. 1

az igényes írás nem 1

az igényes magyar nyelv használata 1

Az internet hatására igen, a mobiltelefon hatására nem. 1

az internet hatására igen, mert a tizenévesek írnak a legocsmányabban a neten, és a leghelytelenebbül is, ami minősíti majd a nyelvtanukat, és nem csak a jegyeiket, de a nyelvhasználatukat is. 1

Az internettel, közösségi portálokkal kapcsolatos kifejezések írásban is megjelennek. Az emotikonok szintén. Informális írásos közlésekben az interneten és sms-ben megszokott rövidítések, fonetikusan leírt szóalakok jobban elterjednek. 1

Az ÍRÁS nem homogén dolog. Tehát a kérdést nem igazán tudom értelmezni. A fentiekhez hasonló rövidítések évszázadok óta léteznek. Most sokat használunk olyan műfajokat, amelyben valamiért szűkös az erőforrás, így jobban előjönnek. De hát a kódexek írói se tettek másként. 1

Az írás nem változik, az íráskép az változhat és a nyelv állandóan változik. Fontos hogy hivatalos szövegekben, írásokban mindenki törekszik a normák betartására. Tehát egy vizsgadolgozatban nem fog senki LOL-t írni, mert nincs értelme a hangos nevetésnek. Más közegben fordul elő az emotikonok használata, egy pihenés időszakában, amikor ismerősökkel beszélgetek, csetelek. 1

Az íráskép nem. Az írás tartalma, a szóhasználat és a helyesírás igen. 1

Azt nem tudom, de az emberek írástudatlansága szörnyű és lehet, hogy a nyelv is leépül ennek hatására 1

Azt nem tudom, hogy változik-e, de a káros hatások egyértelműek. A cset, az sms nagyon nagy mértékben hozzájárul ahhoz, hogy a fiatalok nem tudnak helyesen írni és fogalmazni sem. 1

befolyásolja 1

Bizonyára igen. Főleg a kevésbé iskolázottak körében – ezen a csak érettségivel rendelkezőket is értem, természetesen tisztelet a kivételnek. 1
Bizonyára. A helyesírás például romlik... 1
bizonyos rövídítések átkúsznak a mindennapokba 1
Biztos, de el kell telnie szerintem 10- 20 évnek, hogy az írásban tartósan megmaradjon. 1
Biztos, én csak azt vettem észre, hogy tömörebben fogalmazunk, s többször használunk szmájlikat.
Biztosan egyszerűsödik és ez sokszor lehet jó irány is, de többnyire ront a nyelven. 1
Biztosan hatással van rá. 2
biztosan, hiszen van, aki már furcsának tartja, ha "rendesen", szabályszerűen kell írnia; ha azoknál nem is változik, akik nap mint nap használják a normát írásban (egyetem, szellemi munka), azoknál biztosan gyökeresen más lesz, akik ritkán írnak ellenőrzött szöveget 1
biztosan, ilyen a nyelv, változik 1
biztosan, sok ismerősnél megfigyeltem 1
Biztosan. De hogy ez jó-e... Majd kiderül pár generáció múlva. 1
Biztosan... Ezek mind visszahatnak a nyelvre, hiszen ezekben is a nyelvet használjuk, ami így alakul.
Csak akkor, ha beragad a Caps Lock. 1
csak bizonyos korosztályban, aztán azt is kinőjük amikor már nem használjuk tovább ilyen rendszerességgel a csetet és társait
Csak nagyon lassan 1
de meg mennyire (lasd az ekezetek hianyat:) 1
De még mennyire!:) Tartalmilag is, és külalakra is... 1
Egészen bizonyosan igen. 1
Egészen biztosan változik, bár én ebben elég konzervatívnak tartom magam. 1
egészen biztosan, mert egyre kevesebb írott dolog, és egyre több internetes van jelen. Az internet nagyon hasznos, fontos, de nem kellene ennyire elnyomnia az írott felületeket 1
egy adott generáción belül nem hiszem, de az egymást követő generációknál igen 1
egyelőre nem érezhető számottevően, de igen. sajnos leginkább a helyesírás változik meg 1
egyértelműen igen, szókincsünk csökken, ezért kevésbé árnyalt a beszédünk (és az írásunk) 1
Egyértelműen igen. Ha mást nem is tekintünk, a számítástechnika az írásbeliség rovására terjeszkedik. A fiatalabb generációknak (a mostani 12–15 éves korúak) iszonyú a helyesírása, amin a Word nem segít (arról nem beszélve, hogy egy számítógépes program sem helyettesítheti minden nyelv helyesírási rendszerét tökéletesen – ugyanúgy, ahogyan a Word sem helyettesíti a magyart.) Az emotikonok lassan átveszik az emberi érzések helyét – manapság szinte mindenre van már szmájli.
egyértelműen változik 1
Egyértelműen, de ez szerintem természetes. Akkor is változott az írás, amikor a kézzel írott levelezés beindult. 1
egyértelműen. de pl. egy egyetemi előadáson jól is jön, mikor jegyzetelni kell 1

Egyértelműen. Tudatosan kell használni az internetes nyelvet és így ki lehet védeni a kézírás, vagy a verbalitás elkorcsosulását.ó

egyeseknél igen, ugyanis a rövidített alakokat írják be az érettségibe is. vagy a kézzel írott levelek is tele vannak rövidítéssel. kiváncsi lennék, hogy ezek az emberek hogyan fogalmaznának meg egy hivatalos levelet...leginkább megkérik idősebb szüleiket. meg a sok szmájli 1

Egyre inkább a rövidítés, kódszerű beszélgetés lesz. 1

Elhanyagolja a mai ember az írást. Csak ha muszáj akkor írunk, most már a vak gépelés a domináló és nem a szép írás. 1

Eltűnik 1

Elveszíti árnyaltságát, finomságát: degenerálódik. Ez a mi nyelvünkre különösen tragikus csapást jelent. 1

elveszti jelentőségét 1

Én eléggé hagyományos vagyok, nálam fiatalabbaknál gyakran előfordul, hogy használják a fenti dolgokat kézírásban is, igen. 1

Én nem vettem észre. 1

Én tudom még, mit mikor és hogyan akarok. 1

Erre én is kíváncsi vagyok. Nem tudom. Feltehetőleg igen. Sőt: valószínűleg. A rövidítések és a mosolyszimbólumok mindenképpen beszivárognak a nem hivatalos célú írásbeliségbe. (A kérdést a kézírásra értelmezem.) Viszont a szleng nem új dolog, ráadásul változékony. Van is, volt is, lesz is.

ESETEMBEN NEM JELLEMZŐ,TANÓRÁKON TERMÉSZETESEN SZOKTAK RÖVIDÍTENI, DE NEM AZ INTERNET ÉS MOBILTELEFON HATÁSÁRA. ESETLEG A TANÁROK MEGJEGYZÉSE MELLÉ KERÜL EGY KÉT MOSOLY JEL. DE ALAPVETŐEN NEM GONDOLOM, HOGY A KÉZÍRÁSBAN ÉS FŐLEG A HIVATALOS IRATOKNÁL JELENTKEZNE HATÁSUK 1

Ez csak attól függ, hogy az illető mennyire disztingvál az egyes kommunikációs helyzetek között. Egyes emberek írásbeli kommunikációjában lehetnek az internet és a mobiltelefon hatására létrejövő interferenciák, de ettől még maga az ÍRÁS nem változik. Ha valaki kétnyelvű, attól sem változik az a két nyelv maga, ami a fejében van. 1

Ezt fent leírtam: rövidítések alakulnak ki. 1

Fiatalok körében mindenképp,de akadnak,akik megpróbálják megőrizin az eredetei nyelvet. 1

gondolom igen; a dikjaim szeretnek rövidíteni pl.: megmondtam – +mondtam 1

Ha nem figyelünk oda, igen 1

Ha túlzott mértékben használja valaki az internetet, akkor belekeveredhet, de véleményem szerint nem változik jelentősen...A szlengek megjelennek internet nélkül is. 1

Ha valaki külön tudja választani ezeket, akkor nem. 1

hát majd meglátjuk 1

határozottan igen. sokkal több a rövidítés, az ékezethiány, a kisbetűs írásmód, rövid és tömör fogalmazásmód 1

hivatalos írásban nem, otherként igen 1

Hivatalos írások, dolgozatok esetében semmiképp, a magánjellegű dolgokban már megváltozott,főleg a fiatalok körében. 1
hivatalos levelekben és vizsgadolgozatban nem, a többi helyre beivódik 1
Hogyne! Gaudagodik az új jelekkel, kreatívabb lesz a rövidítések, képek használatával, de a helyesírésra rossz hatással van.
hosszabb időt tekintve biztosan (volt egy osztálytársam, aki a csetes alakokat nyugodt szívvel írta le akár a dolgozatokban is)
igen de az lehet az általános gyorsuló tempótól is 1
Igen – az ember minél rövidebben és hanyagabban, sablonosan és pontatlanul fog kifejezni magát.
igen (organikus megalapozottság l. fMRI vizsgálatok, valamint írástanítás nehézségei a begyakorlási idő csökkenése miatt a számítógép/mobiltelefon javára; a komplex fogalmazás szükségességének lecsökkenése l. internetes űrlapok kitöltése, 160 karakteres smsek, mmsek) – az írás véleményem szerint luxussá válik, újra a kevesek kiváltsága lesz, a kalligráfia jellege kerül előtérbe, a szépen és jól (kézzel) írás társadalmi szintjelző lesz 1
igen a hogy szerintem nem sokáig él kézírásban, teljesen értelmetlen kiírni 1
igen és nagyon is rossz irányba, egyre kevesebben tudnak helyesen írni... 1
igen nagyon, de ez inkább a kisebb korosztályt érinti akiknek 5 évesen mobiljuk van már 1
Igen változik. Édesanyám úgy írt levelet 30 évvel ez előtt egy orosz levelezőtársának, mint Petőfi Aranynak (jó,ez kis túlzás). Én ma úgy írok egy amerikai levelezőtársamnak, mintha az osztálytársammal beszélnék. 1
igen, a fenti fórumok objektív adottságai (angol kifejezések, helyszűke pl.) hatnak az írásra 1
Igen, a férjem például már NEM TUD KÉZZEL ÍRNI!!! 1
Igen, a fiatalok körében látványosabban, romlik a helyesírás 2
Igen, a használt rövidítések, jelek megjelennek az írásban is. 1
igen, a helyesírás is veszít jelentőségéből 1
Igen, a helyesirasunk sinyli meg, egyre gyakrabban kerul a kezembe olyan hivatalos anyag, ami hemzseg a hibaktol. 1
igen, a karakterek csökkentése érdekében rövidítünk – idegen szavakat használunk, amennyiben az közismert és rövidebb, mint a magyar megfelelője 1
Igen, a megváltozott gondolkodási sebesség okozója és következménye is egyszerre a pillanatkommunikációnak, melyet a cyber világ lehetővé tesz. Ez az, ami befolyást gyakorol a hagyományos kultúrtechnikákra, nemcsak az írásra, de az olvasásra is. 1
Igen, a nagy és a kisbetűk elveszítették a szerepüket. 1
Igen, a rövidítések félelmetesen terjednek. Példának okáért ide másolok egy hozzászólást az oldalról, ahol főszerkesztő vagyok. A delikvens tizenhat éves, a téma egy koncertbeszámoló: "Ott voltam^^(a Spongyabobos táskámmal eggyütt:Dszal észre lehetett venni) Nah mikor megtudtam hogy, a kisterembe lesz kicsit Sokkot kaptam o.o,de aztán nem zavart annyira mert így közelebb lehettem az egész banzájhoz.De levegő az nem kell az hülyeség...-.-
A SATEP az fasza volt jól nyomták a srácok:) csak a sok elmebeteg "táncikáló ember" vette el

akedvem attól hogy, énis zúzzak na mind1. Azt hittem szarabb lesz aSuffokatemert nekem eddig nem igazán tetszett,sőt..De most megtetszett:Dlehet egy hölgyemény miattis:$nah de szal megtetszett a Suffokate! És amit nagyon vártam VEIL OF MAYAaz elején elvolt baszva a hangosítás kb. az első két számnál:/ de aztán megoldották:).Egy számot vártam nagyon a Namaste-tahw*-* s annyira jóvolt: O de az összes számuk az volt^^. S Végül a CarnifeXnemtudom mi a baj velük,sztem ultra faszás volt üvölteni a mikibe^^ s ultra közelről látni ahogy Scott "énekel":Dés áh az egész nagyon jóvolt^^ja és lett egy Carnifex-es dobverőm: D!" Hiányzik a központozás, a helyesírást hírből sem ismeri, és annyira szövevényes a rövidítések alkalmazása, hogy gyakorlatilag olvashatatlanná válik az egész. 1
Igen, a sok rövidítés,helytelen szóhasználat, elhamarkodott kijelentések hatására. 1
Igen, a tőlem fiatalabbak gyakran használják levelezésben, beszélgetésben és mobilon is, ami nem feltétlen jó, mert ronthatja a helyesírásukat, nem ösztönöz választékos kifejezésmódra. 1
Igen, abszolút, bár még ez a változás kezdeti szintű, idő kérdése és beépül majd jobban is. 1
igen, abszolúte, nem lesz igény/szükség "jókai regényekre" (direkt kisbetűvel) 1
Igen, alapvetőn. Az emberek szivesen rövidítenek, az sms és az im meg szépen támogatja őket ebben, hiszen általában ez a fajta kommunikáció így vagy úgy szöveghossz-korlátozott. 1
Igen, átszüremlik, de ellenállok:) 1
igen, átvesszük a rövidített formákat a kézírásban is 1
igen, az előző kérdésnél leírtam. Ráadásul rohamosan romlik a helyesírás is, mert amit "tréfásan" helytelenül leírunk, az idővel rögzül, ráadásul arra is hat, aki olvassa. 1
Igen, az emotikonok, szmájlik használatának elterjedése, rövidítések tekintetében, a szavak átalakulnak. 1
igen, bár nem jelentősen 1
Igen, bár szerintem ez csak igényesség kérdése. 1
Igen, beépülhetnek az írott nyelvbe az életképes rövidítések. Gyakran rajzolnak kézzel írott üzenetekbe emotikonokat, érdekes módon 90 fokkal elforgatva is sokszor, mintha a számítógép alfanumerikus kötöttségei megmaradnának kézírásnál is. A helyesírás szintje romolhat, mivel az Interneten kommunikáló fiatalok hibáit nem javítja senki, és egymástól is többnyire helytelen alakokat látnak. 1
igen, csökken a helyesen írók íránya 1
Igen, de azért ez egy kicsit összetett kérdés. 1
igen, de csak korlátozottan, bizonyos fórumokon 1
igen, de ez hagyományosabb írott felületeken illetve műveltebb, formálisabb írott nyelvben kevésbé látszik 1
igen, de fokepp csak az internetes iras 1
igen, de mindig lesz különbség 1
Igen, de nem az én korosztályomban. 1
igen, durvul 1
igen, egyértelműen hat az internet és a mobiltelefon a kézírásra 1

igen, egyértelműen változik	1
igen, egyre kevesebben tudnak elfogadhatóan központozni magyarul és nagyon bosszant, ha ezeket valaki kézírásban, vagy nem megfelelő környezetben használja	1
Igen, egyre szegényesebb és ostobább lesz	1
igen, egyre több rövidítés jön létre, és az emotikonok száma is megugrott	1
Igen, egyre többen jegyzetelnek pl. rögtön számítógépre, ezenkívül pedig nem kell olyan szépen írni, mert beegépelve úgyis olvasható lesz a szöveg	1
Igen, egyszerűsödik, főleg ha ismerősöknek írunk. Hivatalos levélben ezt nem lehet megtenni.	1
igen, elég sokat	1
Igen, elterjednek a különböző idegen (leginkább angol) rövidítések, melyeket lassan minden számítógép előtt ülő ember megért, használni kezd.	1
igen, elterjednek a sz sms/cset rövidítések	1
igen, éppen azokat illetően, amikre rákérdeztél a fentiekben (pl. írásban is emotikon, rövidítés)	1
Igen, érezhetően változik.	1
igen, erősen negatív irányban, amikor a gyerekeknél az a menő, ha minél helytelenebbül írnak, a nickname-jük is pld. Andee Andi helyett...	1
igen, és nem az előnyére	1
igen, és változik a nyelvünk is	1
igen, ezek a kifejezések átszűrődhetnek írásban is	1
Igen, feltétlenül .... és az ravasz, fogós kérdés, hogy ez jó-e vagy rossz:-) Szerintem kötetlenebb formákban nem nagy tragédia, csak tudni kell, mik azok a helyzetek, amikor semmi estre sem megengedhető ez a lazább, nyeglébb stílus.	1
igen, fenn áll a veszélye, hogy ponyolábban fognak fogalmazni a jövő generációjának jelentős része	
igen, folyamatosan	1
igen, főképpen a rövidítések miatt	2
Igen, főleg a fiatalok körében.	1
igen, főleg a tizenévesek helyesírása	1
igen, gyerekek szókincse lecsökken vagy szlengre vált és nem tudnak egy összefüggő kerek fogalmazást írni a helyesírásról, már nem is beszélnék	1
Igen, hacsak az egyén nem kifejezetten aprólékos, a rövidebb kifejezésmód, felületesség, a stilisztikai sokszínűség csökken, sablonosság, praktikusság kerül előtérbe. Csak az internet, és a közvetlen emberi kommunikáció. A hivatalos, protokollárisan szabályozott dolgokban hosszú, nagyon lassan tartó változási folyamatot tudok elképzelni.(30–40 év)	1
Igen, hasonlóan, mint a fentiek.	1
igen, hiszen a fiatalok körében az internet és a mobiltelefon sokkal jobban előtérbe kerül, mint a könyv, papír és a hagyományos levél.	1

Igen, hiszen egyre több rövidítésre törekszenek az emberek, elterjed a szleng a közösségek, csoportok között és az lesz fontos, hogy mindent egyre rövidebben, egyszerűbben tudjunk kifejezni. 1

igen, kevesebbet írunk kézzel, és az irásképünk mindenképpen változik. a nyelvezet pedig az internetes nyelv hatására változik. 1

Igen, konkrétan a kézírás egyre inkább olvashatatlan lesz, hála a ritka használatának. 1

igen, leegyszerűsödik 1

Igen, legalábbis én azt tapasztalom a fiatalabbak körében, hogy a hosszú mássalhangzók helyett csak egyet írnak, az ly-t j-vel helyettesítik és a hosszú magánhangzók helyett rövideket használnak és ezt már kézzel írt formában is alkalmazzák. 1

igen, létrejön egy egyszerűsített, rövidített írásmód is, amelyet a többség ismer, de nem használ minden helyzetben.

igen, liberalizálódott 1

Igen, megfigyelhetően romlik a gyerekek helyesírása. 1

igen, mert a mai fiatalok, tizenévesek már néha a dolgozataikban is használják ezeket a nyelvi formákat, nem tudnak különbséget tenni a formális, hivatalos és a köznapi, netes-csetes alakok közt, nem érzik a stílusbeli különbséget

Igen, mert a növekvő adatmennyiség vonzata a kommunikációs technológiák hatékonyságának fejlődése, mely előbb-utóbb az írásban is megjelenik. 1

igen, mert az emberek rászoknak 1

igen, mert az újabb generáció ezen nőlt föl, és ezeket szokta meg az internetes chatnél, és egyre gyakrabban használja őket az írásban is 1

igen, mert egy kézzel írott levélbe is beleírjuk az interneten megszokott smile-kat, mintha anélkül nem lenne érthető, hogy egy vicces vagy szomorú dologról beszélünk. 1

igen, mert egyszerú és gyors, csupán korosztály változás után, mindenki a netet és a mobilt fogja használni. 1

igen, mert sokan írnak úgy kézzel is, mint cseten pl. gyakori rövidítéseket (h), írásjelek helytelen használata (teljes elhagyása) 1

igen, mivel kevesebbet hasznaljak az emberek, igy nehezebben megy amikor kell. pl hivatalos level megfogalmazasa eleg problemas. 1

Igen, nagyon is rossz irányba. A helyesírás egyre rosszabb, normálisan pedig egyre kevesebben tudják kifejezni magukat. 1

igen, nagyon is. sokkal helytelenebbül írunk és lassan már fel sem tűnik. az anno mindennaposan használt szavakat, manapság előfordul, hogy helyesírási szótárban nézem meg, mivel olyan régen használtam helyesen pl. most is nagyon helytelenül írok, nem használok nagybetűket, mert gyorsan végezni akarok (nem a kérdőív miatt, hanem megszokásból) 1

Igen, nagyon nagy mértékben. Egyre inkább átvesszük az írásban a rövidítéseket, a szlenget, stb.

Igen, nagyon, de ez nem feltétlen jó. 1

igen, nagyon, és drámai a helyzete a helyesírásnak. 1
Igen, negativ iranyban. (ezt magamon is eszreveszem...) 2
Igen, pl. egy változás: sokkal többet írok számítógéppel, mint kézírással (pl. vázlatokat, kivonatokat is gépbe írom). A kézírás során is sok rövidítést használok. 1
igen, pl. probléma a központozás eltűnése 1
Igen, romlik a kifejezőképesség és a helyesírás. 1
Igen, romlik. 2
Igen, rosszabb, hibásabb, fantáziátlanabb, szegényesebb lesz. 1
igen, rövidebbé válik és gyorsabbá 1
igen, rövidítések, emotikonok megjelenése 1
igen, rövidül 1
igen, sajnos a helyesírást sokaknál rontja, illetve a praktikus rövidítések közül van, ami bekerülhet a kézírásba is 1
Igen, sajnos egyre több interneten és sms-ben használt alak kerül be a diákok és valószínűleg a felnőtt társadalom írásaiba is. Ez a tendencia lassanként a nyelv elkorcsosulásához vezet. 1
Igen, sajnos jellemző – főleg a fiatal generációra – hogy hajlamos az internetes nyelvet az írásra is kivetíteni, ami véleményem szerint nem szerencsés, mert lassan elveszítik a helyesírásukat, valamint elfelejtenek szépen fogalmazni! 1
Igen, sok fiatal használja kézírásban is azokat a jeleket, rövidítéseket, amiket az interneten, telefonon is 1
igen, sokkal tömörebbé válik a nyelv 1
igen, szabadosabbá, kevésbé szerkesztetté válik 1
Igen, szegényebb lesz, a helyesírási szabályok felülíródnak. 2
igen, természetesen, rövidítések, jelek megjelenhetnek. 1
Igen, ugyanis az internetes rövidítések az írásban is megjelennek. 1
Igen, változik, egyre több rövidítést használnak, ami nem baj, de az olyan szavak, mint: fürcsi,lávollak, lávcsi, szercsi...na ezek iszonyatosan idegesítőek, ezek tipikusan a különféle közösségi oldalakon fordulnak elő, üzenőfalakon, meg adatlapokon... S sajnos olyan is előfordul, hogy egy-egy diák már nem tud különbséget tenni, s pl. dolgozatban is használ ilyen, vagy hasonló szavakat. 1
Igen, változik. De inkább az internet hatására. 1
Igen, változik. Habár az interneten és mobiltelefonon alkotott jelek egy része speciálisan az ottani kommunikáció igényeihez igazodik, másokat – elsősorban a rövidítéseket – át lehet emelni az informális írott nyelvbe is. 1
Igen, véleményem szerint a helyesírásra kifejezetten negatívan hat. 1
igen, véleményem szerint idővel a kézzel írott levelezésben is megjelennek/sőt akár már meg is jelentek az otherként internetes/mobiltelefonos kommunikáció során gyakran használt rövidítések/emotikonok... 1
Igen,változik bár nem olyan nagy mértékben. 1

igen; a mennyisége mindenképpen – de nyilván hat a szóhasználatra is az internet stb. 1

Igen. Az ésszerű rövidítések megszokottá válnak és beszivárognak az írás minden formájába. A hangulatjelek érzelmi töltést adnak az írott szövegnek, olyat, amit írásban nehéz megfogalmazni.

igen. pl. romlik a helyesírás már az emailekben is, illetve elterjed az ekezet nelkuli szoveg 1

Igen. a bonyolult szavakat több gondolkodással tudom helyesen leírni, mert msn, emil közben nem törődök a helyesírással, illetve a programok ki is javítják. 1

Igen. A helyesírás romlik, mivel az emberek önkéntelenül ráállnak az itt használt normára. 1

Igen. A kézírás elsatnyul, a rövidítések és a tömörítés pedig olyan estekben is alkalmazásra kerül, amikor használatuk teljesen indokolatlan. A szókincs, a kifejezőkészség és a nyelv intelligens használata csökken. 1

Igen. A múltkor belenéztem egy középiskolás füzetébe. Nem kellett volna. 1

Igen. Az emotikonok alkalmazása a leggyakoribb. Az other jellemz?k szerintem inkább a levelezésre nyomják rá a bélyegük.

Igen. Bizonyos emberek meg vannak gyozodve arrol, hogy a kifacsart/helyteleul irt szoalak a helyes.

Igen. Eltűnik a kis-nagybetű használata. Emiatt a diákok még a saját nevüket is kis betűvel írják! Eltűnik a folyamatos kézírás, nyomtatott betűkkel írnak. 1

Igen. Háttérbe szorulnak a levelek, egyszerűsödik a szókincs. 1

Igen. Jelentősen torzul. Vannak fiatalok, aki azért nem szívesen válaszolnak, mert én pontosan leírom a szavakat. Szerintem a tömörség és a torzítás nem szükségszerűen függ össze. 1

Igen. Kevésbé kötött formává válik a kézírás is az internet/mobiltelefon hatására. 1

igen. Kissé csökken a használata. 1

Igen. Középiskolásoknál tapasztaltam, h nagyon rontja a helyesírásukat, képtelenek rövidítés nélkül fogalmazni. s gyakran emotikonokat is használnak. Legjellemzőbb példa: nem tudják, h hogyan használják az idézőjelet. 1

Igen. Leginkább a fiatalok nyelvi kultúrájában lehet ezt lemérni. Nem szeretnek írni. Fárasztja őket egy hosszan megfogalmazandó mondat, nem is tudnak fogalmazni. E mellett a helyesírás botrányosan alakul a “teccik”, “asszem” szavak hatására. 1

Igen. Mindent minél rövidebben próbálunk fogalmazni. Főleg, aki nem tud gépírni, de kell számítógépet használnia.:-) 1

Igen. Mivel nincsenek ellenőrizve a szabályok, ezért sokkal lazábbra veszi mindenki a megfogalmazást. 1

igen. nagyon is. csak meg kel nézni pár érettségi feladatlapot 1

igen. nyelvtanilag elhanyagolódik. 1

Igen. Romlik, gepen nem figyelunk oda a helyesirasra. 1

Igen. Sok rövidítést, illetve emotikont átveszünk és írásban is használunk. 1

Igen. Sokak egyszerűen átemelik a hangulatjeleket, valamint a rövidítéseket az internetről. 1

Igen. Sokan átveszik. Gyerekek dolgozatokban, fogalmazásokban is használják őket. Volt olyan aki teleszmájlizta a fogalmazását.

Igen. Sokat romolhat a helyesírás. Én mindenkit megkérek a rövíditéseken és az angol kifejezéseken kívül, hogy írjanak helyesen, mert tudom hogy berögzül vizuálisan a helytelenül leírt magyar szó. Amúgy sem túl jó a helyesírásom, sajnos és ezért igyekszem odafigyelni rá. 1

Igen. Veszít az információ gazdagságából, pontosságából. 1

Inkább az elektronika hatására, ami nem feltétlenül az internetet jelenti. A számítógépen írás hatására mára minimálisra csökkent a kézzel írás, egyesek már semmit nem írnak kézzel, vannak, akik nem is tudnak. A szóhasználat is bizonyára degradálódik, gyakoribbak a tőmondatok és a rövidítések. De általánosságban nem jelenteném ki az internet romboló hatását, ugyanis ott is megjelennek igényes írások (pl. ÉS, HVG, Litera....), ahogy a nyomtatott sajtóban is megjelennek igénytelen írások (pl. Story, Blikk). 1

képivé válik 1

Kétségtelenül változik az írás a hatásukra, a nálam fiatalabbak közül már sokan használnak például emotikonokat vagy interneten elterjedt rövidítéseket egy gimnáziumi vagy akár egyetemi házidolgozatban. 1

kevesebbet írnak az emberek, a kézírás “kimegy a divatból” 1

kevesebbet írunk 1

kézírás elhanyagolódik 1

kissé torzul, új elemek épülnek bele 1

Lassan igen, de szerintem azért a kettő elkülönül, és csak az érthetőbb/értelmesebb dolgokat veszi át az ÍRÁS. 1

Lassan, de szerintem igen. 1

lehet, hogy egyszerűsödni fog 1

Lehet, hogy sokan kézírással is ugyanúgy írnak, mintha smst írnának, tele rövidítéssel. 1

lehet, nem gondolkoztam még rajta 1

Lehet,de nem biztos 1

lehetséges, bár ameddig annyi bölcsészt képeznek az egyetemeken, a nyelv fennmarad 1

Lehetséges, de erre térjünk vissza 100 év múlva:-) Ha változni fog ilyen irányban, azt nem annyira az infokummunikáció előretörésének, hanem a sztenderd tekintélyvesztésének tulajdonítanám. 1

magánjellegű írásokban igen. 1

Mi nem változik a technikai vívmányok hatására? De ha másban nem,annyiban biztosan,hogy egyre kevesebben írnak hagyományos leveleket. Bár ez a mai rohanó világban szerintem érthető. Egy e-mailt abban a másodpercben megkap a címzett,míg egy levelet csak 1–2 napon belül. 1

milyen értelemben? ha az a kérdés, hogy mondjuk a standard irodalmi vonalba, a vizsgadolgozatokba és a törvénycikkekbe bekerülnek-e a rövidítések, akkor természetesen nem-nem-soha. nem hiszek a közösségi média és a modern információs technológiák nyelvromboló, írásromboló, kommunikációromboló hatásában. 1

Mindenképp. Eleve az írás módja, formája, nyelvtani szerkezete azóta változik, amióta létrejött, ez is hatással van rá, esetleg a gyorsabb lesz a változás, mert sokkal rövidebb idő alatt sokkal több emberhez jut el. 1
Mindenképp.Egyre több rövidítést használunk. 1
Mindenképpen romlik. Főleg a fiatalok helyesírását rontja. 1
Mindenképpen változik valamennyit, de kellő intelligenciával azért megoldható, hogy egy vizsgán ne írja tele az ember emoticonokkal az esszéjét – az irodalmi és tudományos szövegek nem változnak szükségszerűen, a jegyzetelésben pedig segítenek az egyértelmű és megszokott rövidítések. 1
Mindenképpen, főleg az amerikai angolban, ott már jobban beépültek a csetes rövidítések. 1
Mindenképpen, ha nem figyelünk oda a gyerekekre, lehet, hogy a "hogy" szót "h"-ként fogják leírni egy dolgozatban (megtörtént eset), netalán ékezetetk nélkül fognak írni (szintén megtörtént eset).
mindenképpen, sajnos 1
mindenképpen,hiszem egyre inkább kevésbé velős a szöveg,csak lényegre törő,megszokva az 1 sms-ben lehetséges karakter számokat. 1
Mindenképpen: egyre rosszabb lesz a helyesírás és egyre érthetetlenebb lesz az idősebb generációk számára. 1
Mindenképpen. Főleg az emberek helyesírása, hiszen rászoknak a rövidítésre, helytelen nyelvi formákra. 1
Mindenképpen. Pl. sosem használok ékezetes betűket levelezéskor, mert megszoktam az angol billentyűzetet. Hozzám hasonlóan az ismerőseim is egyre inkább mellőzik őket. Macerás ide-oda váltogatni a billentyűkiosztást, mert bizonyos karakterek nem ott vannak a magyarban, ahol lenniük kellene, és ez nagyon zavaró gyors írásnal. a legsűrübben ezeket keverem össze az átkapcsolás miatt pl.:; ' " @ [] \ /? 0 z/s 1
mindig változik, nemcsak az internet és a mobiltelefon hatására 1
nagyon megváltozik, azon kapja magát az ember, hogy alig tud kézzel leírni 1–2 épkézláb mondatot, miután napi több órát netezik, emailezik, számítógépen ír. Nem annyira a mobiltelefon hatására változik. 1
nagyon remélem, hogy igen. az enyém biztos. pl. a mondatot gyakran kisbetűvel kezdem. 1
Nagyon sokat. Minden oldalon lehet találni helyesírási hibákat és ezt átveszik az emberek. 1
nálam nem 1
negatív irányba változik, romlik a helyes írásunk, egyre több a rövidítés a kézzel írott dolgainkban is.
néhány betűszó, rövidítés talán bekerül az írott szövegekbe, vannak kortárs kötlők akik használnak chat vagy sms kifejezéseket, de ezek inkább hangulatfestésként, vagy a kortárs magyar irodalmi nyelv problémáinak a felvétésére irányul véleményem szerint
Néhány esetben változhat.Ha valaki túl sokat csetel akkor esetleg. 1

Nem hiszem, hogy nagymértékben változna. Nyilván egy hivatalos levelet senki nem úgy ír, mint egy sms-t, ahol az a lényeg, hogy minél rövidebb legyen, egy levélnél meg ez nincsen így, ott lehet terjengősebben is írni. 1

nem jelentosen 2

Nem szabad, hogy változzon. 1

Nem tudom, hogy mennyiben változik, de az tény, hogy egyre kevesebbet írnak az emberek kézzel, és ez indukálhat egyfajta változást. 1

Nem változik, csak az emberek felejtik el az édesanyanyelvüket jól használni. 1

Nem változik, legalábbis kevésbé változik, mint mondjuk a szóbeli társaság. 1

Nem, az írásnak megmarad a magasabb presztízse 1

nem, de ritkul 1

nem, mert a mobiltelefon és az internet használata csak kettő a számtalan írottnyelv-használati regiszter közül, tehát egy kiforrott írásbeliséggel rendelkező személynél nem idéz elő változást. Fiatalabbaknál lehet hatással, de ennek is számos együtthatója van, pl. családi háttér, internetezési szokások, internetezéssel töltött idő stb stb stb 1

nem, mert ahogy öltözködünk, ahogy viselkedünk, ugyanúgy a beszéd- és közlésstílusunk is más: operába nagyestélyit veszünk, moziba pedig jó a farmer, üzleti öltözék a szürke/fekete/kék konzervatív kosztüm, randira meg valami kacérabb... 1

nem, nem hiszem, illetve a fiatalok írásbeli nyelvhasználata talán igen (lávcsi és társai) 1

nem,mert a hivatalos életben ugyanúgy azt a nyelvet használjuk, amit általános iskolában tanítottak nekünk. 1

nem. a kérdés nem jó. a helyesírás egy kötött cucc, egy tucat ember rögzíti. sajnos a kérdésben nem válik ek az írás és a helyesírás, pedig a 2 nem 1. én 5 éve írok LY nélkül, hihetetlen sokat, napi több oldalt, de ha váltok, tudok LY-vel írni, csöppet sem romlott a helyesírásom. sztem nem igazolt, h romlik, ha nem az ún. helyesírási norma szerint írunk. többféle stílus van szóban is, normális, h írásban IS, és nem kell mindig betartani az ún. helyesírást. öltönyben sem kell 24 órát lenni... 1

Némiképp igen, de lehet, hogy ezek nem hosszútávú hatások. 1

Nyelvhasználati színtértől függ. 1

Őszintén remélem, hogy NEM! Az írás olyan dolog, ami egyfajta szellemi örökség, és azt ápolni, de főleg tisztelni kell. Én nagyon szeretek írni, kézzel, bár nem írok túl szépen és olvashatóan. Szeretek leveleket írni, vagy a gondolataimat, vagy történeteimet. De ha csak arra gondolok, hogy a teendőimet leírjam, hogy el ne felejtsem, azt is szívesebben írom a fizikai határidőnaplómba, mint a telefon naptárába. 1

persze, a csetelésben ez látszik jól, de nem hiszem, hogy ezek a hivatalos levelekre hatással lennének

persze, de csak a mindennapokban 1

Persze, de ez sokkal jobban érezhető volt 7–8 évvel ezelőtt. Mára azok a szokások beépültek és nehezen fogadják be huzamosabb ideig az újat. Főleg a felgyorsult világnak köszönhetően egyre többen egyre tömörebben fogalmaznak és írnak, ha nem hivatalos dolgokról van szó.

Manapság gyorsan megtalálunk mindent és azt gyorsan meg is akarjuk osztani, mert más is lehet gyorsabb nálunk a szabadságnak köszönhetően. Munkahelyi levelezésben meg az időhiány miatt -> felgyorsult világ. 1
Persze. Lehet küzdeni ellene, de nem érdemes. 4
persze. nyilván vannak olyan új kifejezések, amik az írásba is átjönnek (már most is vannak). a szmájlik is lehet hogy jobban meghonosodnak, pl. idősebb tanáraimtól is kapok már olyan emaileket, amiben szmájlit használnak. ezek kiegészítik az írást, és szemlátomást hasznosak, ezért használja őket szinte mindenki. rövidítések is lehet, hogy megmaradnak (ha egyértelmű, mit rövidítenek, miért is ne?). 1
remélem nem 2
Remélem nem fog változni, mert ezáltal a nyelvünk már nem lesz olyan szép és színes. 1
remélem nem, bár az iskoali dolgozatokban, amiket diákjaimmal iratok rendre előfordulnak a “mienek”, egy re gyakrabban...
remélem, nem; valószínűleg mégis 1
Remeljuk, hogy nem. De biztosan beszivarog 1–2 uj szo, rovidites az irott NYELV-be is. 1
rengeteget változik, ez nem kérdés, főleg a rövidítések és a kiejtés szerinti írásmód kezd elterjedni
ritkább alkalmakkor kerül rá sor (a kézzel írásra értem), általánosságban a nyelv silányul az other platformok használatakor 1
romlik a helyesírás... 1
Romlik. 2
rondább lesz 1
rövidül 1
Saját tapasztalat, hogy az emberek helyesírása egyértelműen rosszabb lesz a számítógépezés hatására.
Sajnálatos módon, igen. 1
sajnos a helyesírás csorbát szenved, illetve hajlamossá válunk a megszokott szlengben használatos szavakat alkalmazni és nehezebb szép kerek nyelvtanilag helyes mondatokat formálni, hiszen sokat rövidítünk 1
Sajnos biztos változik, és akkor ez nem jelent jót bár nem vagyok purista:). Ugyanakkor ez csak a közvetlen kapcsolatokban lehetséges. 1
Sajnos igen és nem jó irányba. Elfelejtünk fogalmazni, mert megszoktuk a Back Space gombot, és meglepődünk, hogy kézírásnál ilyen nincs... Tavaly iskolapadba visszaülve döbbentem rá, hogy szinte elfejeltettem kézzel írni, és tényleg fizikailag megerőltető egy-egy írásbeli vizsga. És kaptam már rosszabb jegyet azért, mert nem tudták elolvasni az írásomat! 1
Sajnos igen, az archaizmusok gyorsabban halnak ki, a szókincs elsivárosodik. 1
sajnos igen, egyre kevesebben tudnak nyelvtanilag helyesen írni 1
Sajnos igen, meglehetősen negatív irányban, a fentebb kifejtettek következtében. 1

Sajnos igen, mert egyre jobban alkalmazkodnak az idősebbek is a nyelvhasználat változásához. 1
sajnos igen, rengeteg helytelen alak terjed el. 1
Sajnos igen. 12
Sajnos igen. A gyerekek jó része nem tud helyesen írni és helyesírási dolgozatokban előkerülnek rövidítések. 1
Sajnos igen. Egyre rosszabb lesz. 1
sajnos nagyon, most már nem is írunk, és ha igen akkor azt is hibásan, nem szépen 1
sajnos sokaknak romolhat a helyesírása 1
Sajnos változik, de én próbálok figyelni arra, hogy én ezeket ne használjam. 1
Sajnos, igen. Ezt úgy értem, ma már nem lehetetlenség, hogy egy blog- vagy twitter bejegyzésekből álló regény jelenjen meg, vagy éppen az SMS-regény. Ami mind szép és jó és hasznos, csak nem nyomtatásban. Szerintem ezzel valamilyen szinten a nyelvünk is elkorcsosul, ami viszont nem jó, mert minden nyelv szép és egyedi a maga nemében és megérdemli, hogy tisztelettel bánjunk velük, legyen az magyar, angol vagy bármely másik nyelv. Szóval lehet, hogy lustaságomnak hála, vagy mert szorít az idő, vagy a karakterszám, használok néha ilyen rövidítéseket, azonban beszédben sohasem és ha olyat írok, akkor írásban sem.
sokak helyesiras romlik,m az interneten nem kell figyelni ra, de pl dolgozatokban nem tudnak mar helyesen irni...
Sokat 1
sokkal több a helyesírási hiba 1
Személytől függ. 1
Szerencsére úgy vettem észre, hogy nem. Az olvasás hiánya az, ami elszínteleníti az írást..... 1
Szeretném hinni, hogy nem. De azt látom, hogy a mostani tizenéves korosztály élőszóban és kézírásban is használja ezeket a nyelvi formákat, s komoly gondot okoz nekik "átváltani". (Például, hogy a teccik, jáccik helyett tetszik, játszik szerepeljen a dolgozatukban.) 1
Szerintem a fiatalok körében változhat, de ők alapjáraton is sok szlenget használnak 1
szerintem egyre rondábban írnak kézzel az emberek. a sajtónyelv lazább, feloldottabb lesz, ahol ez megengedhető (szakfolyóiratban általában nem) 1
Szerintem elkerülhetetlen lesz. Angolban már tudok is erre példát mondani: through --> thru, night --> nite. 1
Szerintem igen, mert sokan átveszik a rövidítéseket, kifejezéseket az internetes nyelvből az írásba.
Szerintem más szövegtípusokat nem befolyásol döntően az SMS-ek és a csetelés helyesírása. 1
Szerintem nem, mindent a maga helyén. 1
Szerintem úgy megszokják a rövidítéseket, helytelen írásokat, hogy a diákok pl. dolgozat írásakor is rosszul írják a szavakat.

Sztem, – na jó, csak vicceltem – szóval szerintem az internet és a mobiltelefon hatása nem fog kiterjedni például a hivatalos levélváltás írásmódjára, viszont kétségtelenül befolyásolja az írást más – hétköznapibb -területeken. 1
Talán a nálunk fiatalabbaknál, akiknek már teljesen természetes volt, hogy létezik ez a nyelv. 1
talán még nem 1
tanítványokén már észrevettem 1
Teljes mértékben, és inkább negatív irányba. 1
Teljesen megváltozik. De azt gondolom az az írás, aminek hivatalosnak kell lennie, az az is marad. Iskolai dolgozatoknál, vagy valamilyen hivatalos ügy intézésénél nem fog, vagyis nem szabadna változnia. 1
Természetesen, egyre inkább teret kapnak a rövidítések, és a szlenget is egyre gyakrabban használjuk.
Természetesen, ez elkerülhetetlen pláne, hogy már általános iskolában internetes forrásokból dolgoznak. 1
Természetesen! Érdemes viszont legalább kétfelé bontani az "Írás"-t! Hivatalos dokumentumokban nem célszerű, és a többség el is kerüli az egyszerűsítések, emotikonok használatát, míg kötetlenebb közlési módba átváltva, ezek majdhogynem stílteremtően hatnak. 1
Természetesen. Az internet és a mobiltelefon hatására meg kell tanulnunk kiszűrni a fontos információkat (az üzenetet) a mondandónkból. Ugyanis ha nincs karaktervisszaszámláló, akkor sokkal inkább szabadjára engedjük a gondolatainkat. 1
Torzul, de a legrosszabb, hogy elfelejtenek írni az emberek. 1
több rövidítés keletkezik 1
Több speciális rövidítés jön be. 1
Úgy gondolom és remélem, hogy nem! A tanítók feladata, hogy ne változzon. 1
Úgy tapasztalom, hogy igen. A sajtó a fent említett módon (az internet kihat a nyomtatott médiumokra is), az egyéni forma pedig (baráti levél, órai jegyzetek stb.) szintén hajlik efelé. 1
Úgy tűnik Igen, de nem tudni mennyire tartósan. 1
Új forma jelent meg, amely természetesen a hivatalos szövegek írására nem hat. 1
Valamely módon igen. 1
valamelyest hatással van rá 1
Valamennyire biztosan, de véleményem szerint még mindig különbséget tudunk tenni a között, hogy kinek is írunk és mit. Hiszen egy hivatalos levélben még a legnagyobb szlenghasználó sem ír bele olyat, amit esetleg cseten használ.
valamennyire igen 2
Valamennyire igen, ami nem feltétlenül jelent pozitív változást, hiszen a rövidítések miatt a helyesírás szabályait nem veszik figyelembe és ez rossz hatással van az emberek helyesírására. 1
Valamennyire igen, de azért nem rohamosan. 1

Valamennyire igen, ugyanakkor az interneten elterjedt rövidítések (tom) és a hasonló megoldások, melyek főként a gyorsabb gépelésre szolgálnak, alig, vagy egyáltalán nem jelennek meg 1
Valamilyen szinten igen. Viszont, aki különbséget tud tenni az internet, a mobiltelefon és akár a hivatalos levél stílusában, annak az írásmódja nem változik. 1
valamilyenm szinten több lesz a rövídítés 1
Valószínűleg fog, legalábbis azoknál a beszélőknél, akik sokat használják a netet meg a mobilt; bár nem tetszik a nyakló nélküli rövidítésözön. 1
valószínűleg igen, de pl. egy hivatalos levél szabályai nem fognak változni irodalmi művekben viszont már előfordul ilyesmi
Változhat. És úgy tűnik, hogy változik is. Messzemenő következtetéseket nem mernék ugyan levonni sem a saját, sem az ismerőseim írásaiból, és azt sem tudnám megmondani, hogy pontosan miben, illetve mennyire változik, de hogy kötetlenebb, élőbeszédszerűbb lett (és lesz, még inkább, úgy sejtem), azt eléggé határozottan állítom. Aztán hogy ebben a változásban mekkora szerepe van az internetnek és a mobiltelefonnak, azt megint csak tippelhetném. De valószínűleg elég nagy. (A cset gyakorlatilag írott szóbeliségnek tekinthető, és így nyilvánvalónak látszik, hogy a szóbeliség irányába mozdítja el az írást is.) 1
Változik, a fiataloknál. Korosztályom nem használja ezeket. 1
Változik, a rövidítések és a helyesírás tekintetében mindenképp. Az is elmondható, hogy hamarabb ragad valaki mobiltelefont, vagy ír egy e-mailt, mint hogy levélpapírt és borítékot vegyen kézbe. Egyszerűbb, olcsóbb és gyorsabb válaszolást tesz lehetővé az internetes, mobiltelefonos kapcsolattartás. 1
változik, a rövidített, kényelmesebbnek tűnő elterjedt kifejezések hamar átkerülnek 1
Változik, azt kell, hogy mondjam negatív irányban. Ugyanazokat az érveket tudom ide is állítani mint a 21-es kérdésben. 1
változik, de csak az annak megfelelő helyzetben lesznek használatosak a megváltozott formák. hivatalos, hagyományos körülmények között még ugyanúgy érvényben lesznek a helyesírás szabályai.
Változik, de ez a változás korlátozott mértékű, a hivatalos levelekben használt írás nem módosul.
Változik, hatékonyabb lesz (nem feltétlenül jó dolog ez persze) és kevésbé szabályozott 1
változik. korcsosul. (láttam már diákújságban a hogy-ot h-val írva) 1
Változik. Először megfigyelhető volt az idegen eredetű szavak helytelen ragozása (kötőjelezés akkor is, amikor nem kellene), majd követte a -val/vel hibás használata. Sajnos ma már annyira elharapódzott, hogy a tévében is megjelent (reklámban: Vanish-sel; pedig helyesen Vanishsel. Legutoljára az RTL Klubon láttam az egyik műsorban a feliratban: Milye van XY-nak? Ezek a hibák 10 éve elképzelhetetlenek lettek volna.) 1
Van erre irányuló "törekvés", pl. a diákság körében, ahol a legnagyobb az internet interaktív funkcióit használók aránya előfordul, hogy elmossák a formális és az informális írásbeliség

különbségeit, ezt azonban az oktatás ellensúlyozza, így számottevő és radikális, gyors változás nincsen, a hosszútávú folyamatok pedig nem észlelhetőek.	
vannak új szóalakok, igen, de mindenki tudja hol használhatja őket és hol nem	1
Véleményem szerint az interneten és az sms-en kívül maximum saját papírlapokra írogathatunk rövidítéseket, smile-kat, stb-t. A világ azért még megköveteli az írásban való megnyilvánulás komolyságát. Az viszont már változtatja az írást, hogy a most felnövő fiatalok már lehet, hogy nem fognak tudni úgy írni, mint ahogyan mi, illetve az elöttünk felnövő generációk, de reményeim szerint az iskolák e téren való megbízhatósága és az értelmes diákok nevelésének célja nem változik.	1

Do you think ORAL CONVERSATIONS are being changed by internet and mobile communication?

Igen. = yes	248
Nem. = no	54
Nem tudom.. / nincs válasz	9
(csak) a szókészlet változik	6
a net elveszi az időt a szóbeli kommunikációtól; nincs idő beszélgetni	3
amennyiben on-line is aktív emberek közt (pl. facebook barátok) folyik	1
angol kifejezések	1
becézgetés egyre gyakoribb (szercsi, lávcsi, cukcsi)	4
beszélgetőpartnertől függ	3
Biztosan, hiszen gyakran én is használom a lol vagy az xD jeleket szóban is. Mivel sokat csetelünk, így egy idő után a szóbeli társalgásban is átvesszük azokat a szavakat, amiket cseten használunk.	1
csak a fiatalok körében	2
csetes kifejezések	1
Egyelőre nem mondhatnánk hogy változik, azzal, hogy sokan használják, inkább csak az internetre való utalás, de nem helyettesítés.	1
Egyes kifejezések, rövidítések átvételével mindenképpen.	1
egyeseknél tuti, elkezdték használni, hogy lol vagy az ilyen rövidítéseket.	1
Egyre több lesz a szleng	1
Egyre több szlengszó jelenik meg a beszédben.	1
Előtérbe kerülnek a rövidítések, és még oldottabb, lazább lesz a szóbeli beszélgetés, mert ezzel az oldottsággal lehet helyettesíteni az oly gyakran használt szmájlikat.	1
elsősorban a fiatalok nyelvhasználatában, az közép és idősebb korúakét nem változtatja meg.	1
enyhen biztos, en nem tapasztalom.	1

érdekes kérdés, valamennyit igen. magam is használok az internetes nyelvhasználatból átvett kifejezéseket (pl. ikszdé, óemgé), illetve szercsi, hizcsi, puszcsi és hasonló kifejezéseket poénból	1
érezhető is a változás: sok idegen szó használata, a szóbeli társalgásban is megjelennek a rövidítések	1
erre is az a válaszom, h igen, és ez viszont hátrányára	1
Ez egy jó kérdés. Talán néhány szóalak lerövidül, nem tudom. Ez azonban nyelvenként is változó, szóval nem hiszem, hogy erre lehetne mindent figyelembe véve válaszolni.	1
Ez is már változott, de véleményem szerint lassabb ütemben fog, ha fog.	1
ez már változik, naponta kapom a válaszokat, hogy nm (nincs mit) és a LOL kifejezést, ami kiborító	1
ez nem annyira veszti el, megmarad	1
ez talán kevésbé, de itt is egyre több angol kifejezés fog elterjedni	1
Ezen még nem gondolkodtam el, egy kicsit biztosan...	1
Ezt kevésbé tartom valószínűnek. (Csetes szavak, kifejezések átkerülhetnek persze a “rendes” szóbeliségbe is, de ez nem túl fontos, meg nem is túl érdekes.)	1
ezt még nem érzékelem. A környezetem elég választékosan beszélget, de bizonyára hat.	1
Fiatalok között bizonyára	1
fiataloknál már megtörtént	8
fiatalos szleng igen; hivatalos, irodalmi nyelv nem változik	4
főleg fiatalabbaknál jelentősen. Talán a szókincset érinti leginkább.	1
Gondolom, mar csak azaltal is, hogy beszelumk ezekrol. Pld: le vagyok merulve, dobok neki egy e-mail-t.	1
Ha arra gondolunk, hogy az egyes generációk milyen kifejezéseket használnak, akkor igen. De ez mindig is úgy volt, hogy az új technikákkal új terminusok jelentek meg. Szükségszer?en jönnek az új szavak, amiket használunk (amiket szükséges használnunk). Ha arra gondolunk, hogy a szóbeli társalgás fontossága megmarad-e, és hogy majd a jöv?ben is ugyanilyen jelent?s lesz-e, mint ma, akkor válaszom az, hogy nem. Szerepe és jelent?sége nem fog megváltozni, mindig fontos lesz, mert ez az, amivel kifejezzük magunkat, amivel képesek vagyunk kapcsolatot teremteni és ápolni egymás között. (Hogy ezt a mai fiatalabb generációk hogyan oldják meg, az már más kérdés...)	1
Ha nem figyelünk oda, igen	1
ha változik akkor nagyon minimálisan	1
hallotam már szóban azt, hogy lol és XD, de nem hiszem hogy nagyon, és tartósan befolyásolná	1
Használjuk a LOL és other szavakat beszéd közben is. Ahelyet, hogy mondjuk egy jót nevetnénk valóban.	1
Idősebb korosztályban, ahova és is tartozom, nem tapasztalom, hogy változtatná.	1
Idővel igen, hiszen már most is ömlik a médiából az a szókincs, amely leginkább a fent említett internetes oldalakon, illetve chat-ben előfordul. Nyilván így akarják megszólítani	

az adott korosztályt, lesznek szavak, amelyek megmaradnak, de nagyobb részük kikopik a nyelvből az évek során. 1
igazi magyar nyelv használata maradjon mg és törekedjünk a _:helyesírásra is. 1
Igen – az ember minél rövidebben és hanyagabban, sablonosan, pontatlanul fog kifejezni magát. A beszélgetőpartner jelenléte kevésbe fontossá válik. 1
igen (kisebb mértékben, mint az írás) 1
Igen és ez sem hoz minden esetben pozitív változást. 1
Igen változik. Az igénytelen közösségek használják a maguk primitív formáikat. Aki igényességre vágyik, az keres más társaságot. Fontos viszont, hogy régen is voltak "bunkók", akik hümmögve, eltorzult kiejtéssel, durván, káromkodással telve szólították meg egymást, vagy egy társaságot aki elsétált elöttük. Ezek ma is megvannak, csupán a "hő" helyett wtf, a "figyeld azt a csajt, szívesen elvinném egy körre" helyett pedig a "szúrnám a fo*óját" kifejezéssel élnek. 1
Igen, a beszédet is butítja az internet. Nincs szókincsük a fiataloknak. Mert a könyv olvasás kimarad az életükből. 1
igen, a divatos kifejezéseket nem csak az interneten használják, hanem szóban is 1
Igen, a legtöbb fiatal szóban is használja az internetes beszédet. 2
Igen, a mondandó megfogalmazásában vannak hiányosságok. 1
igen, a netes nyelvezet beépül a köznyelvbe, a hétköznapok részét képezi, ezért elkerülhetetlen 1
igen, a rövidítések elterjednek 1
Igen, a rövidített formulák átmennek a beszélt nyelvbe is. 1
Igen, a sok rövidítés, helytelen szóhasználat,elhamarkodott kijelentések hatására. 1
Igen, átveszik ezeket a szavakat szóban is. 1
igen, átvesznek rövidítéseket – én nem szeretem, hogy átveszik 1
Igen, az átvett szóalakok a szóbeli nyelvi humorhoz tesznek hozzá. 1
igen, az elterjedt rövidítések szófordulatok átterjednek a beszélt nyelvbe is 1
igen, az is felgyorsul. és kevesebb a szóbeli társalgás mindezek hatására. 1
Igen, az ott használt kifejezések gyakran átkerülnek az élő nyelvbe. 1
igen, azokat a kifejezéseket, amelyeket az interneten olvasunk, használunk, beépül a beszédünkbe, de azért szerintem meg tudjuk különböztetni a kettőt, és el tudjuk választani, azaz ha kell, tudunk "rendesen" beszélni 1
Igen, bár nem annyira erősen, mint az írás terén. Szerintem a szóbeli társalgás (át)alakításában a média játszik nagyobb szerepet. 1
igen, bizonyos rövidítések, illetve angol kifejezések az élő nyelvbe is bekerülnek. 1
Igen, biztosan, sok új szó kerül be a mindennapi szóhasználatba 1
Igen, de a beszélt nyelv változásában ez kevésbé zavar. 2
Igen, de nem a szavak formája, hanem a dinamika és a struktúra. 1
Igen, de nem szeretem hallani. 1
igen, de sokkal kevesebbet, mint az internetes iras 1

Igen, de szerintem csak az általános és középiskolás korosztálynál. 1
Igen, de szerintem nem annyira, mint az írás. Az internetes, csetes nyelv a szóbeliségből alakult ki, de nyilván itt is van kölcsönhatás. 1
igen, de talán nem olyan gyorsan, mint az írás 1
Igen, egyre szegényesebb és ostobább lesz 1
Igen, egyre több csetes rövidítés kerül be a szóbeli társalgába. 1
Igen, egyre több szleng szó kerül be a köznylevbe. 1
Igen, egyszerűsödik. 1
Igen, elég sok rövidítést (pl.:koncert=konci, omg, lol és társaik) becsempészünk mostmár a beszédünkbe, és mivel ezeket szokjuk meg, elfelejtünk ragozni, a mondatok sokszor értelmetlenné, zagyvává válnak. 1
igen, előfordulhatnak rövidítések, torzítások 1
igen, elterjednek a sz sms/cset rövidítések a beszédben is valamint sokkal több lesz az angol szó kifejezés 1
Igen, én azt mondom sajnos változik. Mobil hatására kevésbbé, inkább az internet hatása nagyobb. 1
Igen, erről is azt gondolom, amit az írásnál írtam (írván, bocsi). Azért remélem, a cukinyelv nem lesz kötelező.;-) 1
igen, és nem az előnyére 1
igen, és változik a nyelvünk is 1
Igen, ez is változóban van – szóban is mondunk egy csomó olyan dolgot, amelyet teletűzdelünk számos (az üzenet szempontjából) lényegtelen információval. Interneten és mobiltelefonon ezek korlátozva vannak. 1
igen, ez is. kérdés csak az, melyik milyen gyorsan. 1
Igen, f?leg a fiatalok között. 1
Igen, felgyorsul a beszéd, rövidítünk. Pl.: elköszönéskor. Kevés gesztikulációt, kevésbé választékos szókincset használunk. Az internetes órák helyett olvashatnánk is egy fotelben egy szépirodalmi m?vet, de az interneten is megtehetjük ezt. 1
igen, feltétlen vélemény: lsd. 1el fentebb 1
igen, folyamatosan, de lassabban, mint a chaten 1
igen, főleg a fiatalabbak körében 1
Igen, gyorsul (de ez már a telefon megjelenésével is elkezdődött véleményem szerint) és szlengesedik. 2
igen, kevesbe kifinomult, szines es elegyszerusodik 1
igen, kevesebb lesz a személyes társalgás 1
Igen, lazabb lesz, es nem udvariasabb... 1
Igen, lecsökken sajnos. Sokan szívesebben "beszélgetnek", ismerkednek cseten, msnen, interneten, mint személyesen. Sok embernek ez biztonságot, önbizalmat ad. 1
Igen, méghozzá egyre kevesebbet és azt is lényegre törőbben fogunk beszélgetni. 1
igen, megszünik a személyes kapcsolat és egy telefon is elég, ahhoz hogy beszélgessünk 1

Igen, mert a fiatalabb korosztály kevésbbé tud helyesen és szépen beszélni. 1
igen, mert gyakran használjuk az interneten elterjedt szavakat pl. lol 1
Igen, mert ott is használják ezeket a rövidítéseket, szlenget. 1
Igen, mert sok olyan kifejezést használnak az emberek szóban, amit leírva láttak. Például ilyen az XD is, amit kiejtve is sokszor hallok. 1
igen, mindenképpen beszivárog a társalgásba az interneten használt nyelv 1
Igen, mivel interneten és mobiltelefonon megbeszélt dolgokról tudunk beszélni élőben. 1
Igen, nagy hatással van rá, legfőképpen rontja a minőségét, kevesebb szót használnak az internet hatására, illetve több csúnya szót, mint előtte 1
igen, nagyon, néha érthetetlen ahogy a fiatalok beszélgetnek. 1
igen, néha elejtünk 1-1 csetben használt kifejezést (LOL) 1
igen, oldottabb lett, kevésbé távolságtartó 1
Igen, részben. Több trágár szavat használnak az emberek és felszínesek társalgásukben. Nem tudnak a sok rövidítés és idióta formulák miatt intelligensen társalogni, kicsi a szókincsük, nem tudnak színesen kommunikálni. 1
Igen, rövidebb mondatokat használunk, tömörítjük mondanivalónkat. 1
igen, rövidítések, angol szavak egyre elterjedtebbek és akik nem hazsnálják napi szinten az internetet, azok a szóbeli beszélgetések alkalmával "felsülnek", mivel nem értik a közös kódokat. mindent nagyon gyorsan le akarunk tudni, a mobiltelefon tarifái miatt gyorsabban beszélünk, összefüggéstelenebbül, az internet miatt is. csak itt szleng és szubkulturális kifejezésekkel "gazdagodunk" 1
igen, rövidítések, kifejezések használata 2
igen, sok idegenül hangzó szó betör a köznyelvbe is. kifejezetten jó példa erre a ""lájkolni, "lájkolás", "lájkolnám" és az "ez lájk" kifejezések. 1
igen, sok mindent kimondunk szóban is (lol), amit előtte csak írásban használtunk 1
Igen, sokszor hallottam már, hogy valaki szóban mondja, hogy XD (konkrétan "ikszdé"). Nekem nagyon furcsa, bár fiatal vagyok. Gondolom egyre inkább így lesz. 1
igen, szóban is rendszersen használjuk az otherként internetes/mobiltelefonos kommunikáció során elterjedt kifejezéseket...
Igen, tartalmában is, formájában is. Az internetes szociális élet nem váltja fel, hanem kiegészíti a társas életet, de egy péntek esti összejövé 85 %-ban olyan témákat érint, melyekhez nem sok közünk lenne internet nélkül. Másrészt a mobiltelefonnak és pl a facebook vagy twitter hírfolyamának köszönhetően cenzúrázatlan gondolatokat közlünk ismerőseinkkel, tehát folyamatos tudásunk van a másikról, tehát egy találkozás nem az újdonság erejével és az új információk kicserélésével telik. Ami részben jó, mert nem kell meghallgatni 4 hónap történését monológban, részben rossz, mert aki nincs benne ebben a facebook őrületben az lemarad a laza beszédtémákról. 1
Igen, több hivatkozás esik a már interneten, telefonon megbeszélt dolgokra, ill. egy másik szintről (mélyebb) indul a személyes beszélgetés, ha már előtte vagy közben beszélgettek neten az érintettek. Illetve az internetes félreértések miatt több konfliktus is áttevődhet a

szóbeli társalgásba, ami ha csak ott jelent volna meg a metakommunikáció hatására nem okozott volna problémát.

Igen, több lesz benne a szleng, a rövidítés.	1
Igen, több szófordulat kerül át az élőbeszédbe az internet hatására.	1
Igen, többet hivatkozunk azokra a dolgokra, amelyek az interneten történtek velünk, és ezt értelemszerűen az internet nyelvén mondjuk el.	1
Igen, új szavakkal bővül, pl.: lájkol, attendingel stb.	1
Igen, változik. De inkább az internet hatására.	1
igen,észrevettem például,hogy szivesen mondanak olyan szavakat mint például:láwcsi, puszcsi stb.	1
igen,jelentősen Az emberek szabadabban beszélgetnek,és a nyelvhasználat sem olyan már,-mint kellene.	1
Igen. Az emberek többet kommunikálnak egymással gépen, telefonon keresztül. Ez személytelenebbé teszi a kapcsolatokat.	
Igen. A beszélők leszoknak a választékosságról, az ízes fordulatokról, a szinonímák használatáról.	1
Igen. A fiatalabb generációk esetében mindenképpen, de talán a tinédzserek körében a leginkább.	1
Igen. A fiatalok aktívan használják az internetes kifejezéseket vagy rövidítéseket.	1
Igen. Az internetes nyelvhasználat térthódít a szöbeli közléseknél. Ezt a fiataloknál tapasztalom, de az idősebb korosztály is "kénytelen " erre tendélni, ha a fiatalokkal szót akar érteni.	1
Igen. E tekintetben az előző pontnál kifejtett véleményem él, csak szóbeli formákra és érintkezésre vonatkoztatva.	1
Igen. Egyszerűsödés jellemzi, bár ez nem minden esetben jó. De szerintem a tudatosan és mélyebb gondolkodású embereknek mindig is meglesz az igénye a tartalmas és értelmes beszélgetésekre. Ezeket a formákat inkábbb csak viccelődés és hangulatoldás szempontjából jó használni.	1
Igen. Korábban nem mondtunk olyan mondatokat, hogy "várj egy pillanatot, megnézem a neten", vagy hogy "mingyár felhívlak a mobilomról".	1
igen. sokan rengeteget tudnak csetelni, ám szóban meg se bírnak mukkani	1
Igen. Szerintem érezhetően romlik a szavak kifejezőereje, ezáltal az irónia megragadása is.	1
Igen. Szűkül a szókincs.	1
Igen. Több a becézés, hanyagság.	1
Igen. Trendinek vélt kifejezések hódítanak.	1
Inkább a fiatalabb, tizenéves vagy huszonéves korosztálynál hat szerintem.	1
Inkább a szóbeli kifejezések mentek át az internet és mobiltelefon világának írott formájára, biztos hat visszafelé is, de szóban talán kevéssé feltűnő.	1
Inkább az internet szakszavai kerülnek be, mivel ezeknek sokszor nincs megfelelője.	1
irasban jobban, szoban kevesbe	1

Kevésbé észrevehető, szembetűnő, de nem kizárt. 1

Kevésbé esztétikus, "félvállról vett szöveg darabok dobálása a másik felé" Azt gondolom, annyira elterjedt a mostani civilizációnak ez a "vagány" beszédstílus,hogy már rutinból jönnek ezek a szavak. Érettségin nekem is nagyon oda kellett figyeljek,hogy ne használjak rövidítéseket,amiket már megszoktam. 1

kevésbé m az írás m=mint:) 1

kevésbé szerintem, bár az OMG és LOL kezd átmenni a szóbeli informális beszédbe is 1

Kevésbé szerintem, mint az írott szövegek, a rövidítések eléggé furcsán hangzanak élőbeszédben, ráadásul az emberek törekednek rá, hogy kerüljék őket, mert igénytelenségnek érzik. 1

Kevésbé változik, illetve lassabban, mint az írás. 6

Kevésbé, mint az írásbeliség. 7

Kevesebb a szókincs és nem tiszta a beszéd. 1

kevesebbet találkoznak az emberek, mert telefonon "mindent el lehet intézni", a mobiltelefon miatt fontosabb az, aki telefonál, mint az, aki ott van az ember mellett 1

kicsit igen. Fiatalabbaknál inkább. Pl a szóban használt LOL, amitől az idősebbeket a hideg ráz. 1

kisebb mértékben, de igen 13

Korosztálytól függhet, de szerintem idővel biztos változni fog 1

már szóban is meg ez a 'lávcsi'-zás, szóval ez is valószínűleg kihatással van. 1

Másoknál észrevettem már, hogy beszivárognak a megjegyzéseikbe a LOL és társai, magamon még nem, de lehet, hogy ez is ilyen "más szemében a szálkát"-dolog. 1

Megeshet, bár nem valószínűsítem, hogy az emberek a hogy-ot képesek lesznek lerövidíteni. 1

Minden korosztálynak megvan a saját divatos szókincse, amit az internet befolyásol; lájkolnal, posztolnak, tvítelnek... de ez csak a fiatalokra, azok közül is a kevesebb intelligenciájúakra hat a szóbeli beszédben. 1

Mindenképpen, bár én igyekszem szóban sokkal érthetőbben, magyar kifejezéseket használva társalogni. 1

Mindenképpen. De csak bizonyos körökben. Például tanár-diák beszélgetésben nem. 1

Minimális mennyiségben. Mert tegyük fel az ember azért küld mosolyjelet, hogy kimutassa hogy mosolyog a gép előtt, de ha szóban beszélgetünk akkor ott látja, hogy mosolygunk... Szó, a hang nem cserélhető le könnyen. 1

Minimálisan igen – a netes kifejezések átkerülnek a beszélt nyelvbe. 1

nagyon 1

Nálam nem igazán, de a fiatalabbaknál fontos menőnek lenni, és kötelező olyan szavakat használni, amitől menőbbek, ezek pedig változtatják a nyelvet. De ez mindig is így volt. 1

nálam nem, másoknál nem tudom 1

Negatív irányba változik. Szóban szinte már nem is tudjuk magunkat kifejezni, mivel hozzá vagyunk szokva az internethez és az sms-ek szabad világához, ahol mindent le tudunk írni valahogyan és a smiley-k segítségével még az arckifejezésünket is le tudjuk írni 1

Néha elhangzik élőben is egy LOL valakitől, de megnyugtató módon ehhez gyakran negatívan viszonyul, aki hallja. 1
néhány szó átkerülésétől eltekintve nem tapasztalom, és nem is látom elkerülhetetlennek a változást. bár ma talán az internetes /sms-ben zajló kommunikáció során könnyebben és jobban fejezzük ki érzelmeinket, mint szóban... 1
nem annyira, mint az írott kommunikáció 1
nem feltételn, sokan mondják hogy megölik a személyes kapcsolatot, szerintem ez nem igaz, mert ha a barátom epten van én pedig itthon nem tudunk találkozni, csak így tudjuk megoldani, a szomszéd lánnyal nem neten fogok bezsélgetni valószínű 1
Nem hinném, hogy olyan nagy hatást gyakorolna rá egy két töltelékszó megjelenésén kívül. 1
Nem hinném. Akitől kapok is olyan levélkéket, amiben am, naon, meg ilyes rövídítések vannak, élőszóban mindig végigmondja. 1
Nem hiszem, hogy az internet hatására változna a szóbeli nyelv, a nyelvhasználatunk. Ehhez szerintem nem kell külső tényező. 1
nem igazán. Csak a szlenges rövidítésekben (pl. telcsi). 1
Nem jelentősen. 17
Nem kifejezetten, amit egyre gyakrabban tapasztalok hogy a szokasos generacios szleng mellett a roviditesek itt is elofordulnak, nevezetesen a -csi "becezest" ragasztva minden moge. Ennek kialakulasa szamomra nem tunik egyertelmuen az internet hatasanak, inkabb a mobiltelefon hasznalate. 1
nem láttam ilyet, inkább fordítva, szóbeli rövidítéseket használok sms-ben leírva is 1
Nem számottevően, de a hatása jelen van. Általános kultúráltsági szinttől függ, azaz, ahol sikk eleve szlengesen beszélni, ott sikk lesz ezekkel a beszivárgó elemekkel kommunikálni. 1
Nem tudom megítélni igazán, amennyire látom (hallom a fiatalakat, tizenéveseket), az is romlik 1
nem vagyok benne biztos, hogy közvetlen a kapcsolat... mivel minden (fél-)generáció megváltoztatja a fiatalok nyelvét (márpedig elsősorban rájuk jellemző az extrém internetnyelv), nem tudom, hogy ez is csak része-e a rendes változásoknak, vagy van ok-okozati összefüggés... 1
nem, a helyes és szép beszédmodor mindig is követelmény lesz 1
Nem, ezt nem gondolnám. Persze önmagamból indulok ki. 1
Nem, inkább fordítva: a mobiltelefonálás hasonlít egyre inkább a szóbeli társalgásra, bár a vizuális kapcsolat hiányában soha nem lesz ugyanolyan. 1
Nem, mert a személyes találkozást nem helyettesítheti az elektronikus kommunikáció. pl: kézfogás, ölelés, vállveregetés, puszi, csók, feromonok, 1
Nem. Vagy nagyon kevéssé. De inkább nem. Remélem. 1
Nyilván ezek a nyelvi elemek köznyelbe befértőznek, hiszen sok fiatal ezt használja szóban is, nekik beépül a szókincsükbe, s mivel életük része a cset stb., nem hiszem, hogy elhagynák a kor előrehaladtával, s náluk a beszélt köznapi szókincsbe kerül. Ezeket én nem tartom jó nyelvkincsbővítő szavaknak. Bár egyesek szerint a nyelvújítás során is voltak

merész szóalkotások, és a nyelv nem állandó. Ez igaz. Szerintem azonban ezek a szavak nem bírnak olyan esztétikai minőséggel. 1
Nyílván, bár szerintem ez azért diáknyelv és iskolázottság kérdése is. 1
Olyannyira, hogy lassan a fiatalok nem is beszélgetnek egymással. Nincsenek “csajos dumapartik”. Na ez sem lett valami magyaros. 1
persze, például már használják szóban azt hogy lájkol, dzsoinol egy groupba, képet addol, kommentel stb. szóval beépül az idegen nyelv 1
persze, pl. feljön mint téma. mit értsek szóbeli társalgás változásán? a szóbeli társalgás melyik jellemzőjét?
persze, új kifejezések terén mindenképp. 1
Pl: LOL... 1
remélem, nem 5
Remélem nem, nem vettem észre, az emberek szóban nem ügyelnek a percdíjakra, ezért természetesen beszélnek. 1
Remeljuk, hogy nem. De biztosan beszivarog 1–2 uj szo, rovidites az irott NYELV-be is. 1
részben igen, de ez messze nem olyan mértékű, mint a kézírásra gyakorolt hatás 1
Részben, elkezdik használni a netes kifejezéseket szűkebb környezetben 1
Röviden.Nem . Azt a protokoll, a szervezeti kultúra, és a másik fél is segíthet a régi mederben tartani. 1
rövidítések szóban is (a lol, omg ikszdé(XD) kifejezéseket többször hallottam már) 7
Rövidül és gyorsul. 1
Sajnos biztos vagyok benne, hogy befolyásolja, és változik, az emberek egyre kevesebbet tudnak szóban megnyilvánulni, nem tudnak beszélni, csak nehezen, gondot okoz, hogy szép kerek mondatokban beszéljenek. A stílus is változik, gyakran egyre alpáribb lesz, tanárként sokszor meg kell küzdenem ezzel, hogy ne beszéljenek csúnyán a diákok. Elég nehéz. 1
Sajnos igen. Kétségtelenül beszivárognak a nyelvbe az olyan rövidített megnyilvánulások, mint a LOL. 1
Sajnos változik, talán jobban is, mint írásban, és sokkal rosszabb szerintem, ha szóban mond valaki olyat, mint “asszem”, mintha egy internetes oldalon. 1
sok fiatalnál megszokássá válik a szlengesített rövidített szavak használata, így tehát változik 1
sok technikai kifejezés a beszélgetésben 2
sokan használnak már most rövidített, szlenges nyelvi formát szóban, főleg a tizenévesek. remélem felnőve megváltozik a szóhasználatuk, és ez a szóhasználat csak erre a korosztályra lesz jellemző. 1
Sokban változik.Sok internetes kifejezést használunk a köznyelvben is. 1
sokszor rövid, lényegretörő, sőt akadozó, (ezért kínlódnak sokan a levélírás feladattal nyelvvizsgára készülve) barátnőmnél nincs hatás (hosszú telefonbeszélgetések...) 1
Személyes találkozás helyett akár a szomszéd szobából is kommenteken keresztül kommunikálunk egymással. 1

Személytől függ.	1
Szerencsére még nem beszélünk rövidítésekben, de elhangozhat a LOL, OMG, nemtom, stb.	1
Szerencsére talán a rövidítések a szóbeli beszélgetésben annyira még nem terjedtek el, de a fent említett szavak (LOL, WOW, "ikszdé" azért kezdenek belopódzni a szóbeliségbe is.	1
Szerintem elkerülhetetlen a változás, mivel az internet és a mobil a mindennapi életünk részei.	1
Szerintem igen, mert ha az internetes játékokról vagy weboldalakról van szó, akkor az ezeken alkalmazott szavakkal fejezik ki magukat a beszélget? felek.	1
Szerintem igen. Az embernek így egyre kevesebbet kell pl kölünböző konfliktusokat vagy hivatalos ügyeket személyesen megoldania.	1
Szerintem inkább az internet hatására, de változik. Bár itt is megfigyelhető, hogy a beszélgetőpartnertől függően változtatja az ember a beszédstílusát, tehát idősebbekkel, idegenekkel szemben képes megválni az angol kifejezések használatától, míg a megszokott társaságban külső szemlélő számára szinte érthetetlenül beszél.	
szerintem nem ezek hatására, inkább az idő múlásával változik	1
szerintem sokan átveszik az ott használt kifejezéseket, sokan nem is reflektálnak erre a folyamatra, illetve olyan emberekkel érintkezik, akikkel közösen kialakított, ilyen szavakból is álló "szótárt" használ	1
szlengek száma nő változatos kifejezőképesség lecsökken	1
szóbeli beszédre is ugyanaz hat, mint az írásbelire	1
szóhasználat nem változik	1
Szókincs tekintetében esetleg. Az könnyen lehet.	1
tabutémák eltűnnek az internet hatására	1
Talán annyiban, hogy a technika e két vívmányáról is állandóan szó esik.:-)	1
talán egy kicsit (azt hiszem = asszem) de egyelőre nem számottevő	1
talán gyorsabb és tömörebb lesz, bár mivel az elektromos társalgásnak a beszélt nyelv tömörsége az alapja, nem	
talán igen, ha bizonyos kifejezések átkerülnek a kisebb csoportok nyelvhasználatába	1
talán kevésbé, mint az írás	1
talán legészrevehetőbben a tizenéves korosztály szóhasználatában	1
Talán nem.	2
talán nem?	1
talán	9
Természetesen. Olyan dolgok, amelyek korábban nem voltak szükségesek a mindennapi életben, de hasznosak voltak az elektronikus kommunikációban, most populárissá, bevetté válnak akarva, akaratlanul is.	1
több betűszó keletkezik. A hangulatjelek szerintem nehezebben használhatók a köznyelvben, pl a:), (L) nehezen változtatható (rövid) szóvá.	1
Több szleng kifejezést használunk.	1
tömörítés, pongyola fogalmazás	1

Úgy gondolom, hogy igen. Az interneten használt rövidítések összevont szóalakok a beszédben is egyre népszerűek.	1
Úgy gondolom, hogy jelentős a kölcsönhatás az írásbeli és a szóbeli nyelv között, így az írásban vagy szóban helytelenül használt alakok lassanként a másik nyelvhasználatra is átterjednek.	1
ugyan az a válaszom, mint az előző kérdésnél	1
új szavak keletkezhetnek ezáltal, a legjobb példa erre a facebookról a lájkolni ige, ami mostanra már az ismerőseim között igencsak elterjedt	1
új szavakkal bővül	1
Valamennyire átvesznek szavakat az emberek, de nem mindenki, én pl. nem, és nem is szeretem a buta rövidítéseket, kicsinyítéseket... bár már kezdem megszokni, és hat is rám.	1
Valamennyire biztosan változik, de szerintem nem annyira markánsan, mint az írott nyelv. Hallottam már tinédzserektől, hogy "lávollak" és "lol".	1
Valamennyire igen, az angol mozaikszók belekerülhetnek, de közel se annyira, mint az írás.	1
valamennyire igen, de talán nem annyira, mint az írás	1
valószínű, de sokkal kevésbé, mint az írás	1
Valószínűleg igen, de csak kis mértékben, a szavak szintjén. És nem csak új szavak jelennek meg, hanem a nyelvjárási és a köznyelvi elemek keveredése is felgyorsul (talán).	1
valószínűleg, gyorsabb, kapkodóbb, lényegretörőbb talán a rohanó kor következményeképp.	1
Valószínűnek tartom, hogy idővel az is változni fog.	1
valszeg	1
változik, de nem annak hatására	1
Változik, már most mondják, hogy XD és LOL és egyebeket	1
Változik, mert másképp beszélünk neten, mint telefonon, egyrészt függ attól hogy kivel barát munkatárs, családtag. Neten kicsit másképp van,mert nem figyelünk annyira viszont telefonnál csak a lényeget sűrítjük bele a beszélgetésbe.	1
Változik, mert meghonosodnak korábban csak interneten használt új kifejezések, de ez a változás kisebb mértékű, mint az írást érintő.	1
Változik, minden változik. Mindennek van előnye és hátránya. Itt a hátrány, hogy lehet hogy egy délutáni személyes találkozó helyett a cseten beszélem meg a barátnőmmel a történéseket, az előny, hogy olyanokkal is megbeszélhetem a történéseket, akikkel a távolság miatt nem tehetném.	1
Változott, véleményem szerint elég kis mértékben, hiszen itt számos nem verbális kifejezőeszköz is rendelkezésre áll, nem csak a puszta karakterek.	1
vannak, akik kimondják, hogy LOL, ez nekem nem annyira tetszik	1

Annex 12

Dynamic and graphic emoticons (MSN, Skype, collected: 2011–2012)

Emoticon colours

Facial expression emoticons and gesture emoticons

"Occupation" emoticons and style emoticons

Action emoticons

Textual emoticons

Logo emoticons (Skype)

Sadness, love, agression

Emoticons in phatic function: saying hello and sending good wishes

Punctuation mark emoticons

DOI 10.1515/9783110499117-018

References

Aczél, Petra (2012). *Médiaretorika*. Budapest: Magyar Mercurius.

Aczél, Petra (2013). Netoric: An Approach to New Media Rhetoric. In: van Belle, Hilde et al. (eds.): *Verbal and Visual Rhetoric in a Media World*. Amsterdam: Amsterdam University Press, 311–329.

Aczél, Petra (2014). Expressivity and Emotion in Visionary Rhetoric. In: Benedek, András & Nyíri, Kristóf (eds.): *The Power of the Image: Emotion, Expression, Explanation*. Frankfurt am Main et al.: Peter Lang Verlag, 61–72.

Aczél, Petra (2015). Ingenious Rhetoric: The Visual Secret of Rhetoricality. In: Benedek, András & Nyíri, Kristóf (eds.): *Beyond Words: Pictures, Parables, Paradoxes*. Frankfurt: Peter Lang Verlag, 165–177.

A. Jászó, Anna (ed.) (1991). *A magyar nyelv könyve*. Budapest: Trezor.

Adamik, Tamás; A. Jászó, Anna & Aczél, Petra (2004). *Retorika*. Budapest: Osiris Kiadó.

Ágel, Vilmos & Hennig, Mathilde (eds.) (2007). *Zugänge zur Grammatik der gesprochenen Sprache*. Tübingen: Max Niemeyer Verlag.

Amaghlobeli, Natia (2012). Linguistic Features of Typographic Emoticons in SMS Discourse. *Theory and Practice in Language Studies2/2*, 348–354.

Androutsopoulos, Jannis K. & Ziegler, Evelyn (2003). Regionalismen in einer Chat-Gemeinschaft. In: Androutsopoulos, Jannis K. & Ziegler, Evelyn (eds.): *„Standardfragen". Soziolinguistische Perspektiven auf Sprachgeschichte, Sprachkontakt und Sprachvariation*. Frankfurt am Main: Peter Lang, 251–279.

Aprile, Marcello (2005). *Dalle parole ai dizionari*. Bologna: Il Mulino, 56–67.

Aronoff, Mark (1985). Orthography and linguistic theory: the syntactic basis of Masoratic Hebrew Punctuation. *Language* (Baltimore) 61, 28–72.

Auer, Peter & Günthner, Susanne (2003). Die Entstehung von Diskursmarkern im Deutschen – ein Fall von Grammatikalisierung? *InLiSt – Interaction and Linguistic Structures* 38.http://kops.uni-konstanz.de/bitstream/handle/123456789/3832/Inlist38.pdf?sequence=1 [23. 03. 2016]

B. Gergely, Piroska (1997). Az Erdélyi magyar szótörténeti tár használhatósága a beszélt nyelv történeti vizsgálatában. *Magyar Nyelvjárások* XXXIX, 65–74.

Balaskó, Mária; Balázs, Géza & Kovács, László (eds.) (2010). *Hálózatkutatás. Hálózatok a társadalomban és a nyelvben*. Segédkönyvek a nyelvészet tanulmányozásához 103. Budapest: Tinta Könyvkiadó.

Balázs, Géza & Bódi, Zoltán (eds.) (2005). *Az internetkorszak kommunikációja. Tanulmányok.* Budapest: Gondolat–Infonia.

Balázs, Géza; H. Varga, Gyula & Veszelszki, Ágnes (eds.) (2005). *A magyar szemiotika negyedfél évtized után*. Budapest–Eger: Líceum Kiadó.

Balázs, Géza & Veszelszki, Ágnes (eds.) (2016). *Generációk nyelve. Tanulmánykötet*. Budapest: Mai Magyar Nyelvi Tanszék – Inter – Magyar Szemiotikai Társaság.

Balázs, Géza (ed.) (2002). *Informatikai technológia és nyelvhasználat*. Válogatás a Nemzeti Kulturális Örökség Minisztériuma anyanyelvi pályázatából. Budapest: Trezor.

Balázs, Géza (ed.) (2011). *Nyelvészetről mindenkinek. 77 nyelvészeti összefoglaló*. Budapest: Inter.

Balázs, Géza (1998a). Külső nyelvészet az informatika korában. *Magyar Nyelvőr*, 324–337.

Balázs, Géza (1998b). *Magyar nyelvkultúra az ezredfordulón*. Budapest: A–Z Kiadó.

DOI 10.1515/9783110499117-019

Balázs, Géza (1999). Kommunikációs létformák és átcsapások. Nyelvi-szemiotikai megjegyzések az ezredvégen. *Magyar Nyelv*, 138–154.
Balázs, Géza (2003). „Minden házfalat cseréljetek sms-falra". Sms-fal mint elektronikus graffiti. *Magyar Nyelvőr*, 144–159.
Balázs, Géza (2004). Választási sms-ek folklorisztikai-szövegtani vizsgálata. *Magyar Nyelvőr*, 36–53.
Balázs, Géza (2005a). Az internetkorszak kommunikációja. In: Balázs, Géza & Bódi, Zoltán (eds.): *Az internetkorszak kommunikációja. Tanulmányok*. Budapest: Gondolat–Infonia, 25–57.
Balázs, Géza (2005b). Az új média új műfaja, az sms-hír – nyelvészeti megközelítésben. *Magyar Nyelvőr*, 129–150.
Balázs, Géza (2006). Az sms-folklór – a minimálfolklór nyelvi képe I. *Magyar Nyelvőr*, 439–456.
Balázs, Géza (2007a). Az sms-folklór – a minimálfolklór nyelvi képe II. *Magyar Nyelvőr*, 48–62.
Balázs, Géza (2007b). Szövegantropológia. Szövegek többirányú megközelítése. Szombathely: Berzsenyi Dániel Főiskola; Budapest: Inter Kultúra-, Nyelv- és Médiakutató Központ.
Balázs, Géza (2009). Az informatika hatása a nyelvre. In: Bárdosi, Vilmos (ed.): *Quo vadis philologia temporum nostrorum? Korunk civilizációjának nyelvi képe*. Budapest: Tinta Könyvkiadó, 23–42.
Balázs, Géza (2011). *Sms-nyelv és -folklór*. Budapest: Magyar Szemiotikai Társaság, Inter Kft., Prae.hu.
Balázs, Géza (2015). Netfolklór – intermedialitás és terjedés. *Replika*, 2015/1–2, 171–185.
Balogh, Andrea (2012). Kazuárok és hardcore-ok: Angol szavak a számítógépes játékokban. In: *A sokszínű alkalmazott nyelvészet*. Budapest: Tinta Könyvkiadó, 40–43.
Balogh, Andrea (2014). Geeknyelvtan? A számítógépes játékok nyelvéről. In: Lengyel, Klára & Veszelszki, Ágnes (eds.): *Tudomány, technolektus, terminológia*. Budapest: Éghajlat Könyvkiadó, 345–352.
Bańczerowski, Janusz (2000). *A nyelv és a nyelvi kommunikáció alapkérdései*. Nyomárkay, István (ed.). Budapest: ELTE Szláv és Balti Filológiai Intézet, Lengyel Filológiai Tanszék.
Bańczerowski, Janusz (2008). *A világ nyelvi képe. A világkép mint a valóság metaképe a nyelvben és a nyelvhasználatban*. Segédkönyvek a nyelvészet tanulmányozásához 86. Budapest: Tinta Könyvkiadó.
Barabási, Albert-László (2005). *Behálózva – A hálózatok csodálatos világa a sejtektől a világhálóig*. http://www.mindentudas.hu/barabasialbertlaszlo/20051010barabasi.html [retrieved: 20. 12. 2006; checked: 20. 09. 2016, not available]
Barabási, Albert-László (2002). *Linked: The New Science of Networks*. Perseus Books.
Barabási, Albert-László (2016). *Network Science*. Cambridge University Press.
Bard, Alexander & Söderqvist, Jan (2002). *Netocracy – The New Power Elite and Life After Capitalism*. London: BookHouse Publishing.
Barnes, John A. (1954). *Class and Committees in a Norwegian Island Parish, „Human Relations"*. Buckingham: Open University Press.
Bárthází, Eszter (2008). Manipuláció, valamint manipulációra alkalmas nyelvhasználati eszközök a sajtóreklámokban. *Magyar Nyelv*, 443–463.
Baudrillard, Jean (1987). *Das Andere Selbst*. Wien: Böhlau.
Bauer, András & Berács, József (1998). *Bevezetés a marketingbe*. Budapest: Külkereskedelmi Oktatási és Továbbképző Központ.
Becher, Johannes & Becher, Viktor (2011). Gegen ein Anti-Wikipedia-Dogma an Hochschulen. *Academics.de*. http://www.academics.de/wissenschaft/

gegen_ein_anti_wikipedia_dogma_an_hochschulen_44842.html [retrieved: 17. 02. 2011; checked: 20. 09. 2016]
Beck, Mihály & Beck, Mihályné (1999). A rövidítések természetrajza. *Magyar Nyelv*, 459–463.
Behaghel, Otto (1927/1967). Geschriebenes Deutsch und gesprochenes Deutsch. In: Behaghel, Otto: *Von deutscher Sprache. Aufsätze, Vorträge und Plaudereien*. Lahr. Reprint: Wiesbaden: Sandig, 11–34.
Beißwenger, Michael (ed.) (2001). *Chat-Kommunikation. Sprache, Interaktion, Sozialität & Identität in synchroner computervermittelter Kommunikation. Perspektiven auf ein interdisziplinäres Forschungsfeld.* Stuttgart: Ibidem.
Bencédy, József; Fábián, Pál; Rácz, Endre & Velcsov, Mártonné (1974). *A mai magyar nyelv*. Budapest: Tankönyvkiadó.
Bencze, Lóránt (1992). Deixis és referencia. In: Kozocsa, Sándor Géza (ed.): *Emlékkönyv Fábián Pál hetvenedik születésnapjára*. Budapest: ELTE, 41–49.
Benczik, Vilmos (2001). *Nyelv, írás, irodalom kommunikációelméleti megközelítésben*. Budapest: Trezor Kiadó.
Benczik, Vilmos (2002). A hang mágiájától a hang forradalmáig. In: Nyíri, Kristóf (ed.): *Mobilközösség – mobilmegismerés. Tanulmányok*. Budapest: MTA Filozófiai Kutatóintézete, 91–107.
Benedek, András & Nyíri, Kristóf (eds.) (2011). *Images in Language: Metaphors and Metamorphoses*. Series Visual Learning, vol. 1. Frankfurt am Main: Peter Lang.
Benedek, András & Nyíri, Kristóf (eds.) (2012). *The Iconic Turn in Education*. Series Visual Learning, vol. 2. Frankfurt am Main: Peter Lang.
Benedek, András & Nyíri, Kristóf (eds.) (2013). *How to Do Things with Pictures: Skill, Practice, Performance*. Series Visual Learning, vol. 3. Frankfurt am Main: Peter Lang.
Benedek, András & Nyíri, Kristóf (eds.) (2014). *The Power of the Image: Emotion, Expression, Explanation*. Series Visual Learning, vol. 4. Frankfurt am Main: Peter Lang.
Benedek, András & Nyíri, Kristóf (eds.) (2015). *Beyond Words: Pictures, Parable, Paradoxes*. Series Visual Learning, vol. 5. Frankfurt am Main: Peter Lang.
Benedek, András & Veszelszki, Ágnes (eds.) (2016). *In the Beginning Was the Image. The Omnipresence of Pictures. Time, Truth, Tradition*. Frankfurt/M.: Peter Lang.
Berencsi Erika (1996). Abszurd humor – nyelvi humor. Manuscript. Kossuth Lajos Tudományegyetem. http://www.mek.iif.hu/porta/szint/tarsad/nyelvtud/abszurd/abszurd.htm [retrieved: 03. 06. 2011; checked: 20. 09. 2016]
Beutner, Yvonne (2002). *E-Mall Kommunikation. Eine Analyse*. Stuttgart: ibidem Verlag.
Biber, Douglas (1988). *Variation accross speech and writing*. Cambridge et al.: Cambridge University Press.
Bieswanger, Markus (2007). 2 abbrevi8 or not 2 abbrevi8. A contrastive analysis of different space and time-saving strategies in English and German text messages. In: Floyd, S. et al. (eds.): *Texas Linguistic Forum 50*. Austin: Texas Linguistic Forum, 1–12.
Blog, napló, irodalom. *Korunk*, March 2008.
Bloomfield, Leonard (1933). *Language*. New York: Henry Holt.
Bódi, Zoltán & Veszelszki, Ágnes (2006). *Emotikonok. Érzelemkifejezés az internetes kommunikációban*. Budapest: Magyar Szemiotikai Társaság.
Bódi, Zoltán (1997). Cybernyelv, avagy a nemzetközi számítástechnikai hálózatok nyelvi világa. In: Fercsik, Erzsébet & T. Somogyi, Magda (eds.): *Oktatási tapasztalatok – Kutatási eredmények. Tanulmányok az anyanyelv használatának és oktatásának kérdéseiről*. Budapest: Korona Nova, 156–172.

Bódi, Zoltán (1998). Internetes kommunikáció – beszédkommunikáció. In: Gósy, Mária (ed.): *Beszédkutatás'98. Beszéd, spontán beszéd, beszédkommunikáció.* Budapest: MTA Nyelvtudományi Intézete, 178–188.
Bódi, Zoltán (1999). *Az elektronikus levelezés és az internet nyelvhasználatáról.* Presentation on the Conference IX. Magyar Alkalmazott Nyelvészeti Kongresszus. Veszprém University, Veszprém, 8–10. 04. 1999.
Bódi, Zoltán (2000). Az informatikai nyelvhasználat jelei és kommunikációs konfliktusai. In: Balázs, Géza & Voigt, Vilmos (eds.): *Jeles jogok – jogos jelek. Nyelvi jogok – társadalmi konfliktusok.* Budapest: Magyar Szemiotikai Társaság, 197–208.
Bódi, Zoltán (2004a). *A világháló nyelve. Internetezők és internetes nyelvhasználat a magyar társadalomban.* Budapest: Gondolat Kiadó.
Bódi, Zoltán (2004b). Az írás és a beszéd viszonya az internetes interakcióban. *Magyar Nyelvőr*, 286–294.
Bódi, Zoltán (2005). Szimbolikus írásbeliség az internetes interakcióban. In: Balázs, Géza & Bódi, Zoltán (eds.): *Az internetkorszak kommunikációja. Tanulmányok.* Budapest: Gondolat-Infonia, 195–212.
Bódi, Zoltán (2011). *Infoszótár. Informatikai fogalmak eredete, magyarázata és használata.* Budapest: Tinta Könyvkiadó.
Boehm, Gottfried (ed.) (1994). *Was ist ein Bild?* München: Wilhelm Fink Verlag.
Boellstorff, Tom (2013). Making big data, in theory. *First Monday*, 18/10, 7 October 2013. Online: http://firstmonday.org/ojs/index.php/fm/article/view/4869/3750 [retrieved: 21. 01. 2014; checked: 20. 09. 2016]
Bőgel, György (ed.) (2006). *Blogvilág. Egy műfaj születése.* Budapest: HVG Kiadó.
Bonka, Boneva; Kraut, Robert & Frohlich, David (2003). E-Mail und interpersonale Beziehungen. Das Geschlecht macht den Unterschied. In: Höflich, Joachim R. & Gebhardt, Julian (eds.): *Vermittlungskulturen im Wandel. Brief, E-mail, SMS.* Frankfurt am Main – Berlin – Bern – Bruxelles – New York – Oxford – Wien: Lang, 147–172.
Boyd, Danah M. & Ellison, Nicole B. (2007). Social network sites: Definition, history, and scholarship. *Journal of Computer-Mediated Communication* 13/1: article 11.
Brammerts, Helmut (1996). Leitfaden für das Sprachenlernen im Tandem. Tandem per Internet und das International E-Mail Tandem Network. In: Brammerts, Helmut & Little, David (eds.): *Leitfaden für das Sprachenlernen im Tandem über das Internet.* Bochum: Brockmeyer, 1–14.
Brammerts, Helmut (1999). Fremdsprachenlernen mit neuen Technologien: Didaktische Überlegungen. In: Donath, Reinhard (ed.): *Internet und Multimedia in der Erwachsenenbildung.* Stuttgart: Klett, 32–43.
Brubaker, Jed R. & Vertesi, Janet (2010). *Death and the Social Network.* Presentation. CHI 2010 Workshop on HCI at the End of Life: Understanding Death, Dying, and the Digital, Atlanta, GA, USA. http://www.jedbrubaker.com/wp-content/uploads/2010/02/brubaker-vertesi-death-sns.pdf [retrieved: 21. 01. 2014; checked: 20. 09. 2016]
Bußman, Hadumod (ed.) (2002). *Lexikon der Sprachwissenschaft.* Stuttgart: Alfred Kröner Verlag.
Bybee, Joan (2003). Mechanisms of Change in Grammaticalization: The Role of Frequency. In: Joseph, Brian D. & Janda, Richard D. (eds.): *The Handbook of Historical Linguistics.* Malden – Oxford – Melbourne – Berlin: Blackwell, 602–623.
Chafe, Wallace & Danielewitz, Jane (1987). Properties of spoken and written language. In: Holowitz, Rosalind & Samuels, Jay S. (eds.): *Comprehending oral and written language.* San Diego et al.: Academic Press, 83–113.

Cheal, Daniel (1987). Showing them you love them: Gift giving and the dialectic of intimacy. *The Sociological Review* 36, 150–169.

Cohen, David (2015). SurePayroll: Infographic: Hashing Out Hashtags. *Adweek*. 23. 04. 2015. http://www.adweek.com/socialtimes/infographic-surepayroll-hashtags/619169 [retrieved: 14. 10. 2015; checked: 14. 10. 2016]

Croft, William (2001). *Radical Construction Grammar*. Oxford: University Press.

Condliffe, Jamie (2013). Will There Ever Be More Dead People Than Living Ones on Facebook? *Gizmodo*, 10. 31. 2013. http://gizmodo.com/will-there-ever-be-more-dead-people-than-living-ones-on-1455816826 [21. 01. 2014]

Crystal, David (2001). *Language and the Internet*. Cambridge: University Press.

Crystal, David (2003). *A nyelv enciklopédiája*. Budapest: Osiris Kiadó.

Crystal, David (2008). *Txtng. The gr8 db8*. Oxford et al.: Oxford Univ. Press

Csala, Bertalan (2009). Morfondírozások a blogról a közhely és az újítások kapcsán. In: Bárdosi, Vilmos (ed.): *Quo vadis philologia temporum nostrorum? Korunk civilizációjának nyelvi képe*. Budapest: Tinta Könyvkiadó, 77–86.

Csépe, Valéria (2002). Gyermekek a mobil információs társadalomban. Kognitív előnyök és hátrányok. In: Nyíri, Kristóf (ed.): *Mobilközösség – mobilmegismerés. Tanulmányok*. Budapest: MTA Filozófiai Kutatóintézete, 167–177.

Daneš, František (1970). One instance of Prague school methodology: Functional analysis of utterance and text. In: Garvin, P. L. (ed.): *Method and Theory in Linguistics*. The Hague: Mouton, 132–146.

Daneš, František (1974). Functional sentence perspective and the organization of the text. In: Daneš, František (ed.): *Papers on Functional Sentence Perspective*. Janua Linguarum, Series Minor, 147. The Hague: Mouton, 106–128.

Daróczi, Ildikó (2016). Kontrastive Aspekte des Neologismus 'Selfie'. *Argumentum* 12, 56–86.

Dawkins, Richard (1976). *The Selfish Gene*. Oxford: University Press.

Dawkins, Richard (1993). *Viruses of the mind. Dennett and his critics: Demystifying Mind*. Cambridge (Mass.): Blackwell.

Deli, Zsuzsa (2005). *Nemzetközi marketing*. Budapest: Műegyetemi Kiadó.

Dennett, Daniel C. (1990). Memes and the exploitation of imagination. *The Journal of Aesthetics and Art Criticism* 48, 127–135.

Deppermann, Arnulf; Fiehler, Reinhard & Spranz-Fogasy, Thomas (eds.) (2006). *Grammatik und Interaktion. Untersuchungen zum Zusammenhang von grammatischen Strukturen und Gesprächsprozessen*. Radolfzell: Verlag für Gesprächsforschung.

Deppermann, Arnulf; Fiehler, Reinhard & Spranz-Fogasy, Thomas (2006). Zur Einführung: Grammatik und Interaktion. In: Deppermann, Arnulf; Fiehler, Reinhard & Spranz-Fogasy, Thomas (eds.): *Grammatik und Interaktion. Untersuchungen zum Zusammenhang von grammatischen Strukturen und Gesprächsprozessen*. Radolfzell: Verlag für Gesprächsforschung, 5–10.

Deppermann, Arnulf (2006). Construction Grammar – Eine Grammatik für die Interaktion? In: Deppermann, Arnulf; Fiehler, Reinhard & Spranz-Fogasy, Thomas (Hrsg.): *Grammatik und Interaktion. Untersuchungen zum Zusammenhang von grammatischen Strukturen und Gesprächsprozessen*. Radolfzell: Verlag für Gesprächsforschung, 43–66.

Dér, Csilla Ilona (2008). *Grammatikalizáció*. Nyelvtudományi Értekezések 158. Budapest: Akadémiai Kiadó.

Dér, Csilla Ilona (2009). Mik is a diskurzusjelölők? In: Keszler, Borbála & Tátrai, Szilárd (eds.): *Diskurzus a grammatikában – grammatika a diskurzusban*. Segédkönyvek a nyelvészet tanulmányozásához 88. Budapest: Tinta Könyvkiadó, 293–303.

Dér, Csilla (2013a). Grammaticalization: A specific type of semantic, categorical, and prosodic change. *Berliner Beiträge zur Hungarologie* 18: 160–179.

Dér, Csilla Ilona (2013b). Mi számít grammatikalizációnak? Kísérlet a fogalom újradefiniálására. In: *A nyelvtörténeti kutatások újabb eredményei VII*. Szeged: Szegedi Tudományegyetem Magyar Nyelvészeti Tanszék, 17–28.

Di Meola, Claudio (2000). *Die Grammatikalisierung deutscher Präpositionen*. Tübingen: Stauffenburg.

Diewald, Gabriele (1997). *Grammatikalisierung. Eine Einführung in Sein und Werden grammatischer Formen. Germanistische Arbeitshefte*. Tübingen: Max Niemeyer.

Dittmann, Miguel (2001). *Sprachverwendung im Internet. Untersuchungen des Internet Relay Chat (IRC) in Deutschland und Frankreich*. Sarlat: Éditions Indoles.

Domokos, Mariann & Vargha, Katalin (2015). Elektronikus választási folklór 2014. *Replika* 2015/1–2, 141–169.

Domokos, Mariann (2014). Methodological Issues in Electronic Folklore. *Slovensky Národopis* 62, 283–295.

Domonkosi, Ágnes (2012). Önstilizáció, stílusstratégia, átértékelődő stílusminták. In: *A stílus szociokulturális tényezői*. Budapest: ELTE, 191–221.

Domonkosi, Ágnes (2013). Self-stylization, style strategy, stylistic patterns re-evaluated: the main sytlistic features of a personal blog. *Studia Linguistica Hungarica* 28, 99–120.

Dömötör, Adrienne (2009). Idézőjelben mondom. In: Keszler, Borbála & Tátrai, Szilárd (eds.): *Diskurzus a grammatikában – grammatika a diskurzusban*. Segédkönyvek a nyelvészet tanulmányozásához 88. Budapest: Tinta Könyvkiadó, 312–321.

Döring, Nicola (1997). Kommunikation im Internet. Neun theoretische Ansätze. In: Batinic, Bernad (ed.): *Internet für Psychologen*. Göttingen: Hogrefe, 267–298.

Döring, Nicola (2002). "Kurzm. wird gesendet" – Abkürzungen und Akronyme in der SMS-Kommunikation. *Muttersprache. Vierteljahresschrift für deutsche Sprache*, 2.

Dunbar, Robin I. M. (2002). Vannak-e kognitív korlátai az e-világnak? In: Nyíri, Kristóf (ed.): *Mobilközösség – mobilmegismerés. Tanulmányok*. Budapest: MTA Filozófiai Kutatóintézete, 55–66.

Dürscheid, Christa (1999). Zwischen Mündlichkeit und Schriftlichkeit: die Kommunikation im Internet. *Papiere zur Linguistik* 60/1, 17–30.

Dürscheid, Christa (2002). E-Mail und SMS – ein Vergleich. In: Ziegler, Arne & Dürscheid, Christa (eds.): *Kommunikationsform E-Mail*. Tübingen: Stauffenburg. 93–114.

Dürscheid, Christa (2005). E-Mail – verändert sie das Schreiben? In: Siever, Torsten; Schlobinski, Peter & Runkehl, Jens (eds.): *Websprache.net. Sprache und Kommunikation im Internet*. Berlin – New York: Walter de Gruyter, 85–97.

Dürscheid, Christa (2006). *Einführung in die Schriftlinguistik. 3.*, überarbeitete und ergänzte Auflage. Göttingen: Vandenhoek&Ruprecht.

Dürscheid, Christa (i.p.) (2016). Nähe, Distanz und neue Medien. In: Hennig, Mathilde & Feilke, Helmuth (eds.): *Zur Karriere von Nähe und Distanz* [zur Publikation angenommen].

É. Kiss, Katalin (1979). A mondatátszövődésről. In: Szathmári, István & Várkonyi, István (eds.): *A szövegtan a kutatásban és az oktatásban*. Kaposvár, 93–105.

É. Kiss, Katalin (1983). *A magyar mondatszerkezet generatív leírása*. NyTudÉrt 116. Budapest: Akadémiai Kiadó.

Eckardt, Regine (2006). *Meaning Change in Grammaticalization. An Enquiry into Semantic Reanalysis*. Oxford: University Press.

Eckkramer, Eva Martha & Eder, Hildegrund Maria (2000). *(Cyber)Diskurs zwischen Konvention und Revolution. Eine multilinguale textlinguistische Analyse von Gebrauchstextsorten im realen und virtuellen Raum.* Frankfurt am Main et al.: Peter Lang.

Eisenberg, Peter (2007). Sollen Grammatiken die gesprochene Sprache beschreiben? In: Ágel, Vilmos & Hennig, Mathilde (eds.): *Zugänge zur Grammatik der gesprochenen Sprache.* Tübingen: Max Niemeyer Verlag, 276–295.

Érsok, Nikoletta Ágnes (2003). Írva csevegés – virtuális írásbeliség. *Magyar Nyelvőr*, 99–104.

Érsok, Nikoletta Ágnes (2004). Sömös, sumus, írj vissza. *Magyar Nyelvőr*, 294–313.

Érsok, Nikoletta Ágnes (2006). Szóbeliség és/vagy írásbeliség. *Magyar Nyelvőr*, 165–176.

Érsok, Nikoletta Ágnes (2007). *Az internetes kommunikáció műfajai (különös tekintettel az interaktív magánéleti műfajokra).* PhD dissertation. Manuscript. Budapest: ELTE.

Fábián, Pál (1998). A magyar nyelvművelés története. In: Fábián, Pál & Tátrai, Szilárd (eds.): *Nyelvművelés.* Budapest: Nemzeti Tankönyvkiadó, 9–70.

Fábián, Zsuzsanna (2009). Neologizmusok és szótárak. In: Bárdosi, Vilmos (ed.): *Quo vadis philologia temporum nostrorum? Korunk civilizációjának nyelvi képe.* Budapest: Tinta Könyvkiadó, 105–120.

Fahlman, Scott E. (n.d.): *Smiley Lore.* http://www-2.cs.cmu.edu/~sef/sefSmiley.htm [retrieved: 03. 06. 2011; checked: 08. 10. 2016]

Faragó, László (1931). Hozzászólás a Nyugat reklám-ankétjához. *Nyugat* 1931/4. Disputa.

Fazakas, Noémi (2015). Language Myths and Ideologies in the New Media. *Revue Internationale D'études En Langues Modernes Appliquées / International Review Of Studies In Applied Modern Languages 8*, 102–113.

Feilke, Helmuth; Kappest, Klaus-Peter & Knobloch, Clemens (eds.) (2001). *Grammatikalisierung, Spracherwerb und Schriftlichkeit.* Tübingen: Max Niemeyer.

Ferrara, Kathleen; Brunner, Hans & Whittemore, Greg (1991). Interactive written discourse as an emergent register. *Written Communication* 8/1, 8–34.

Fehér, Katalin (2012). Az internettől az új médiáig. *Századvég* 66, 3–18.

Fiehler, Reinhard ; Barden, Birgit; Elstermann, Mechthild & Kraft, Barbara (2004). *Eigenschaften gesprochener Sprache.* Tübingen: Gunter Narr Verlag.

Fiehler, Reinhard (2000). Gesprochene Sprache – gibt's die? In: Ágel, Vilmos & Herzog, Andreas (eds.): *Jahrbuch der ungarischen Germanistik 2000.* Budapest – Bonn. 93–104.

Fiehler, Reinhard (2005). Gesprochene Sprache. In: *Duden. Die Grammatik.* 7. Auflage. Mannheim: Dudenverlag, 1175–256.

Fiehler, Reinhard (2006). Was gehört in eine Grammatik gesprochener Sprache? Erfahrungen beim Schreiben eines Kapitels der neuen DUDEN-Grammatik. In: Deppermann, Arnulf; Fiehler, Reinhard & Spranz-Fogasy, Thomas (eds.): *Grammatik und Interaktion. Untersuchungen zum Zusammenhang von grammatischen Strukturen und Gesprächsprozessen.* Radolfzell: Verlag für Gesprächsforschung, 21–42.

Fiehler, Reinhard (2007). Thesen zur Struktur einer Grammatik der gesprochenen Sprache. In: Ágel, Vilmos & Hennig, Mathilde (eds.): *Zugänge zur Grammatik der gesprochenen Sprache.* Tübingen: Max Niemeyer Verlag, 297–314.

Firbas, Jan (1964). On defining the theme in functional sentence analysis. *Travaux linguistiques de Prague* 1, 267–280.

Fillmore, Charles J. (1981). Pragmatics and the description of discourse. In: Cole, Peter (ed.): *Radical pragmatics.* New York: Academic Press, 143–166.

Fix, Tina (2001). *Generation @ im Chat. Hintergrund und explorative Motivstudie zur jugendlichen Netzkommunikation.* München: KoPäd Verlag.

Flusser, Vilém (2004). *A technikai képek mindensége [univerzuma] felé*. Budapest: Kijárat Kiadó.

Főző, Attila László (2006). Szinkrón és aszinkrón kommunikáció IKT alapú oktatási projektekben. *Új Pedagógiai Szemle* 2006/1, 104–112.

Frehner, Carmen (2008). *Email – SMS – MMS. The Linguistic Creativity of Asynchronous Discourse in the New Media Age*. Bern et al.: Peter Lang.

Frindte, Wolfgang & Köhler, Thomas (eds.) (1999). *Kommunikation im Internet*. Frankfurt am Main et al.: Lang.

Frindte, Wolfgang (1999a). Neue Medien – neue Wirklichkeiten. Szenarios. In: Frindte, Wolfgang & Köhler, Thomas (eds.): *Kommunikation im Internet*. Frankfurt am Main et al.: Lang, 11–22.

Frindte, Wolfgang (1999b). Dialoge in Netzstrukturen: Medienphilosophisches. In: Frindte, Wolfgang & Köhler, Thomas (eds.): *Kommunikation im Internet*. Frankfurt am Main et al.: Lang, 23–50.

Fülöp, Géza (1996). *Az információ*. Budapest: ELTE Könyvtártudományi-Informatikai Tanszék.

Gagyi, Benedek (2008). Blogirodalom. *Korunk*. Blog, napló, irodalom. March 2008, 25–30.

Gellert, Christian Fürchtegott (1989). Gedanken von einem guten deutschen Briefe, an den Herrn F. H. v. W. In: Witte, Bernd (ed.): *Gesammelte Schriften. Kritische, kommentierte Ausgabe*. Band 4. Berlin – New York.

Giles, Jim (2005). Special Report: Internet encyclopaedias go head to head. *Nature* 438, 900–901.

Giuliani, Luca (2006). Macht und Ohnmacht der Bilder. Eine frisch gewaschene Schürze und die gemordeten Mamelucken. In: Maar, Christa & Burda, Hubert (eds.): *Iconic worlds. Neue Bilderwelten und Wissensräume*. Köln: DuMont, 185–204.

Glózer, Rita (2015). Internetes paródiavideók és ifjúsági médiahasználat. *Replika* 90–91, 117–139.

Gonda, Zsuzsa (2015). *Digitális szövegek olvasásának típusai és stratégiái*. Bölcsészet- és Művészetpedagógiai Kiadványok. Budapest: Eötvös Loránd Tudományegyetem.

Gósy, Mária (1999). Hangzó és írott anyanyelvünk. In: *Magyar Nyelvjárások. A Kossuth Lajos Tudományegyetem Magyar Nyelvtudományi Tanszékének évkönyve, XXXVI*. Debrecen: Kossuth Lajos Tudományegyetem, 3–24.

Gottsched, Johann Christoph (1973). Von poetischen Sendschreiben oder Briefen. In: *Ausgewählte Werke*. Band 6/2. Berlin – New York.

Gülich, Elisabeth (1986). Textsorten in der Kommunikationspraxis. In: Kallmeyer, Werner (ed.): *Kommunikationstypologie. Handlungsmuster, Textsorten, Situationstypen*. Düsseldorf: Schwann, 15–46.

Günther, Harmut (1988). *Schriftliche Sprache. Strukturen geschriebener Wörter und ihre Verarbeitung beim Lesen*. Tübingen: Max Niemeyer Verlag.

Günther, Klaus B. & Günther, Harmut (eds.) (1983). *Schrift, Schreiben, Schriftlichkeit. Arbeiten zur Struktur, Funktion und Entwicklung schriftlicher Sprache*. Tübingen: Max Niemeyer Verlag.

Gyarmathy, Dorottya (2013). Hungarikum a nyelvben. In: Bárdosi, Vilmos (ed.): *Reáliák. A lexikológiától a frazeológiáig*. Budapest: Tinta Könyvkiadó, 229–238.

Gyuricza, Katalin (2016). Az intertextualitás vizsgálata nyelvészeti szempontból. *Magyar Nyelvőr* 140/1, 71–95.

Gyuris, Beáta (2003). The Semantics of Contrastive Topics in Hungarian. Doctoral dissertation. Budapest: Eötvös Loránd University.

Haas, Christina; Takayoshi, Pamela; Carr, Brandon; Hudson, Kimberley & Pollock, Ross (2011). Young People's Everyday Literacies: The Language Features of Instant Messaging. *Research in the Teaching of English* 45/4, n. p.
Hardaker, Claire (2010). Trolling in asynchronous computer-mediated communication: From user discussions to academic definitions. *Journal of Politeness Research* 6, 215–242.
H. Kietzmann, Jan & Hermkens, Kristopher (2011). Social media? Get serious! Understanding the functional building blocks of social media. *Business Horizons* 54, 241–251.
H. Molnár, Ilona (1961). Egy sajátos határozó alakú mondatrészről. In: Sulán, Béla (ed.): *Nyelvtani tanulmányok*. Budapest: Tankönyvkiadó, 143–154.
H. Molnár, Ilona (1968). *Módosító szók és módosító mondatrészletek a mai magyar nyelvben*. Nyelvtudományi Értekezések 60. Budapest: Akadémiai Kiadó.
Haader, Lea (2000). A tárgyi mellékmondat. In: Keszler, Borbála (ed.): *Magyar grammatika*. Budapest: Nemzeti Tankönyvkiadó, 487–490.
Haader, Lea (2003). Az összetett mondat. In: Kiss, Jenő & Pusztai, Ferenc (eds.): *Magyar nyelvtörténet*. Budapest: Osiris, 500–560.
Haader, Lea (2008). Az alárendelt mondatok változási irányairól. In: Haader, Lea & Horváth, László (eds.): *Tanulmányok a középmagyar kor mondattana köréből*. Segédkönyvek a nyelvészet tanulmányozásához 89. Budapest: Tinta Könyvkiadó, 77–87.
Haase, Martin; Huber, Michael; Krumeich, Alexander & Rehm, Georg (1997). Internetkommunikation und Sprachwandel. In: Weingarten, Rüdiger (ed.): *Sprachwandel durch Computer*. Opladen: Westdeutscher Verlag, 51–85.
Harnad, Stevan (2006a). Back to the oral tradition. In: Origgi, Gloria (ed.): *text-e. Text in the Age of the Internet*. Houndmills – New York: Palgrave Macmillan, 15–16.
Harnad, Stevan (2006b). Skyreading and Skywriting for researchers: a post-Gutenberg anomaly and how to resolve it. In: Origgi, Gloria (ed.): *text-e. Text in the Age of the Internet*. Houndmills – New York: Palgrave Macmillan, 46–61.
Haspelmath, Martin (1999). Why is grammaticalization irreversible? *Linguistics* 37/6, 1043–1068.
Heijnk, Stefan (2002). *Texten fürs Web. Grundlagen und Praxiswissen für Online-Redakteure*. Heidelberg: dpunkt.verlag.
Heine, Bernd & Kuteva, Tania (2002). *World Lexicon of Grammaticalization*. Cambridge: University Press.
Heine, Bernd; Claudi, Ulrika & Hünnemeyer, Friederike (1991). *Grammaticalization: a conceptual framework*. Chicago–London: University of Chicago Press.
Hodel, Jan (2010). Recherche: Google – and Far Beyond. In: Gasteiner, Martin & Haber, Peter (eds.): *Digitale Arbeitstechniken für die Geistes- und Kulturwissenschaften*. Wien – Köln – Weimar: Böhlau, 25–37.
Hoffmann, Ludger (1998). *Grammatik der gesprochenen Sprache*. Heidelberg: Groos.
Höflich, Joachim R. (2003a). Mediatisierung des Alltags und der Wandel von Vermittlungskulturen. In: Höflich, Joachim R. & Gebhardt, Julian (eds.): *Vermittlungskulturen im Wandel. Brief, E-mail, SMS*. Frankfurt am Main et al.: Lang, 7–20.
Höflich, Joachim R. (2003b). Vermittlungskulturen im Wandel. Brief, E-mail, SMS. In: Höflich, Joachim R. & Gebhardt, Julian (eds.): *Vermittlungskulturen im Wandel. Brief, E-mail, SMS*. Frankfurt am Main et al.: Lang, 39–61.
Hoppál, Mihály (1984). Szent Antal-lánc, avagy a szerencse lánclevele. *Világosság* XXV/12, 769–776.

Hopper, Paul J. & Traugott, Elizabeth Closs (1993). *Grammaticalization*. Cambridge: University Press.

Hortoványi, Judit (2015). A média elemeinek hatása 14–16 éves serdülők önkifejező rajzain. *Psychiatria Hungarica* 30, 192–200.

Horváth, Dóra (2014). Az alma megosztása és feldolgozása: Kép, szöveg és mém az új generáció változó információ fogyasztásában. In: Lehota, József (ed.): *Tomcsányi Pál akadémikus 90 éves*. Budapest: MTA Agrár-közgazdasági Bizottság Agrármarketing Albizottság, 93–109.

Horváth, Dóra & Mitev, Ariel (2016a). Értékesek-e a mémek? Internetes mémérték a felhasználók véleményén keresztül. *Marketing és Menedzsment* 50/1, 25–35.

Horváth, Dóra & Mitev, Ariel (2016b). Internet Memes: The Meteorites of the Online World: Spontaneous Online Content with Corporate Relevance. In: De Pelsmacker, Patrick (ed.): *Advertising in New Formats and Media: Current Research and Implications for Marketers*. Emerald Group Publishing Ltd., 243–262.

Horváth, Dóra; Mitev, Ariel & Veszelszki, Ágnes (2013). Egy kiállítás mémei: a mémek megosztásának és interpretálásának fogyasztói élményvilága. In: Király, Éva (ed.): *„Kiterjesztett" marketing*. Budapest: Budapesti Gazdasági Főiskola, 140–164.

Horváth, Dóra (2015). Memes at an Exhibition: Consumer Interpretations Of Internet Memes. In: Verlegh, Peeter (ed.): *Advances in Advertising Research*. Vol. VI. Wiesbaden: Springer Fachmedien, 51–62.

Horváth, János (1911). A Nyugat magyartalanságairól. *Magyar Nyelv* 7, 61–74.

Hrachovec, Herbert (2002). Közvetett-közvetített jelenlét. In: Nyíri, Kristóf (ed.): *Mobilközösség – mobilmegismerés. Tanulmányok*. Budapest: MTA Filozófiai Kutatóintézete, 115–125.

Hug, Theo (2010). *Media competency and visual literacy – Conceptual considerations*. Presentation in the Visual Learning Lab, Budapest, 6. October 2010. Abstract: http://www.vll.mpt.bme.hu/images/stories/PDF/hug_abs.pdf [17. 05. 2011]

Imo, Wolfgang (2007a). *Construction Grammar und Gesprochene-Sprache-Forschung. Konstruktionen mit zehn matrixsatzfähigen Verben im gesprochenen Deutsch*. Tübingen: Max Niemeyer Verlag.

Imo, Wolfgang (2007b). Zur Anwendung der Construction Grammar auf die gesprochene Sprache – der Fall „ich mein(e)". In: Ágel, Vilmos & Hennig, Mathilde (eds.): *Zugänge zur Grammatik der gesprochenen Sprache*. Tübingen: Max Niemeyer Verlag, 3–34.

Jakusné Harmos, Éva (2005). A kereskedelmi és a politikai propaganda nyelvi eszközei. *Magyar Nyelvőr*, 419–436.

Juhász, Dezső (1991). A módosítószók. In: Benkő, Loránd (ed.): *A magyar nyelv történeti nyelvtana I*. Budapest: Akadémiai Kiadó, 501–514.

Juhász, Valéria (2010). Az azonnali üzenetküldések diskurzusszerveződésének sajátosságai. In: Kukorelli, Katalin (ed.): *A tartalom és forma harmóniájának kommunikációja*. Dunaújváros: Dunaújvárosi Főiskola, 62–74.

Juhász, Valéria (2011). *IM az EMC-ben*. Presentation in the Series A Gutenberg-galaxistól a Google-galaxisig. ELTE BTK, Budapest, 04. 05. 2011.

Kádár, Edit (2002). A relatív állítások és az ún. témaismétlő névmások. In: Andor, József; Benkes, Zsuzsa & Bókay, Antal (eds.): *Szöveg az egész világ. Petőfi S. János 70. születésnapjára*. Budapest: Tinta Könyvkiadó, 302–312.

Kálmán, László & Trón, Viktor (2005). *Bevezetés a nyelvtudományba*. Budapest: Tinta Könyvkiadó.

Kálmán, László & Trón, Viktor (2007). *Bevezetés a nyelvtudományba*. Second edition. Budapest: Tinta Könyvkiadó.

Kamer, Nimrod (2013). Brace yourselves for the proliferation of the 'finger hashtag'. *Wired*, 06. 03. 2013. http://www.wired.co.uk/news/archive/2013-03/06/hashtags [retrieved: 09. 05. 2015; checked: 07. 10. 2016; webpage content has changed]

Kaplan, Andreas M. & Haenlein, Michael (2010). Users of the world, unite! The challenges and opportunities of social media. *Business Horizons* 53/1, 61.

Karácsony, András (2002). Individualitás a nomádok földjén. Pillanatképek. In: Nyíri, Kristóf (ed.): *Mobilközösség – mobilmegismerés. Tanulmányok*. Budapest: MTA Filozófiai Kutatóintézete, 127–138.

Karinthy, Frigyes (1929). *Minden másképpen van. Ötvenkét vasárnap*. Budapest: Athenaeum Irodalmi és Nyomdai Rt.

Károly, Sándor (1958). *Az értelmező és az értelmezői mondat a magyarban*. NytudÉrt. 16. Budapest: Akadémiai Kiadó.

Károly, Sándor (1970). *Általános és magyar jelentéstan*. Budapest: Akadémiai Kiadó.

Kassai, Ilona (1993). Gyorsult-e a magyar beszéd tempója az elmúlt 100–120 évben? In: Gósy, Mária (ed.): *Beszédkutatás '93*. Budapest: MTA Nyelvtudományi Intézet, 62–69.

Kelle, Bernhard (2000). Regionale Varietäten in Internet-Chat als Wegbereiter einer regionalen Schriftlichkeit. *Deutsche Sprache* 2000/4, 357–371.

Keszler, Borbála (ed.) (2000). *Magyar grammatika*. Budapest: Nemzeti Tankönyvkiadó.

Keszler, Borbála (1977). Az egyszerű és az összetett mondat határsávja. In: Rácz, Endre & Szathmári, István (eds.): *Tanulmányok a mai magyar nyelv mondattana köréből*. Budapest: Nemzeti Tankönyvkiadó, 111–133.

Keszler, Borbála (1983). Kötetlen beszélgetések mondat- és szövegtani vizsgálata. In: Rácz, Endre & Szathmári, István (eds.): *Tanulmányok a mai magyar nyelv szövegtana köréből*. Budapest: Tankönyvkiadó, 164–202.

Keszler, Borbála (1985). Über die Verwendung der Füllwörter. Annales Univ. Sci. Budapest. *Sectio Linguistica* XVIII, 11–26.

Keszler, Borbála (1986). *A magyar írásjelhasználat és grammatikai alapjai*. Dissertation (for the degree „kandidátus"). Manuscript.

Keszler, Borbála (1992). A közbevetések grammatikai és mondatfonetikai sajátosságai. In: Keszler, Borbála (ed.): *Újabb fejezetek a magyar leíró nyelvtan köréből*. Budapest: Tankönyvkiadó, 82–88.

Keszler, Borbála (1998). Bevezető szók és kifejezések. In: Zoltán, András (ed.): *Nyelv – stílus – irodalom. Emlékkönyv Péter Mihály 70. születésnapjára*. Budapest: [n. d.], 290–298.

Keszler, Borbála (2000a). A szóképzés. In: Keszler, Borbála (ed.): *Magyar grammatika*. Budapest: Nemzeti Tankönyvkiadó, 307–320.

Keszler, Borbála (2000b). A szóképzés. *Magyar Nyelvőr*, 59–73.

Keszler, Borbála (2004). *Írásjeltan. Az írásjelhasználat szabályai, problémái és története*. Budapest: Nemzeti Tankönyvkiadó.

Keszler, Borbála (2006). *Írásjel-használati gyakorlókönyv. Az írásjelhasználat szabályai, problémái és története*. Budapest: Nemzeti Tankönyvkiadó.

Keszler, Borbála (2009). A nyelvújítás és a magyar orvosi nyelv. In: Bárdosi, Vilmos (ed.): *Quo vadis philologia temporum nostrorum? Korunk civilizációjának nyelvi képe*. Budapest: Tinta Könyvkiadó, 189–198.

Keszler, Borbála (2011). *A meggyőzés eszközei a régi magyar receptekben*. Manuscript.

Kilian, Jörg (2001). T@stentöne. Geschriebene Umgangssprache in computervermittelter Kommunikation. Historisch-kritische Ergänzungen zu einem neuen Feld der linguistischen Forschung. In: Beißwenger, Michael (ed.): *Chat-Kommunikation. Sprache, Interaktion,*

Sozialität & Identität in synchroner computervermittelter Kommunikation. Perspektiven auf ein interdisziplinäres Forschungsfeld. Stuttgart: Ibidem, 55–78.

Kis, Tamás (1991). *A magyar katonai szleng szótára (1980–1990)*. Debrecen: Kossuth Egyetemi Kiadó.

Kiss, Jenő (1995/2002). *Társadalom és nyelvhasználat. Szociolingvisztikai alapfogalmak*. Budapest: Nemzeti Tankönyvkiadó.

Kiss, Mariann (2005). *Alapmarketing*. Budapest: Aula Kiadó.

Knoop, Ulrich (1983). Zum Status der Schriftlichkeit in der Sprache der Neuzeit. In: Günther, Klaus B. & Günther, Harmut (eds.): *Schrift, Schreiben, Schriftlichkeit. Arbeiten zur Struktur, Funktion und Entwicklung schriftlicher Sprache*. Tübingen: Max Niemeyer Verlag, 159–167.

Knorr, Dagmar & Jakobs, Eva-Maria (eds.) (1997). *Textproduktion in elektronischen Umgebungen*. Frankfurt am Main et al.: Peter Lang.

Koch, Peter & Oesterreicher, Wulf (1985). Sprache der Nähe – Sprache der Distanz. Mündlichkeit und Schriftlichkeit im Spannungsfeld von Sprachtheorie und Sprachgeschichte. In: *Romanistisches Jahrbuch* 36, 15–43.

Koch, Peter & Oesterreicher, Wulf (1994). Schriftlichkeit und Sprache. In: Günther, Hartmut & Ludwig, Otto (eds.): *Schrift und Schriftlichkeit. Ein interdisziplinäres Handbuch internationaler Forschung*. 1. Halbband. Berlin – New York: Walter de Gruyter, 587–604.

Köhler, Thomas (1999). Methodik und Methodologie der Forschung zur Internet-Kommunikation. In: Frindte, Wolfgang & Köhler, Thomas (eds.): Kommunikation im Internet. Frankfurt am Main et al.: Lang, 183–192.

Kovács, Kristóf; Krajcsi, Attila & Pléh, Csaba (2002). Mobilhasználat, időgazdálkodás és extraverzió. In: Nyíri, Kristóf (ed.): *Mobilközösség – mobilmegismerés. Tanulmányok*. Budapest: MTA Filozófiai Kutatóintézete. 179–200.

Kovalovszky, Miklós (1977). *Nyelvfejlődés – nyelvhelyesség*. Budapest: Akadémiai Kiadó.

Krauthamer, Helene Seltzer (1999). *Spoken language interference patterns in written English*. New York: Peter Lang.

Krishnamurthy, Balachander & Wills, Craig E. (2008). Characterizing privacy in online social networks. Proceeding WOSN '08 *Proceedings of the first workshop on Online social networks*. New York: ACM, 37–42.

Kroker, Michael (2001). „LiDuMiNo?“. *Wirtschaftswoche* 24, 7. 6. 2001, 145–147.

Krotz, Friedrich (2001). *Mediatisierung kommunikativen Handelns. Der Wandel von Alltag und sozialen Beziehungen, Kultur und Gesellschaft durch Medien*. Wiesbaden: VS.

Kugler, Nóra (2000a). A partikula. In: Keszler, Borbála (ed.): *Magyar grammatika*. Budapest: Nemzeti Tankönyvkiadó, 275–281.

Kugler, Nóra (2000b). A tagadószó. In: Keszler, Borbála (ed.): *Magyar grammatika*. Budapest: Nemzeti Tankönyvkiadó, 289–291.

Kugler, Nóra (2015). *Megfigyelés és következtetés a nyelvi tevékenységben*. Budapest: Tinta Könyvkiadó.

Kuna, Ágnes (2007). Az orvosi receptről. Beszélt nyelvi elemek a 16–17. századi receptekben. *Magyar Nyelvőr*, 470–479.

Kupper, Sabine (2006). *Anglizismen in deutschen Werbeanzeigen. Eine empirische Studie zur stilistischen und ökonomischen Motivation von Anglizismen*. Frankfurt am Main et al.: Peter Lang.

Laczházi, Gyula (2004): *Az emotikonok és a koraújkor érzelemkifejező technikái*. http://magyar-irodalom.elte.hu/arianna/irat/emoticon.html [published: 08. 04. 2004; retrieved: 17. 01. 2011; checked: 07. 10. 2016]

Laczkó, Krisztina (2001). Névmás és referencia. *Magyar Nyelvőr* 125, 103–107.
Laczkó, Krisztina (2003). A mutató névmások funkcionális vizsgálata. *Magyar Nyelvőr* 127, 314–325.
Laczkó, Krisztina (2004). A névmási rendszer funkcionális keretben I. *Magyar Nyelvőr* 128, 469–479.
Laczkó, Krisztina (2005). A névmási rendszer funkcionális keretben II. *Magyar Nyelvőr* 129, 78–88.
Laczkó, Mária (2009). Középiskolai tanulók beszédének temporális jellemzői. *Magyar Nyelvőr*, 447–467.
Ladányi, Mária (1999). Produktivitás a szóképzésben: a természetes morfológia elveinek alkalmazhatósága a magyarra. *Magyar Nyelv*, 166–179.
Ladányi, Mária (2007). *Produktivitás és analógia a szóképzésben: elvek és esetek*. Budapest: Tinta Könyvkiadó.
Lakoff, Robin Tolmach (1982). Some of my favorite writers are literate: the mingling of oral and literate strategies in written communication. In: Tannen, Deborah (ed.): *Spoken and written language: exploring orality and literacy*. Norwood, New Jersey: Ablex Publishing Corporation, 239–260.
Langacker, Ronald W. (1991). *Foundations of cognitive grammar*. Vol. 2. Stanford: Stanford University Press.
Lanstyák, István (2009). *A magyar beszélt nyelv sajátosságai*. Bratislava: STIMUL, Comenius Egyetem Bölcsészettudományi Karának Magyar Nyelv és Irodalom Tanszéke.
Lauer, Bernhard (2001). Die 40 besten SMS-Praxistipps. *Telefon Magazin. Der Ratgeber für Telekommunikation* 5, 18–22.
Lehmann, Christian (1982). *Thoughts on grammaticalization: A programmatic sketch*. Arbeiten des Kölner Universalien-Projekts 48. Cologne: Universität zu Köln, Institut für Sprachwissenschaft.
Lehmann, Christian (1985). Grammaticalization: synchronic variation and diachronic change. *Lingua e Stile* 20, 303–318.
Lehmann, Christian (1991). Grammaticalization and related changes in contemporary German. In: Traugott, Elizabeth C. & Heine, B. (eds.): *Approaches to Grammaticalization*. Vol. II: Focus on Types of Grammatical Markers. Amsterdam – Philadelphia: Benjamins, 493–535.
Lehmann, Christian (1995). *Thoughts on Grammaticalization. LINCOM Studies on Theoretical Linguistics 01*. München – Newcastle: LINCOM EUROPA.
Lehmann, Christian (2002). New reflections on grammaticalization and lexicalization. In: Wischer, Ilse & Diewald, Gabriele (eds.): *New Reflections on Grammaticalization*. Amsterdam – Philadelphia: John Benjamins Publishing Company, 1–18.
Lengyel, Klára (2000). A szóösszetétel. A ritkább szóalkotási módok. In: Keszler, Borbála (ed.): *Magyar grammatika*. Budapest: Nemzeti Tankönyvkiadó, 321–345.
Levinson, Stephen C. (1983/1992). *Pragmatics*. Cambridge: University Press.
Levinson, Stephen C. (1994). Discourse Deixis. In: Asher, R. E. (ed.): *The Encyclopedia of Language and Linguistics*. Pergamon, 846.
Löffler, Heinrich (1985). *Germanistische Soziolinguistik*. Grundlagen der Germanistik, 28. Berlin: Erich Schmidt Verlag.
Ludwig, Otto (1983). Einige Vorschläge zur Begrifflichkeit und Terminologie von Untersuchungen im Bereich der Schriftlichkeit. In: Klaus, B. & Günther, Harmut (eds.): *Schrift, Schreiben, Schriftlichkeit. Arbeiten zur Struktur, Funktion und Entwicklung schriftlicher Sprache*. Tübingen: Max Niemeyer Verlag, 1–15.

Lyons, John (1977/1989). *Semantics*. Vol. 2. Cambridge: Cambridge University Press.
Maar, Christa (2006). Iconic worlds. Bilderwelten nach dem iconic turn. In: Maar, Christa & Burda, Hubert (eds.): *Iconic worlds. Neue Bilderwelten und Wissensräume*. Köln: DuMont, 11–14.
Makaronizmus. In: *Magyar néprajzi lexikon*. http://mek.oszk.hu/02100/02115/html/3-1516.html [03. 06. 2011]
Markó, Alexandra (2003). *Asszem – avagy a módosítószóvá/partikulává válás egy lehetősége a szinkróniában?* Manuscript. Budapest: ELTE.
Márku, Anita (2015). Researching the internet language use as a modern trend in applied linguistics. In: *Modern trends in foreign language teaching and applied linguistics*. Beregszász/Berhovo: II. Rákóczi Ferenc Kárpátaljai Magyar Főiskola, 156–167.
Marton, Lívia (2015). MZ/X! Jelkezz!:O In: Bárdos, Jenő (ed.): *Nyelvpedagógia 2015. Az anyanyelv és az idegen nyelvek tanításának elmélete és gyakorlata*. Eger: Líceum Kiadó, 81–91.
Mathesius, Vilém (1915). O passivu u moderní anglištine. *Sborníkfilologicky* 5, 198–220.
McCarthy, Michael (1998). *Spoken language and applied linguistics*. Cambridge: Cambridge University Press.
McCloud, Cory (2006). The immaterial book. In: Origgi, Gloria (ed.): *text-e. Text in the Age of the Internet*. Houndmills – New York: Palgrave Macmillan, 14–5.
McCrindle, Marc & Wolfinger, Emily (2010). Az XYZ ábécéje. A nemzedékek meghatározása. *Korunk*, November 2010, 13–18.
McLuhan, Marshall (1962). *The Gutenberg Galaxy. The Making of Typographic Man*. University of Toronto Press.
McLuhan, Marshall (1964). *Understanding Media: the Extensions of Man*. New York: McGraw – Hill.
Meier-Seethaler, Carola (1997). *Gefühl und Urteilskraft. Ein Plädoyer für die emotionale Vernunft*. München: CH Beck.
Meillet, Anton (1912). L'evolution des formes grammaticales. In: Meillet, Anton (ed.): *Linguistique Historique et Linguistique Générale*. Paris: Champion, 131–148.
Meinel, Christoph & Sack, Harald (2004). *www. Kommunikation, Internetworking, Web-Technologien*. Springer.
Minya, Károly (2003). *Mai magyar nyelvújítás – szókészletünk módosulása a neologizmusok tükrében a rendszerváltozástól az ezredfordulóig*. Budapest: Tinta Könyvkiadó.
Molnár, Anna (2002). *Die Grammatikalisierung deutscher Modalpartikeln*. Fallstudien. MetaLinguistica 12. Frankfurt am Main et al.: Peter Lang.
Móricz, Éva & Téglássy, Tamás (1999). *Kreatív tervezés a reklámban*. Budapest – Tatabánya: Budapesti Közgazdaságtudományi Egyetem – Modern Üzleti Tudományok Főiskolája.
Móricz, Kálmán (1999). A számítógép és a helyesírás. *Édes Anyanyelvünk* XXI/5, 10.
Nádas, Tímea (2007). Tekintettel egy határozóra – élőnyelvi szempontból. *Magyar Nyelvőr*, 107–117.
Nagy, J. Béla (1963). Kettő meg kettő az négy. – Kettő meg kettő négy. *Magyar Nyelvőr* 87, 384–386.
Nagy, J. Béla (1965). Nem igaz, hogy az az, az csak gaz. *Magyar Nyelv* 61, 352–354.
Nelson, Ted (1974). *Computer lib. Dream machines*. Washington: Tempus Books. (Revised edition: 1987.)
Neuland, Eva (ed.) (2012). *Sprache der Generationen*. Mannheim–Zürich: Dudenverlag.

Nickisch, Reinhard (2003). Der Brief – historische Betrachtungen. In: Höflich, Joachim R. & Gebhardt, Julian (eds.): *Vermittlungskulturen im Wandel. Brief, E-mail, SMS.* Frankfurt am Main et al.: Lang, 63–74.

Nyíri, Kristóf & Szécsi, Gábor (eds.) (1998). *Szóbeliség és írásbeliség. A kommunikációs technológiák története Homérosztól Heideggerig.* Budapest: Áron Kiadó.

Nyíri, Kristóf (2000). *A multimedialitás ismeretfilozófiája.* Uniworld Közhasznú Egyesület. Kultúrák közötti kommunikáció szabadegyeteme. http://www.phil-inst.hu/uniworld/kkk/mm/mm.htm [retrieved: 03. 06. 2011; checked: 07. 10. 2016, not available]

Nyíri, Kristóf (2001). Képjelentés és mobilkommunikáció. Vázlat. In: Nyíri, Kristóf (ed.): *A 21. századi kommunikáció útjai. Tanulmányok.* Budapest: MTA Filozófiai Kutatóintézete, 59–79.

Nyíri, Kristóf (2003). A gondolkodás képelmélete. In: Neumer, Katalin (ed.): *Kép, beszéd, írás.* Budapest: Gondolat Kiadói Kör, 264–278.

Nyíri, Kristóf (2004). *Vernetztes Wissen. Philosophie im Zeitalter des Internets.* Wien: Passagen Verlag.

Nyíri, Kristóf (2006). Mobiltárs a szélessáv sodrában. *Magyar Tudomány* 2006/7. http://www.matud.iif.hu/06jul/12.html [retrieved: 17. 06. 2011; checked: 07. 10. 2016, not available]

Nyomárkay, István (1998). A nyelvhasználat udvariassági stratégiái. *Magyar Nyelvőr*, 277–283.

Ogilvy, David (1992). *Ogilvy a reklámról.* Budapest: Park Könyvkiadó.

Olson, David R. (1977). From utterance to text: the bias of language in speech and writing. *Harvard Educational Review* 47, 257–281.

Onea Gáspár, Edgar (2006). *Sprache und Schrift aus handlungstheoretischer Perspektive.* Studia Linguistica Germanica 81. Berlin – New York: Walter de Gruyter.

Ong, Walter J. (1982). *Orality and literacy. The technologizing of the Word.* London – New York: Methuen.

Origgi, Gloria (ed.) (2006). *text-e. Text in the Age of the Internet.* Houndmills – New York: Palgrave Macmillan.

Origgi, Gloria (2006). Conclusion: How the internet has changed our cultural and scholarly practices. In: Origgi, Gloria (ed.): *text-e. Text in the Age of the Internet.* Houndmills – New York: Palgrave Macmillan, 197–204.

Orwell, George (2000). *1984.* Budapest: Európa Könyvkiadó.

Papp-Váry, Árpád (2007). Márkaépítés mint a modern marketing egyik kulcseleme. In: Svéhlik, Csaba (ed.): *Marketing a 21. században.* Mór: KHEOPS Automobil-Kutató Intézet, 87–138.

Parker, Ashley (2011). Twitter's Secret Handshake. *New York Times*, June 10, 2011. http://www.nytimes.com/2011/06/12/fashion/hashtags-a-new-way-for-tweets-cultural-studies.html [14. 10. 2014]

Pease, Allan (1989). *Testbeszéd. Gondolatolvasás gesztusokból.* Budapest: Park Kiadó.

Peters, Isabella (2009). *Folksonomies. Indexing and Retrieval in Web 2.0.* Wolfgang G. Stock (ed.). Knowledge & Information. Studies in Information Science. Berlin: De Gruyter Saur.

Petőfi, S. János (1995). *A hipertextuális irodalom a perszonal computer elterjedt alkalmazásának korszakában.* http://www.jgytf.u-szeged.hu/~vass/szemm082.htm [09. 05. 2010]

Petykó, Márton (2012a). A blog műfaji jellemzőinek korpuszalapú vizsgálata. *Magyar Nyelvőr* 136/1, 45–72.

Petykó, Márton (2012b). Az írott beszélt nyelvtől a spontán írott nyelv felé. In: *A sokszínű alkalmazott nyelvészet.* Budapest: Tinta Könyvkiadó, 44–52.

Petykó, Márton (2013). Az internetes troll mint identitás kialakítása politikai blogok diskurzusaiban. *Magyar Nyelvőr* 137/3, 274–313.

Plato (1925). *Plato in Twelve Volumes*. Vol. 9. Translated by Harold N. Fowler. Cambridge, MA: Harvard University Press; London: William Heinemann Ltd.

Pléh, Csaba (2010). A számítógép és a személyiség metaforáinak átalakulása. *Alföld*, March 2010.

Pólya, Tamás (2012). Goffman online. Mediatizált dramaturgiai szerepek és közönségek az új mediális (digitális) személyközi kommunikációban. *Századvég* 66, 19–50.

Popcorn, Faith & Marigold, Lys (2001). *Éva-marketing. A nőket megcélzó marketing alapelvei*. Budapest: Geomédia Szakkönyvek.

Porkoláb, Ádám (2015). Az iskolában elsajátított írott nyelvi norma presztízse az internetes kommunikációban. *Iskolakultúra* 2015/11, 75–89.

Pscheida, Daniela (2007). *Internetkompetenz von Erwachsenen. Medienpraxis der Generationen*. Hamburg: Krämer.

Pusztai, Ferenc (ed.) (2003). *Magyar értelmező kéziszótár*. Second edition. Budapest: Akadémiai Kiadó.

Rambukkana, Nathan (ed.) (2015). *Hashtag Publics. The Power and Politics of Discursive Networks*. Frankfurt am Main et al.: Peter Lang.

Róka, Jolán (1998). A reklámról történeti távlatokban. In: Szathmári, István (ed.): *Stilisztika és gyakorlat*. Budapest: Nemzeti Tankönyvkiadó, 351–373.

Róka, Jolán (2003). *Kommunikációtan. Fejezetek a kommunikáció elméletéből és gyakorlatából*. Budapest: Századvég Kiadó, 62–96.

Röttgers, Kurt (2009). The pornographic turn – or: the loss of decency. In: Nyíri, Kristóf (ed.): *Engagement and exposure. Mobile Communication and the ethics of social networking*. Vienna: Passagen Verlag, 87–91.

Runkehl, Jens; Schlobinski, Peter & Siever, Torsten (1998). *Sprache und Kommunikation im Internet. Überblick und Analysen*. Opladen/Wiesbaden: Westdeutscher Verlag.

Runkehl, Jens (2005). Text-Bild-Konstellationen. In: Siever, Torsten; Schlobinski, Peter & Runkehl, Jens (eds.): *Websprache.net. Sprache und Kommunikation im Internet*. Berlin – New York: Walter de Gruyter, 202–218.

Sandig, Barbara (1972). *Zur Differenzierung gebrauchssprachlicher Textsorten im Deutschen*. Athenäum.

Sapir, Edward (1921). *Language. An Introduction to the Study of Speech*. New York: Harcourt Brace & Company.

Schank, Gerd & Schoenthal, Gisela (1976). *Gesprochene Sprache. Eine Einführung in Forschungsansätze und Analysemethoden*. Tübingen: Gruyter.

Scherer, Hans Siegfried (1984). *Sprechen im situativen Kontext. Theorie und Praxis der Analyse spontanen Sprachgebrauchs*. Romanica et comparatistica. Sprach- und literaturwissenschaftliche Studien 3. Tübingen: Stauffenberg Verlag.

Schirm, Anita (2002). Nyelvhasználatunk az informatika korában. In: Balázs, Géza (ed.): *Informatikai technológia és nyelvhasználat. Válogatás a Nemzeti Kulturális Örökség Minisztériuma anyanyelvi pályázatából*. Budapest: Trezor, 7–44.

Schirm, Anita (2009). Partikula és/vagy diskurzusjelölő? In: Keszler, Borbála & Tátrai, Szilárd (eds.): *Diskurzus a grammatikában – grammatika a diskurzusban*. Segédkönyvek a nyelvészet tanulmányozásához 88. Budapest: Tinta Könyvkiadó, 304–311.

Schirm, Anita (2010). *A diskurzusjelölők funkciói: a hát, az -e és a vajon elemek története és jelenkori szinkrón státusa alapján*. Doctoral dissertation. Manuscript.

Schlobinski, Peter (ed.) (1997). *Syntax des gesprochenen Deutsch*. Opladen: Westdeutscher Verlag.

Schlobinski, Peter (ed.) (2006). *Von *hdl* bis *cul8r*. Sprache und Kommunikation in den neuen Medien.* Mannheim et al.: Dudenverlag.
Schlobinski, Peter (2000). Von Chatten im Cyberspace. In: Eichhoff-Cyrus, Karin & Hoberg, Rudolf (eds.): *Die deutsche Sprache zur Jahrtausendwende. Sprachkultur oder Sprachverfall?* Mannheim: Dudenverlag.
Schlobinski, Peter (2005). Sprache und internetbasierte Kommunikation – Voraussetzungen und Perspektiven. In: Siever, Torsten; Schlobinski, Peter & Runkehl, Jens (eds.): *Websprache.net. Sprache und Kommunikation im Internet.* Berlin – New York: Walter de Gruyter, 1–14.
Schlobinski, Peter (2009). *Von HDL bis DUBIDODO – (K)ein Wörterbuch zur SMS.* Mannheim–Leipzig–Wien–Zürich: Duden.
Schmidt, Miriam (2001). *SMS easy. Die besten Emoticons, Sprüche und Kürzel.* München.
Schmitz, Ulrich (1995). *Neue Medien und Gegenwartssprache. Lagebericht und Problemskizze.* Osnabrücker Beiträge zur Sprachtheorie (OBST) 50. Osnabrück.
Schmitz, Ulrich (2003). Text-Bild-Metamorphosen in Medien um 2000. In: Schmitz, Ulrich & Wenzel, Horst (eds.): *Wissen und neue Medien.* Berlin, 241–263.
Schneider, Daniel; Sperling, Sebastian; Schell, Geraldine; Hemmer, Katharina; Glauer, Ramiro & Silberhorn, Daniel (2005). *Instant Messaging – Neue Räume im Cyberspace. Nutzertypen, Gebrauchsweisen, Motive, Regeln.* München: Verlag Reinhard Fischer.
Schwitalla, Johannes (1997). *Gesprochenes Deutsch. Eine Einführung.* Berlin: Erich Schmidt Verlag.
Schwitalla, Johannes (2001). Gesprochene-Sprache-Forschung und ihre Entwicklung zu einer Gesprächsanalyse. In: Brinker, Hans (ed.): *Text- und Gesprächslinguistik. Ein internationales Handbuch zeitgenössischer Forschung. (Linguistics of text and conversation.)* Handbücher zur Sprach- und Kommunikationswissenschaft 16. Halbband 2. Berlin: Walter de Gruyter, 896–903.
Schwitalla, Johannes (2003). *Gesprochenes Deutsch. Eine Einführung.* Second edition. Berlin: Erich Schmidt Verlag.
Scotti Morgana, Silvia (1981). *Le parole nuove.* Bologna: Zanichelli.
Sebestyén, Árpád (1988). A belső nyelvtípusok néhány kérdéséről. In: Kiss, Jenő & Szűts, László (eds.): *A magyar nyelv rétegződése.* Budapest: Akadémiai Kiadó, 108–119.
Selting, Margret (2007). 'Grammatik des gesprochenen Deutsch' im Rahmen der Interaktionalen Linguistik. In: Ágel, Vilmos & Hennig, Mathilde (eds.): *Zugänge zur Grammatik der gesprochenen Sprache.* Tübingen: Max Niemeyer Verlag, 99–135.
Shin, Jiwon (2008). Morality and Internet Behavior: A Study of the Internet Troll and its Relation with Morality on the Internet. *Technology and Teacher Education Annual Review 5*, 28–34.
Sieber, Peter (1998). *Parlando in Texten. Zur Veränderung kommunikativer Grundmuster in der Schriftlichkeit.* Tübingen: de Gruyter.
Síklaki, István (1994). A meggyőzés pszichológiája. Scientia Humana.
Sille, István (2009). *Illem, etikett, protokoll.* Budapest: Akadémiai Kiadó.
Simon, Viktória (2014). Modern kódkeveredés: az internetes nyelvhasználat hatásai fiatalok írásbeli fogalmazásaiban. *Anyanyelv-pedagógia*, Paper 497.
Sperber, Dan (1996). *Explaining culture: a naturalistic approach.* Oxford: Blackwell.
Sperber, Dan & Wilson, Deirdre (1996). *Relevance. Communication and Cognition.* Second Edition. Oxford – Cambridge: Blackwell.

Steger, Hugo (1965–66/1967). Gesprochene Sprache. Zu ihrer Typik und Terminologie. In: Moder, Hugo et al. (eds.): *Satz und Wort im heutigen Deutsch. Jahrbuch 1965/66 des Instituts für deutsche Sprache*. Düsseldorf, 259–291.

Storrer, Angelika (1997). Vom Text zum Hypertext. Die Produktion von Hypertexten auf der Basis traditioneller wissenschaftlicher Texte. In: Knorr, Dagmar & Jakobs, Eva-Maria (eds.): *Textproduktion in elektronischen Umgebungen*. Frankfurt am Main et al.: Peter Lang, 121–139.

Storrer, Angelika (2000). Schriftverkehr auf der Datenautobahn: Besonderheiten der schriftlichen Kommunikation im Internet. In: Voß, G. G.; Holly, W. & Boehnke, K. (eds.): *Neue Medien im Alltag: Begriffsbestimmungen eines interdisziplinären Forschungsfeldes*. Opladen: Leske + Budrich, 153–177.

Storrer, Angelika (2001). Sprachliche Besonderheiten getippter Gespräche: Sprecherwechsel und sprachliches Zeigen in der Chat-Kommunikation. In: Beißwenger, Michael (ed.): *Chat-Kommunikation. Sprache, Interaktion, Sozialität & Identität in synchroner computervermittelter Kommunikation. Perspektiven auf ein interdisziplinäres Forschungsfeld.* Stuttgart: Ibidem, 3–24.

Sz. Tóth, László (2013). Digitális kommunikáció az iskolában – a Netszótár mint híd: Netszótár. @-tól a Zukbergnetig. *Anyanyelv-pedagógia*, Paper 448.

Szalamin, Edit (1978). A mai magyar beszélt nyelv mondattanának kérdéséhez. *Magyar Nyelv*, 293–303.

Szalamin, Edit (1988). Az ún. témaismétlő névmások kérdéséhez. In: Kontra, Miklós (ed.): *Beszélt nyelvi tanulmányok*. Linguistica, Ser. A, Studia et Dissertationes I. Budapest, 91–101.

Szathmári, István (2004). *Stilisztikai lexikon: stilisztikai fogalmak magyarázata szépirodalmi példákkal szemléltetve*. Budapest: Tinta Könyvkiadó.

Szépe, György (1997). Az Internet-korszak nyelvészete. *Modern Nyelvoktatás*, 1997/1–2, 76–89.

Tannen, Deborah (ed.) (1982). *Spoken and written language: exploring orality and literacy*. Norwood, New Jersey: Ablex Publishing Corporation.

Tátrai, Szilárd (2000). Az elbeszélő „én" nyelvi jelöltsége. Kísérlet a perszonális narráció szövegtani megközelítésére. *Magyar Nyelvőr* 124, 226–238.

Tátrai, Szilárd (2005). *A nézőpont szerepe a narratív megértésben*. Általános Nyelvészeti Tanulmányok XXI. Budapest: Akadémiai Kiadó, 207–229.

Teplan, Annegret (2005). *Kommunikation im Chat. Grundlagen, Instrumente, Bewertung*. Berlin: Verlag Dr. Müller.

Terdiman, Daniel (2005). Study: Wikipedia as accurate as Britannica. *CNET News*. http://news.cnet.com/Study-Wikipedia-as-accurate-as-Britannica/2100-1038_3-5997332.html [03. 06. 2011]

Thaler, Verena (2003). *Chat-Kommunikation im Spannungsfeld zwischen Oralität und Literalität*. Berlin: VWF.

Ting, Hiram; Winnie Wong Poh Ming; Ernest Cyril de Run & Sally Lau Yin Choo (2015). Beliefs about the Use of Instagram: An Exploratory Study. *International Journal of Business and Innovation* 2/2, 15–31.

Tokaji, Zsolt & Adamkó, Fanni (2006). *Szerelmesms*. Budapest: Korona Kiadó.

Tolcsvai Nagy, Gábor (1999). Térjelölés a magyar nyelvben. *Magyar Nyelv* 95, 154–165.

Tolcsvai Nagy, Gábor (2001). *A magyar nyelv szövegtana*. Budapest: Nemzeti Tankönyvkiadó.

Tolcsvai Nagy, Gábor (2004). A hálózat nyelve. In: Tolcsvai Nagy, Gábor: *Nyelv, érték, közösség*. Budapest: Gondolat Kiadó, 188–205.

Tompa, József (ed.) (1961–62). *A mai magyar nyelv rendszere*. Volume II. Budapest: Akadémiai Kiadó.
Tompa, József (1964). Az az néha fölösleges. *Magyar Nyelv* 60, 359–363.
Traugott, Elizabeth Closs (1988). Pragmatic strengthening and grammaticalization. *Berkeley Linguistics Society* 14, 406–416.
Traugott, Elizabeth C. (1997). *The discourse connective after all: A historical pragmatic account.* http://www.stanford.edu/~traugott/papers/after_all.pdf [not available]
Traugott, Elizabeth C. (2003). Constructions in Grammaticalization. In: Joseph, B. D. & Janda, R. D. (eds.): *The Handbook of Historical Linguistics*. Malden – Oxford – Melbourne – Berlin: Blackwell Publishing, 624–647.
Trudgill, Peter (1997). *Bevezetés a nyelv és társadalom tanulmányozásába*. Szeged: JGYTF Kiadó.
Vallduví, Enric (1990). *The Informational Component*. Doctoral dissertation. University of Pennsylvania, Philadelphia.
Vallent, Brigitta (2008). Beszélt nyelvi hatások a középiskolások fogalmazásaiban. *Magyar Nyelvőr*, 189–205.
Verschueren, Jef (1995). Pragmatic perspective. In: Verschueren, Jef; Östman, Jan-Ola & Blommaert, Jan (eds.): *Handbook of Pragmatics*. Amsterdam, Philadelphia: John Benjamins, 1–19.
Verschueren, Jef (1999). *Understanding Pragmatics*. London – New York – Sydney –Auckland: Arnold.
Vértes O., András (1986). Érzelmi világunk és a nyelv történeti változásának kölcsönhatása. *Magyar Nyelv*, 11–29.
Vértes O., András (1989). *Gedanken über die geschichtliche Veränderung des Sprechetempos.* Proceedings of the Speech Research '89. International Conference. MFF/HPP, 21, 381–384.
Veszelszki, Ágnes (2013a). A digilektus hatása az írásbeli és a szóbeli kommunikációra egy kérdőíves vizsgálat alapján. *Magyar Nyelvőr* 2013/3, 248–274.
Veszelszki, Ágnes (2017). Verbal and visual aggression in trolling. In: Benedek, András & Veszelszki, Ágnes (eds.): Visual Learning: Virtual – Visual – Veridical (working title). Frankfurt a. M.: Peter Lang (in press).
Veszelszki, Ágnes & Parapatics, Andrea (2014). A részvételtől a részvétig. A halál megjelenése és gyászmunka a közösségi oldalakon. *Magyar Nyelvőr*, 179–199.
Veszelszki, Ágnes & Parapatics, Andrea (2016). From Cooperation to Compassion: Death and Bereavement on Social Networking Websites. In: Knautz, Kathrin & Baran, Katsiaryna S. (eds): *Facets of Facebook. Use and Users*. De Gruyter Saur, 174–211.
Veszelszki, Ágnes (ed.) (2012). *Netszótár. @-tól a Zukbergnetig*. Budapest: ELTE Eötvös Kiadó.
Veszelszki, Ágnes (2002). Üzi a mobon? Az SMS nyelvhasználata. In: Balázs, Géza (ed.): *Informatikai technológia és nyelvhasználat. Tanulmánykötet*. Budapest: Trezor Kiadó, 239–247.
Veszelszki, Ágnes (2005a). Az SMS köszönőformulái. In: Balázs, Géza & Bódi, Zoltán (eds.): *Az internetkorszak kommunikációja. Tanulmányok*. Budapest: Gondolat–Infonia, 284–295.
Veszelszki, Ágnes (2005b). Dinamikus emotikonok. In: Balázs, Géza; Gyula H. Varga & Veszelszki Ágnes (eds.): *A magyar szemiotika negyedfél évtized után*. Budapest – Eger: Líceum Kiadó, 273–286.
Veszelszki, Ágnes (2005c). Írásjelek és szimbólumok az SMS-ekben. *Magyar Nyelvőr*, 111–116.
Veszelszki, Ágnes (2006). *Az sms nyelvhasználatának extralingvális és intralingvális vizsgálata*. MA thesis. Manuscript. Budapest: ELTE BTK.
Veszelszki, Ágnes (2007). Asszem, nemtom, h +fejt7őek-e nkd az sms-röv. – avagy az sms-beli rövidítésekről. *Kommunikáció, Média, Gazdaság* V/1, 29–51.

Veszelszki, Ágnes (2008). Egy beszélt nyelvi jellemző, a *témaismétlő névmás* csevegésszövegekben. *Magyar Nyelvőr*, 235–244.
Veszelszki, Ágnes (2009). Képírás vagy képes írás? Az infokommunikációs technológia hatása a felső tagozatosok írására. In: Balázs, Géza & H. Varga, Gyula (eds.): *Ikonikus fordulat a kultúrában*. Budapest: Magyar Szemiotikai Társaság; Eger: Líceum Kiadó, 309–333.
Veszelszki, Ágnes (2010a). Digilektus a lektusok rendszerében. In: Illés-Molnár, Márta; Kaló, Zsuzsa; Klein, Laura & Parapatics, Andrea (eds.): *Félúton 5. Az ELTE BTK Nyelvtudományi Doktori Iskolájának konferenciája*. Budapest: ELTE BTK Nyelvtudományi Doktori Iskola, 199–215.
Veszelszki, Ágnes (2010b). Úton – a digilektus és a dialóguslevelek. In: Balázs, Géza & H. Varga, Gyula (eds.): *Az utazás szemiotikája*. Budapest: Magyar Szemiotikai Társaság; Eger: Líceum Kiadó, 286–298.
Veszelszki, Ágnes (2010c). Grammatikalizáció, különös tekintettel az asszem-re. In: Bárdosi, Vilmos (ed.): *A világ nyelvi képe*. Budapest: Tinta Könyvkiadó, 249–265.
Veszelszki, Ágnes (2011a). Digitális kommunikáció és nyelvtudomány. In: Balázs, Géza (ed.): *Nyelvészetről mindenkinek. 77 nyelvészeti összefoglaló*. Budapest: Inter, 56–60.
Veszelszki, Ágnes (2011b). Szóbeliség és írásbeliség. In: Balázs, Géza (ed.): *Nyelvészetről mindenkinek. 77 nyelvészeti összefoglaló*. Budapest: Inter, 327–330.
Veszelszki, Ágnes (2011c). Verbális agresszió az sms-ekben. In: Balázs, Géza: *Sms-nyelv és -folklór*. Budapest: Magyar Szemiotikai Társaság, Inter Kft., Prae.hu, 98–106.
Veszelszki, Ágnes (2013b). Promiscuity of Images. Memes from an English-Hungarian contrastive perspective. In: Benedek, András & Nyíri, Kristóf (eds.): *How To Do Things With Pictures: Skill, Practice, Performance*. Frankfurt: Peter Lang, 115–127.
Veszelszki, Ágnes (2014). Marketolingvisztika és a marketing szaknyelve. In: Veszelszki, Ágnes & Lengyel, Klára (eds.): *Tudomány, technolektus, terminológia. A tudományok, szakmák nyelve*. Budapest: Éghajlat Könyvkiadó, 321–328.
Veszelszki, Ágnes (2015a). Megfontolások a digitális kommunikáció pragmatikai leírásához. In: Bárdosi, Vilmos (ed.): *A nyelvi pragmatika kérdései szinkrón és diakrón megközelítésben*. Budapest: Tinta Könyvkiadó, 239–250.
Veszelszki, Ágnes (2015b). Érzelemkifejezési módok a digitális kommunikációban: emotikonok és reakciógif-ek. *Magyar Nyelvőr*, 74–85.
Veszelszki, Ágnes (2015c). Emoticons vs. Reaction-Gifs. Non-Verbal Communication on the Internet from the Aspects of Visuality, Verbality and Time. In: Benedek, András & Nyíri, Kristóf (eds.): *Beyond Words. Pictures, Parables, Paradoxes*. Series Visual Learning, vol. 5. Frankfurt: Peter Lang, 131–145.
Veszelszki, Ágnes (2015d). Konfliktuskezelés a közösségi médiában. Esettanulmány a Trollfoci vs. Notts County példáján. *Médiakutató* 2015/2, 39–51.
Veszelszki, Ágnes (2016a). A netnyelvészet terminológiai és kronológiai áttekintése. In: Balázs, Géza (ed.): *Jelentés a magyar nyelvről, 2010–2015*. Budapest: Inter, 56–73.
Veszelszki, Ágnes (2016b). A hashtag mint új frazeologizmus? In: Bárdosi, Vilmos (ed.): *A nyelvi pragmatika kérdései szinkrón és diakrón megközelítésben*. Budapest: Modern Filológiai Társaság – Tinta Könyvkiadó, 147–158.
Veszelszki, Ágnes (2016c). #time, #truth, #tradition. An image-text relationship on Instagram: the hashtag. In: Benedek, András & Veszelszki, Ágnes (eds.): *Time, truth, tradition*. Frankfurt: Peter Lang.
Veszelszki, Ágnes (2016d). Intergenerációs különbségek a digitális kommunikációban. Esettanulmány 65 év feletti nyelvhasználókkal. In: Balázs, Géza & Veszelszki, Ágnes (eds.):

Generációk nyelve. Tanulmánykötet. Budapest: ELTE BTK Mai Magyar Nyelvi Tanszék – Inter. – MSZT, 233–248.

Wacha, Imre (1992). A nyelvi rétegződés kérdései. In: Kemény, Gábor (ed.): *Normatudat – nyelvi norma*. Linguistica. Series A. Studia et dissertationes 8. Budapest: Nyelvtudományi Intézet, 49–105.

Waltereit, Richard (2002). *The Rise of Discourse Markers in Italian: A Specific Type of Language Change*. https://www.academia.edu/2521630/The_rise_of_discourse_markers_in_Italian_A_specific_type_of_language_change [08. 04. 2017]

Wardhaugh, Ronald (1995). *Szociolingvisztika*. Budapest: Osiris Századvég.

Wegener, Heide (1998). Zur Grammatikalisierung von Modalpartikeln. In: Barz, Irmhild & Öhlschläger, Günther (eds.): *Zwischen Grammatik und Lexikon*. Tübingen: Max Niemeyer Verlag, 37–55.

Weller, Katrin; Meckel, Martin Sebastian & Stahl, Matthias (eds.) (2013). Social Media. Recherche Spezial 2013/1. Gesis Leibniz-Institut für Sozial-Wissenschaften. http://www.gesis.org/sowiport/fileadmin/user_upload/pdf_recherche_spezial/rs_13_01_socialmedia.pdf [10. 08. 2013]

Wirth, Uwe (2005). Chatten. Plaudern mit anderen Mitteln. In: Siever, Torsten; Schlobinski, Peter & Runkehl, Jens (eds.): *Websprache.net. Sprache und Kommunikation im Internet*. Berlin – New York: Walter de Gruyter, 67–84.

Zachar, Viktor (2013). Fordítási stratégiák a neologizmusokkal kapcsolatban. In: Bárdosi, Vilmos (ed.): *Reáliák a lexikológiától a frazeológiáig*. Budapest: Tinta Könyvkiadó, 141–151.

Zachar, Viktor (2014). A plágiumügyekkel kapcsolatos neologizmusok a magyar és a német nyelvben. In: Bárdosi, Vilmos (ed.): *Szövegalkotó gyakorlatok, nyelvteremtő praktikák*. Budapest: Tinta Könyvkiadó, 227–236.

Ziegler, Arne & Dürscheid, Christa (eds.) (2002). *Kommunikationsform E-Mail*. Tübingen: Stauffenburg.

Ziegler, Arne (2002). E-Mail – Textsorte oder Kommunikationsform? In: Ziegler, Arne & Dürscheid, Christa (eds.): *Kommunikationsform E-Mail*. Tübingen: Stauffenburg, 9–32.

Zimányi, Árpád (2014). A korszerű infokommunikációs technológiák hatása az anyanyelvi nevelésre. In: *Tanulmányok a tankönyvkutatás feladatairól, módszereiről*. Bratislava: Terra vydavateľstvo, 220–231.

Zimányi, Árpád (2005). SMS-szövegek elemzése. In: Balázs, Géza & Bódi, Zoltán (eds.): *Az internetkorszak kommunikációja. Tanulmányok*. Budapest: Gondolat–Infonia, 269–283.

Zimányi, Árpád (2011). *Nyelvtani viszonyok a neten*. Presentation in the series A Gutenberg-galaxistól a Google-galaxisig. ELTE, Budapest. 27. April 2011.

Zlinszky, Aladár (1925). Az eufémizmusról. *Magyar Nyelv*, 8–9.

Zsigó Anna (2010). *A csetelés mint új nyelvi jelenség*. ELTE BTK. Manuscript.

W1 = Ön lett az év embere a Time szerint. http://origo.hu/techbazis/internet/20061218onlett.html [uploaded: 18. 12. 2006; retrieved: 03. 06. 2011; checked: 07. 10. 2016]

W2 = The First Smiley :-). http://research.microsoft.com/~mbj/Smiley/Smiley.html [retrieved: 03. 06. 2011; checked: 07. 10. 2016]

W3 = A digitális kor véleményvezérei. BrandTrend, 09. 11. 2010. http://www.brandtrend.hu/hir/4683/a_digitalis_kor_velemenyvezerei.html [retrieved: 03. 06. 2011; checked: 07. 10. 2016; not available]

W4 = IBM (2013). What is big data? Bringing big data to the enterprise. IBM. Online: http://www.ibm.com/big-data/us/en/ [checked: 07. 10. 2016; webpage content has changed]

W5 = Blogging + video = vlogging. http://www.wired.com/entertainment/music/news/2005/07/68171 [retrieved: 14. 01. 2009; checked: 07. 10. 2016; not available]
W6 = http://research.microsoft.com/~mbj/Smiley/Smiley.html [retrieved: 18. 05. 2011; checked: 07. 10. 2016]
W7 = http://www-2.cs.cmu.edu/~sef/sefSmiley.htm [retrieved: 17. 06. 2011; checked: 07. 10. 2016]
W8 = http://www.askoxford.com/betterwriting/emoticon /?view=uk [retrieved: 14. 09. 2005; checked: 07. 10. 2016; webpage content has changed]
W9 = http://toplike.eu/like/100262 [retrieved: 24. 05. 2011; checked: 07. 10. 2016; not available]
W10 = Hungarian Code of Ethics for Advertising 1997. http://www.hullamvadasz.hu/ftpterulet/jog/magyar_reklamek.htm [retrieved: 26. 10. 2007; checked: 07. 10. 2016; not available]
W11 = Most popular meme characters. http://knowyourmeme.com/memes/popular [checked: 07. 10. 2016; content has changed].

Index

DOI 10.1515/9783110499117-020